Lecture Notes in Networks and Systems   802

The series "Lecture Notes in Networks and Systems" publishes the latest developments in Networks and Systems—quickly, informally and with high quality. Original research reported in proceedings and post-proceedings represents the core of LNNS.

Volumes published in LNNS embrace all aspects and subfields of, as well as new challenges in, Networks and Systems.

The series contains proceedings and edited volumes in systems and networks, spanning the areas of Cyber-Physical Systems, Autonomous Systems, Sensor Networks, Control Systems, Energy Systems, Automotive Systems, Biological Systems, Vehicular Networking and Connected Vehicles, Aerospace Systems, Automation, Manufacturing, Smart Grids, Nonlinear Systems, Power Systems, Robotics, Social Systems, Economic Systems and other. Of particular value to both the contributors and the readership are the short publication timeframe and the world-wide distribution and exposure which enable both a wide and rapid dissemination of research output.

The series covers the theory, applications, and perspectives on the state of the art and future developments relevant to systems and networks, decision making, control, complex processes and related areas, as embedded in the fields of interdisciplinary and applied sciences, engineering, computer science, physics, economics, social, and life sciences, as well as the paradigms and methodologies behind them.

Indexed by SCOPUS, INSPEC, WTI Frankfurt eG, zbMATH, SCImago.

All books published in the series are submitted for consideration in Web of Science.

For proposals from Asia please contact Aninda Bose (aninda.bose@springer.com).

Alvaro Rocha · Hojjat Adeli ·
Gintautas Dzemyda · Fernando Moreira ·
Valentina Colla
Editors

Information Systems and Technologies

WorldCIST 2023, Volume 4

Editors
Alvaro Rocha
ISEG
Universidade de Lisboa
Lisbon, Cávado, Portugal

Hojjat Adeli
College of Engineering
The Ohio State University
Columbus, OH, USA

Gintautas Dzemyda
Institute of Data Science and Digital
Technologies
Vilnius University
Vilnius, Lithuania

Fernando Moreira
DCT
Universidade Portucalense
Porto, Portugal

Valentina Colla
TeCIP Institute
Scuola Superiore Sant'Anna
Pisa, Italy

ISSN 2367-3370   ISSN 2367-3389 (electronic)
Lecture Notes in Networks and Systems
ISBN 978-3-031-45650-3   ISBN 978-3-031-45651-0 (eBook)
https://doi.org/10.1007/978-3-031-45651-0

This Springer imprint is published by the registered company Springer Nature Switzerland AG
The registered company address is: Gewerbestrasse 11, 6330 Cham, Switzerland

Paper in this product is recyclable.

Preface

This book contains a selection of papers accepted for presentation and discussion at the 2023 World Conference on Information Systems and Technologies (WorldCIST'23). This conference had the scientific support of the Sant'Anna School of Advanced Studies, Pisa, University of Calabria, Information and Technology Management Association (ITMA), IEEE Systems, Man, and Cybernetics Society (IEEE SMC), Iberian Association for Information Systems and Technologies (AISTI), and Global Institute for IT Management (GIIM). It took place in Pisa city, Italy, 4–6 April 2023.

The World Conference on Information Systems and Technologies (WorldCIST) is a global forum for researchers and practitioners to present and discuss recent results and innovations, current trends, professional experiences and challenges of modern Information Systems and Technologies research, technological development, and applications. One of its main aims is to strengthen the drive toward a holistic symbiosis between academy, society, and industry. WorldCIST'23 was built on the successes of: WorldCIST'13 held at Olhão, Algarve, Portugal; WorldCIST'14 held at Funchal, Madeira, Portugal; WorldCIST'15 held at São Miguel, Azores, Portugal; WorldCIST'16 held at Recife, Pernambuco, Brazil; WorldCIST'17 held at Porto Santo, Madeira, Portugal; WorldCIST'18 held at Naples, Italy; WorldCIST'19 held at La Toja, Spain; WorldCIST'20 held at Budva, Montenegro; WorldCIST'21 held at Terceira Island, Portugal; and WorldCIST'22, which took place online at Budva, Montenegro.

The Program Committee of WorldCIST'23 was composed of a multidisciplinary group of 339 experts and those who are intimately concerned with Information Systems and Technologies. They have had the responsibility for evaluating, in a 'blind review' process, and the papers received for each of the main themes proposed for the Conference were: A) Information and Knowledge Management; B) Organizational Models and Information Systems; C) Software and Systems Modeling; D) Software Systems, Architectures, Applications, and Tools; E) Multimedia Systems and Applications; F) Computer Networks, Mobility, and Pervasive Systems; G) Intelligent and Decision Support Systems; H) Big Data Analytics and Applications; I) Human-Computer Interaction; J) Ethics, Computers & Security; K) Health Informatics; L) Information Technologies in Education; M) Information Technologies in Radiocommunications; and N) Technologies for Biomedical Applications.

The conference also included workshop sessions taking place in parallel with the conference ones. Workshop sessions covered themes such as: Novel Computational Paradigms, Methods, and Approaches in Bioinformatics; Artificial Intelligence for Technology Transfer; Blockchain and Distributed Ledger Technology (DLT) in Business; Enabling Software Engineering Practices Via Latest Development's Trends; Information Systems and Technologies for the Steel Sector; Information Systems and Technologies for Digital Cultural Heritage and Tourism; Recent Advances in Deep Learning Methods and Evolutionary Computing for Health Care; Data Mining and Machine Learning in Smart Cities; Digital Marketing and Communication, Technologies, and Applications;

Digital Transformation and Artificial Intelligence; and Open Learning and Inclusive Education Through Information and Communication Technology.

WorldCIST'23 and its workshops received about 400 contributions from 53 countries around the world. The papers accepted for oral presentation and discussion at the conference are published by Springer (this book) in four volumes and will be submitted for indexing by WoS, Scopus, EI-Compendex, DBLP, and/or Google Scholar, among others. Extended versions of selected best papers will be published in special or regular issues of leading and relevant journals, mainly JCR/SCI/SSCI and Scopus/EI-Compendex indexed journals.

We acknowledge all of those that contributed to the staging of WorldCIST'23 (authors, committees, workshop organizers, and sponsors). We deeply appreciate their involvement and support that was crucial for the success of WorldCIST'23.

April 2023

Alvaro Rocha
Hojjat Adeli
Gintautas Dzemyda
Fernando Moreira
Valentina Colla

Organization

Honorary Chair

Hojjat Adeli   The Ohio State University, USA

General Chair

Álvaro Rocha   ISEG, University of Lisbon, Portugal

Co-chairs

Gintautas Dzemyda   Vilnius University, Lithuania
Sandra Costanzo   University of Calabria, Italy

Workshops Chair

Fernando Moreira   Portucalense University, Portugal

Local Organizing Committee

Valentina Cola (Chair)   Scuola Superiore Sant'Anna—TeCIP Institute,
                                         Italy
Marco Vannucci   Scuola Superiore Sant'Anna—TeCIP Institute,
                                         Italy
Vincenzo Iannino   Scuola Superiore Sant'Anna—TeCIP Institute,
                                         Italy
Stefano Dettori   Scuola Superiore Sant'Anna—TeCIP Institute,
                                         Italy

Advisory Committee

Ana Maria Correia (Chair)   University of Sheffield, UK
Brandon Randolph-Seng   Texas A&M University, USA

Chris Kimble	KEDGE Business School & MRM, UM2, Montpellier, France
Damian Niwiński	University of Warsaw, Poland
Florin Gheorghe Filip	Romanian Academy, Romania
Janusz Kacprzyk	Polish Academy of Sciences, Poland
João Tavares	University of Porto, Portugal
Jon Hall	The Open University, UK
John MacIntyre	University of Sunderland, UK
Karl Stroetmann	Empirica Communication & Technology Research, Germany
Majed Al-Mashari	King Saud University, Saudi Arabia
Miguel-Angel Sicilia	University of Alcalá, Spain
Mirjana Ivanovic	University of Novi Sad, Serbia
Paulo Novais	University of Minho, Portugal
Wim Van Grembergen	University of Antwerp, Belgium
Mirjana Ivanovic	University of Novi Sad, Serbia
Reza Langari	Texas A&M University, USA
Wim Van Grembergen	University of Antwerp, Belgium

Program Committee

Abderrahmane Ez-Zahout	Mohammed V University, Morocco
Adriana Gradim	University of Aveiro, Portugal
Adriana Peña Pérez Negrón	Universidad de Guadalajara, Mexico
Adriani Besimi	South East European University, Macedonia
Agostinho Sousa Pinto	Polythecnic of Porto, Portugal
Ahmed El Oualkadi	Abdelmalek Essaadi University, Morocco
Akex Rabasa	University Miguel Hernandez, Spain
Alba Córdoba-Cabús	University of Malaga, Spain
Alberto Freitas	FMUP, University of Porto, Portugal
Aleksandra Labus	University of Belgrade, Serbia
Alessio De Santo	HE-ARC, Switzerland
Alexandru Vulpe	University Politehnica of Bucharest, Romania
Ali Idri	ENSIAS, University Mohamed V, Morocco
Alicia García-Holgado	University of Salamanca, Spain
Amélia Badica	Universti of Craiova, Romania
Amélia Cristina Ferreira Silva	Polytechnic of Porto, Portugal
Amit Shelef	Sapir Academic College, Israel
Alanio de Lima	UFC, Brazil
Almir Souza Silva Neto	IFMA, Brazil
Álvaro López-Martín	University of Malaga, Spain

Ana Carla Amaro	Universidade de Aveiro, Portugal
Ana Isabel Martins	University of Aveiro, Portugal
Anabela Tereso	University of Minho, Portugal
Anabela Gomes	University of Coimbra, Portugal
Anacleto Correia	CINAV, Portugal
Andrew Brosnan	University College Cork, Ireland
Andjela Draganic	University of Montenegro, Montenegro
Aneta Polewko-Klim	University of Białystok, Institute of Informatics, Poland
Aneta Poniszewska-Maranda	Lodz University of Technology, Poland
Angeles Quezada	Instituto Tecnologico de Tijuana, Mexico
Anis Tissaoui	University of Jendouba, Tunisia
Ankur Singh Bist	KIET, India
Ann Svensson	University West, Sweden
Anna Gawrońska	Poznański Instytut Technologiczny, Poland
Antoni Oliver	University of the Balearic Islands, Spain
Antonio Jiménez-Martín	Universidad Politécnica de Madrid, Spain
Aroon Abbu	Bell and Howell, USA
Arslan Enikeev	Kazan Federal University, Russia
Beatriz Berrios Aguayo	University of Jaen, Spain
Benedita Malheiro	Polytechnic of Porto, ISEP, Portugal
Bertil Marques	Polytechnic of Porto, ISEP, Portugal
Boris Shishkov	ULSIT/IMI-BAS/IICREST, Bulgaria
Borja Bordel	Universidad Politécnica de Madrid, Spain
Branko Perisic	Faculty of Technical Sciences, Serbia
Carla Pinto	Polytechnic of Porto, ISEP, Portugal
Carlos Balsa	Polythecnic of Bragança, Portugal
Carlos Rompante Cunha	Polytechnic of Bragança, Portugal
Catarina Reis	Polytechnic of Leiria, Portugal
Célio Gonçalo Marques	Polythenic of Tomar, Portugal
Cengiz Acarturk	Middle East Technical University, Turkey
Cesar Collazos	Universidad del Cauca, Colombia
Christine Gruber	K1-MET, Austria
Christophe Guyeux	Universite de Bourgogne Franche Comté, France
Christophe Soares	University Fernando Pessoa, Portugal
Christos Bouras	University of Patras, Greece
Christos Chrysoulas	London South Bank University, UK
Christos Chrysoulas	Edinburgh Napier University, UK
Ciro Martins	University of Aveiro, Portugal
Claudio Sapateiro	Polytechnic of Setúbal, Portugal
Cosmin Striletchi	Technical University of Cluj-Napoca, Romania
Costin Badica	University of Craiova, Romania

Cristian García Bauza	PLADEMA-UNICEN-CONICET, Argentina
Cristina Caridade	Polytechnic of Coimbra, Portugal
David Cortés-Polo	University of Extremadura, Spain
David Kelly	University College London, UK
Daria Bylieva	Peter the Great St.Petersburg Polytechnic University, Russia
Dayana Spagnuelo	Vrije Universiteit Amsterdam, Netherlands
Dhouha Jaziri	University of Sousse, Tunisia
Dmitry Frolov	HSE University, Russia
Dulce Mourato	ISTEC - Higher Advanced Technologies Institute Lisbon, Portugal
Edita Butrime	Lithuanian University of Health Sciences, Lithuania
Edna Dias Canedo	University of Brasilia, Brazil
Egils Ginters	Riga Technical University, Latvia
Ekaterina Isaeva	Perm State University, Russia
Eliana Leite	University of Minho, Portugal
Enrique Pelaez	ESPOL University, Ecuador
Eriks Sneiders	Stockholm University, Sweden
Esperança Amengual	Universitat de les Illes Balears, Spain
Esteban Castellanos	ESPE, Ecuador
Fatima Azzahra Amazal	Ibn Zohr University, Morocco
Fernando Bobillo	University of Zaragoza, Spain
Fernando Molina-Granja	National University of Chimborazo, Ecuador
Fernando Moreira	Portucalense University, Portugal
Fernando Ribeiro	Polytechnic Castelo Branco, Portugal
Filipe Caldeira	Polythecnic of Viseu, Portugal
Filipe Portela	University of Minho, Portugal
Filippo Neri	University of Naples, Italy
Firat Bestepe	Republic of Turkey Ministry of Development, Turkey
Francesco Bianconi	Università degli Studi di Perugia, Italy
Francisco García-Peñalvo	University of Salamanca, Spain
Francisco Valverde	Universidad Central del Ecuador, Ecuador
Frederico Branco	University of Trás-os-Montes e Alto Douro, Portugal
Galim Vakhitov	Kazan Federal University, Russia
Gayo Diallo	University of Bordeaux, France
Gema Bello-Orgaz	Universidad Politecnica de Madrid, Spain
George Suciu	BEIA Consult International, Romania
Ghani Albaali	Princess Sumaya University for Technology, Jordan

Gian Piero Zarri	University Paris-Sorbonne, France
Giovanni Buonanno	University of Calabria, Italy
Gonçalo Paiva Dias	University of Aveiro, Portugal
Goreti Marreiros	ISEP/GECAD, Portugal
Graciela Lara López	University of Guadalajara, Mexico
Habiba Drias	University of Science and Technology Houari Boumediene, Algeria
Hafed Zarzour	University of Souk Ahras, Algeria
Haji Gul	City University of Science and Information Technology, Pakistan
Hakima Benali Mellah	Cerist, Algeria
Hamid Alasadi	Basra University, Iraq
Hatem Ben Sta	University of Tunis at El Manar, Tunisia
Hector Fernando Gomez Alvarado	Universidad Tecnica de Ambato, Ecuador
Hector Menendez	King's College London, UK
Hélder Gomes	University of Aveiro, Portugal
Helia Guerra	University of the Azores, Portugal
Henrique da Mota Silveira	University of Campinas (UNICAMP), Brazil
Henrique S. Mamede	University Aberta, Portugal
Henrique Vicente	University of Évora, Portugal
Hicham Gueddah	University Mohammed V in Rabat, Morocco
Hing Kai Chan	University of Nottingham Ningbo China, China
Igor Aguilar Alonso	Universidad Nacional Tecnológica de Lima Sur, Peru
Inês Domingues	University of Coimbra, Portugal
Isabel Lopes	Polytechnic of Bragança, Portugal
Isabel Pedrosa	Coimbra Business School - ISCAC, Portugal
Isaías Martins	University of Leon, Spain
Issam Moghrabi	Gulf University for Science and Technology, Kuwait
Ivan Armuelles Voinov	University of Panama, Panama
Ivan Dunđer	University of Zabreb, Croatia
Ivone Amorim	University of Porto, Portugal
Jaime Diaz	University of La Frontera, Chile
Jan Egger	IKIM, Germany
Jan Kubicek	Technical University of Ostrava, Czech Republic
Jeimi Cano	Universidad de los Andes, Colombia
Jesús Gallardo Casero	University of Zaragoza, Spain
Jezreel Mejia	CIMAT, Unidad Zacatecas, Mexico
Jikai Li	The College of New Jersey, USA
Jinzhi Lu	KTH-Royal Institute of Technology, Sweden
Joao Carlos Silva	IPCA, Portugal

João Manuel R. S. Tavares	University of Porto, FEUP, Portugal
João Paulo Pereira	Polytechnic of Bragança, Portugal
João Reis	University of Aveiro, Portugal
João Reis	University of Lisbon, Portugal
João Rodrigues	University of the Algarve, Portugal
João Vidal Carvalho	Polythecnic of Coimbra, Portugal
Joaquin Nicolas Ros	University of Murcia, Spain
John W. Castro	University de Atacama, Chile
Jorge Barbosa	Polythecnic of Coimbra, Portugal
Jorge Buele	Technical University of Ambato, Ecuador
Jorge Gomes	University of Lisbon, Portugal
Jorge Oliveira e Sá	University of Minho, Portugal
José Braga de Vasconcelos	Universidade Lusófona, Portugal
Jose M Parente de Oliveira	Aeronautics Institute of Technology, Brazil
José Machado	University of Minho, Portugal
José Paulo Lousado	Polythecnic of Viseu, Portugal
Jose Quiroga	University of Oviedo, Spain
Jose Silvestre Silva	Academia Militar, Portugal
Jose Torres	Universidty Fernando Pessoa, Portugal
Juan M. Santos	University of Vigo, Spain
Juan Manuel Carrillo de Gea	University of Murcia, Spain
Juan Pablo Damato	UNCPBA-CONICET, Argentina
Kalinka Kaloyanova	Sofia University, Bulgaria
Kamran Shaukat	The University of Newcastle, Australia
Karima Moumane	ENSIAS, Morocco
Katerina Zdravkova	University Ss. Cyril and Methodius, North Macedonia
Khawla Tadist	Marocco
Khalid Benali	LORIA—University of Lorraine, France
Khalid Nafil	Mohammed V University in Rabat, Morocco
Korhan Gunel	Adnan Menderes University, Turkey
Krzysztof Wolk	Polish-Japanese Academy of Information Technology, Poland
Kuan Yew Wong	Universiti Teknologi Malaysia (UTM), Malaysia
Kwanghoon Kim	Kyonggi University, South Korea
Laila Cheikhi	Mohammed V University in Rabat, Morocco
Laura Varela-Candamio	Universidade da Coruña, Spain
Laurentiu Boicescu	E.T.T.I. U.P.B., Romania
Lbtissam Abnane	ENSIAS, Morocco
Lia-Anca Hangan	Technical University of Cluj-Napoca, Romania
Ligia Martinez	CECAR, Colombia
Lila Rao-Graham	University of the West Indies, Jamaica

Łukasz Tomczyk	Pedagogical University of Cracow, Poland
Luis Alvarez Sabucedo	University of Vigo, Spain
Luís Filipe Barbosa	University of Trás-os-Montes e Alto Douro
Luis Mendes Gomes	University of the Azores, Portugal
Luis Pinto Ferreira	Polytechnic of Porto, Portugal
Luis Roseiro	Polytechnic of Coimbra, Portugal
Luis Silva Rodrigues	Polythencic of Porto, Portugal
Mahdieh Zakizadeh	MOP, Iran
Maksim Goman	JKU, Austria
Manal el Bajta	ENSIAS, Morocco
Manuel Antonio Fernández-Villacañas Marín	Technical University of Madrid, Spain
Manuel Ignacio Ayala Chauvin	University Indoamerica, Ecuador
Manuel Silva	Polytechnic of Porto and INESC TEC, Portugal
Manuel Tupia	Pontifical Catholic University of Peru, Peru
Manuel Au-Yong-Oliveira	University of Aveiro, Portugal
Marcelo Mendonça Teixeira	Universidade de Pernambuco, Brazil
Marciele Bernardes	University of Minho, Brazil
Marco Ronchetti	Universita' di Trento, Italy
Mareca María PIlar	Universidad Politécnica de Madrid, Spain
Marek Kvet	Zilinska Univerzita v Ziline, Slovakia
Maria João Ferreira	Universidade Portucalense, Portugal
Maria José Sousa	University of Coimbra, Portugal
María Teresa García-Álvarez	University of A Coruna, Spain
Maria Sokhn	University of Applied Sciences of Western Switzerland, Switzerland
Marijana Despotovic-Zrakic	Faculty Organizational Science, Serbia
Marilio Cardoso	Polythecnic of Porto, Portugal
Mário Antunes	Polythecnic of Leiria & CRACS INESC TEC, Portugal
Marisa Maximiano	Polytechnic Institute of Leiria, Portugal
Marisol Garcia-Valls	Polytechnic University of Valencia, Spain
Maristela Holanda	University of Brasilia, Brazil
Marius Vochin	E.T.T.I. U.P.B., Romania
Martin Henkel	Stockholm University, Sweden
Martín López Nores	University of Vigo, Spain
Martin Zelm	INTEROP-VLab, Belgium
Mazyar Zand	MOP, Iran
Mawloud Mosbah	University 20 Août 1955 of Skikda, Algeria
Michal Adamczak	Poznan School of Logistics, Poland
Michal Kvet	University of Zilina, Slovakia
Miguel Garcia	University of Oviedo, Spain

Miguel Melo	INESC TEC, Portugal
Mihai Lungu	University of Craiova, Romania
Mircea Georgescu	Al. I. Cuza University of Iasi, Romania
Mirna Muñoz	Centro de Investigación en Matemáticas A.C., Mexico
Mohamed Hosni	ENSIAS, Morocco
Monica Leba	University of Petrosani, Romania
Nadesda Abbas	UBO, Chile
Narjes Benameur	Laboratory of Biophysics and Medical Technologies of Tunis, Tunisia
Natalia Grafeeva	Saint Petersburg University, Russia
Natalia Miloslavskaya	National Research Nuclear University MEPhI, Russia
Naveed Ahmed	University of Sharjah, United Arab Emirates
Neeraj Gupta	KIET group of institutions Ghaziabad, India
Nelson Rocha	University of Aveiro, Portugal
Nikola S. Nikolov	University of Limerick, Ireland
Nicolas de Araujo Moreira	Federal University of Ceara, Brazil
Nikolai Prokopyev	Kazan Federal University, Russia
Niranjan S. K.	JSS Science and Technology University, India
Noemi Emanuela Cazzaniga	Politecnico di Milano, Italy
Noureddine Kerzazi	Polytechnique Montréal, Canada
Nuno Melão	Polytechnic of Viseu, Portugal
Nuno Octávio Fernandes	Polytechnic of Castelo Branco, Portugal
Nuno Pombo	University of Beira Interior, Portugal
Olga Kurasova	Vilnius University, Lithuania
Olimpiu Stoicuta	University of Petrosani, Romania
Patricia Zachman	Universidad Nacional del Chaco Austral, Argentina
Paula Serdeira Azevedo	University of Algarve, Portugal
Paula Dias	Polytechnic of Guarda, Portugal
Paulo Alejandro Quezada Sarmiento	University of the Basque Country, Spain
Paulo Maio	Polytechnic of Porto, ISEP, Portugal
Paulvanna Nayaki Marimuthu	Kuwait University, Kuwait
Paweł Karczmarek	The John Paul II Catholic University of Lublin, Poland
Pedro Rangel Henriques	University of Minho, Portugal
Pedro Sobral	University Fernando Pessoa, Portugal
Pedro Sousa	University of Minho, Portugal
Philipp Jordan	University of Hawaii at Manoa, USA
Piotr Kulczycki	Systems Research Institute, Polish Academy of Sciences, Poland

Prabhat Mahanti	University of New Brunswick, Canada
Rabia Azzi	Bordeaux University, France
Radu-Emil Precup	Politehnica University of Timisoara, Romania
Rafael Caldeirinha	Polytechnic of Leiria, Portugal
Raghuraman Rangarajan	Sequoia AT, Portugal
Raiani Ali	Hamad Bin Khalifa University, Qatar
Ramadan Elaiess	University of Benghazi, Libya
Ramayah T.	Universiti Sains Malaysia, Malaysia
Ramazy Mahmoudi	University of Monastir, Tunisia
Ramiro Gonçalves	University of Trás-os-Montes e Alto Douro & INESC TEC, Portugal
Ramon Alcarria	Universidad Politécnica de Madrid, Spain
Ramon Fabregat Gesa	University of Girona, Spain
Ramy Rahimi	Chungnam National University, South Korea
Reiko Hishiyama	Waseda University, Japan
Renata Maria Maracho	Federal University of Minas Gerais, Brazil
Renato Toasa	Israel Technological University, Ecuador
Reyes Juárez Ramírez	Universidad Autonoma de Baja California, Mexico
Rocío González-Sánchez	Rey Juan Carlos University, Spain
Rodrigo Franklin Frogeri	University Center of Minas Gerais South, Brazil
Ruben Pereira	ISCTE, Portugal
Rui Alexandre Castanho	WSB University, Poland
Rui S. Moreira	UFP & INESC TEC & LIACC, Portugal
Rustam Burnashev	Kazan Federal University, Russia
Saeed Salah	Al-Quds University, Palestine
Said Achchab	Mohammed V University in Rabat, Morocco
Sajid Anwar	Institute of Management Sciences Peshawar, Pakistan
Sami Habib	Kuwait University, Kuwait
Samuel Sepulveda	University of La Frontera, Chile
Snadra Costanzo	University of Calabria, Italy
Sandra Patricia Cano Mazuera	University of San Buenaventura Cali, Colombia
Sassi Sassi	FSJEGJ, Tunisia
Seppo Sirkemaa	University of Turku, Finland
Shahnawaz Talpur	Mehran University of Engineering & Technology Jamshoro, Pakistan
Silviu Vert	Politehnica University of Timisoara, Romania
Simona Mirela Riurean	University of Petrosani, Romania
Slawomir Zolkiewski	Silesian University of Technology, Poland
Solange Rito Lima	University of Minho, Portugal
Sonia Morgado	ISCPSI, Portugal

Sonia Sobral	Portucalense University, Portugal
Sorin Zoican	Polytechnic University of Bucharest, Romania
Souraya Hamida	Batna 2 University, Algeria
Stalin Figueroa	University of Alcala, Spain
Sümeyya Ilkin	Kocaeli University, Turkey
Syed Asim Ali	University of Karachi, Pakistan
Syed Nasirin	Universiti Malaysia Sabah, Malaysia
Tatiana Antipova	Institute of Certified Specialists, Russia
Tatianna Rosal	Universtiy of Trás-os-Montes e Alto Douro, Portugal
Tero Kokkonen	JAMK University of Applied Sciences, Finland
The Thanh Van	HCMC University of Food Industry, Vietnam
Thomas Weber	EPFL, Switzerland
Timothy Asiedu	TIM Technology Services Ltd., Ghana
Tom Sander	New College of Humanities, Germany
Tomaž Klobučar	Jozef Stefan Institute, Slovenia
Toshihiko Kato	University of Electro-communications, Japan
Tuomo Sipola	Jamk University of Applied Sciences, Finland
Tzung-Pei Hong	National University of Kaohsiung, Taiwan
Valentim Realinho	Polythecnic of Portalegre, Portugal
Valentina Colla	Scuola Superiore Sant'Anna, Italy
Valerio Stallone	ZHAW, Switzerland
Vicenzo Iannino	Scuola Superiore Sant'Anna, Italy
Vitor Gonçalves	Polythecnic of Bragança, Portugal
Victor Alves	University of Minho, Portugal
Victor Georgiev	Kazan Federal University, Russia
Victor Hugo Medina Garcia	Universidad Distrital Francisco José de Caldas, Colombia
Victor Kaptelinin	Umeå University, Sweden
Viktor Medvedev	Vilnius University, Lithuania
Vincenza Carchiolo	University of Catania, Italy
Waqas Bangyal	University of Gujrat, Pakistan
Wolf Zimmermann	Martin Luther University Halle-Wittenberg, Germany
Yadira Quiñonez	Autonomous University of Sinaloa, Mexico
Yair Wiseman	Bar-Ilan University, Israel
Yassine Drias	University of Algiers, Algeria
Yuhua Li	Cardiff University, UK
Yuwei Lin	University of Roehampton, UK
Zbigniew Suraj	University of Rzeszow, Poland
Zorica Bogdanovic	University of Belgrade, Serbia

Contents

Digital Transformation and Artificial Intelligence

Digital Marketing and Communication, Technologies, and Applications

Data Mining and Machine Learning in Smart Cities

**Information Systems and Technologies for Digital Cultural Heritage
and Tourism**

Information Systems and Technologies for the Steel Sector

**Novel Computational Paradigms, Methods and Approaches in
Bioinformatics**

**Open Learning and Inclusive Education Through Information and
Communication Technology**

Digital Transformation and Artificial Intelligence

Polarization and Similarly of News in Portugal and the Philippines

Joao T. Aparicio[1(✉)], Thomas J. Tiam-Lee[2], and Carlos J. Costa[3(✉)]

[1] INESC-ID, Instituto Superior Técnico, Universidade de Lisboa, Rua Alves Redol 9, 1000-029 Lisbon, Portugal
joao.aparicio@tecnico.ulisboa.pt
[2] De La Salle University, 2401 Taft Avenue, Malate, 1004 Manila, Philippines
thomas.tiam-lee@dlsu.edu.ph
[3] Advance/CSG, ISEG (Lisbon School of Economics and Management),
Universidade de Lisboa, Lisbon, Portugal
cjcosta@iseg.ulisboa.pt

Abstract. The structure of social media platforms allows us to extract meaningful information regarding the dynamics and relationships of real-world actors. In this paper, we used Twitter data to better understand the landscape of news organizations in Portugal and the Philippines through network science techniques. Our findings showed similarities and differences between the two countries. For instance, we found higher levels of similarity in tweet content across news outlets in Portugal. However, in terms of retweeter behavior, we found more polarization on the Philippine news outlets.

Keywords: Twitter · SNA · Network Communities · News · Behaviour · Network Science

1 Introduction

The role of the internet and social media in news consumption and dissemination has grown in recent years. According to the Reuters Institute Digital News Report, consumption of traditional news media (e.g. TV, print) has declined further in almost all markets globally in 2022, while online and social media are on the rise [20]. In the United States, social media has outpaced print newspaper as a news source according to a study by Pew Research Center [23]. In order to keep up with this trend, substantial efforts have been done by traditional news outlets to establish and extend their online presence, especially in social media [14,19,25].

The presence of news organizations in online social media platforms provides interesting opportunities to explore, analyze, and visualize the landscape of news consumption and dissemination. The inherent structure of social media platforms not only allows for data to be extracted more easily; it can also be used to infer relationships between different actors (news outlets, consumers) within the space

A. Rocha et al. (Eds.): WorldCIST 2023, LNNS 802, pp. 3–12, 2024.
https://doi.org/10.1007/978-3-031-45651-0_1

[6]. Therefore, exploring the social media footprint of news organizations can help reveal interesting insights that would otherwise be difficult to extract.

In this paper, we use data and network science techniques to answer a set of research questions pertaining to news organizations in two geographically distant countries: Portugal and the Philippines. Our goal is to better understand the social media network of news organizations and reveal similarities and differences between the two countries. To do this, we extracted and analyzed data from the official Twitter accounts of major news organizations from Portugal and the Philippines. We chose Twitter for its established user base in both countries, as well as the ease of pulling data through its public API. Moreover, Twitter is known to be a go-to resource for reporters due to its low cost and high efficiency [8]. In line with this, polarization, centrality, and exposure to various forms of propaganda and misinformation on Twitter may be used to describe networks on the platform [17]. However, the principles presented in this paper can be extended to other social media platforms as well.

We chose the comparison of news landscape on Twitter on both these countries because they are quite far apart in terms of news trust [20]. Where in the Philippines the overall trust scores is 37% and in Portugal is much higher 61%. Not only that but the perception of political polarization in news is double in the Philippines with 32% of people thinking that outlets are politically far apart. In Portugal that number is 16% [20]. Since the authors are from both the countries mentioned there is wider perception of the underlying cultural differences between these.

There a need to further explore the differences in trust and polarization perspectives across the countries [11]. Using Twitter data we hypothesize that the relationship between the follows and retweets in news related Twitter accounts may also mirror the polarization and relevance aspects. We measure these using network models of follow and retweet behavior and network clustering of news accounts.

The objective of this paper is to grasp the dynamics of news organizations in Twitter, namely on Portugal and the Philippines. We identify the particular questions which can help us better understand the online news ecosystem and verify common connotations within each region:

- How do news organizations influence each other within Twitter, through follow and similarity?
- What is the relationship between the number of followers and follow behavior with other news accounts?
- Can we identify patterns within news organizations in terms of common retweeters?

2   Related Work

We used Scopus to further understand current trends in Twitter news research. We performed a query using the expression `Twitter AND Network AND News`.

As a result, we obtained 1896 documents. In Vosviewer, we selected co-occurrences. The words have to occur at least five times. Then, 670 words were obtained. However, then we filtered some words that occurred many times but were not related to the task at hand, such as "Facebook", "social media", "social media networks" and "social networking". The result is six groups. Current trends in Twitter show three significant clusters related to Twitter analysis. The keywords are based on: on the left, (in green) Machine learning focused research. The data science approach is in the middle (in red), and the social science approach is on the right (in blue) (Fig. 1).

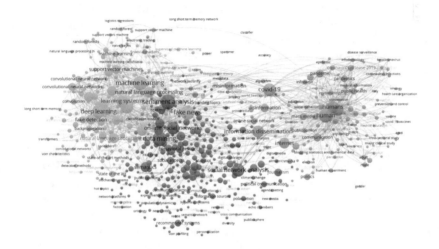

Fig. 1. Trends in Twitter news network analysis

Twitter is known to be a go-to resource for reporters due to its low cost and high efficiency [8]. In line with this, polarization, centrality, and exposure to various forms of propaganda and misinformation on Twitter may be used to describe networks on the platform [17]. Several researchers analysed the behaviour of twitter users in the context of news outlets [3,13,18,22,24]. For example Hayat et al. [18] analyse if the demographics of Twitter followers of Al-Jazeera may be associated with a specific political choice. Soares et al. [24] studied the Brazilian political context and analysed the polarization of the tweets related to presidential candidates. Papadopoulou et al. [21] have used network representation to trace misinformation in social networks like Twitter. His research is mainly focused on the tools design for further usage by researchers and journalists. Unfortunately, there is no focus on understanding patterns inherent in the data, and the visualizations implemented are unsuitable due to unnecessary complexity and uninformative representations.

3   Materials and Methods

To understand follow behaviour between different news outlets we build a network using the follower-followed relationship between each pair of news account. Afterwards we applied a community finding algorithm to better generalize similar behaviour with groups. This algorithm maximizes intra-cluster modularity,

$$Q = \sum_{c=1}^{n} \left[\frac{L_c}{m} - \gamma \left(\frac{k_c^{in} k_c^{out}}{2\,m} \right)^2 \right] \tag{1}$$

where, L_c are the links within a community, c is the number of communities, m is the total number of edges, γ is the resolution parameter, $k_c^{in} k_c^{out}$ are the in and out degrees of the nodes in a community individually summed and then multiplied. The optimiziation was then defined by Clauset et al. [10].

We also relate it to the number of followers each news account has with the number of other news outlets who are followers.

To understand the influence criteria we calculate the similarity of the last 2000 tweets of each news outlet with every other. Using the timestamp of each tweet we can understand who is influenced by who. The similarity is calculated after removing stop words of each of the native languages and stemming afterwards. The tweet similarity is calculated based on TF-IDF to get the pairwise cosine similarity between tweets. The calculation is similar to what Fócil-Arias et al. [15] presented. We then built a network outlet similarity using only tweets with at least 80% similar. Were the Nodes are news accounts and the edges mean that there are tweets with over 80% similarity. The weight of said tweets is the number of tweets that have high similarity between each pair of accounts.

Using the retweeter accounts of each news outlet we map a network of retweeters and outlets to understand if the behaviour is common around each news outlet. To do this, we identified the set of tweets whose retweets are at least a given threshold. We used a threshold of 100 for the Philippine news outlets and a threshold of 10 for the Portuguese news outlets. The discrepancy is to adjust for the difference in Twitter population between the two countries, which is 1:10 [20]. From this, we identify accounts who have often retweeted these set of tweets and generate a graph which visualizes the common retweeters and the news outlets that they retweet from.

The data was extracted using the Twitter API. The selected accounts were based on the available accounts from the news outlets studied in the 2022 Reuters Institute for the study of Journalism report [20]. From each of the acconts, the most recent 2000 tweets were extracted. The employs the POST-DS methodology [12] and is focused on a prescriptive data analytics task [7].

4   Results

The proposed methodology was applied to the extracted Twitter data. However, the data extracted was task specific.

4.1   How Do News Organizations Influence Each Other Within Twitter, Through follow and Similarity?

Using the set of previous 2000 tweets we generated a pairwise distance with temporal constraints to generate the following graphs (Fig. 2).

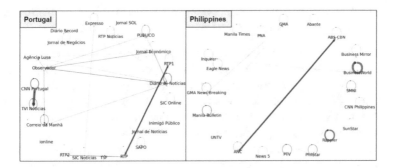

Fig. 2. Similarity of Tweets from each news outlet in the Philippines and Portugal. The edge weights represent similarity and the arrows point to the outlet with the later posts.

The groups of newspapers were clustered according to the community finding algorithm described. Using this method we can see that the accounts mainly influence other accounts within the same brand, like [RTP, RTP1, RTP2], [SIC Noticias, Expresso, Inimigo Público] and [TVI Noticias, CNN Portugal]. Other patterns we see are: Observador being heavily influenced by 6 other news outlets. This is expected since the name of the newspaper literally translates to "Observer". We also see that a sizable amount of the outlets is pulling news from Agência Lusa.

Given the nature of the analysis we opted for a full pairwise similarity of the tweets to also understand how self similar each outlet is. The weight of the eliptical edges tells us just that. Outlets like Diario de notícias, TVI Noticias and CNN Portugal are quite self similar, or in other words, repetitive. While others like TSF Rádio, Inimigo Público, Diario Record, Jornal Sol and ionline are less self similar. We can also observe that Agência Lusa is influenced by Jornal de Noticias.

Unlike the Portuguese scenario, the tweets from the Twitter accounts of Philippine news outlets did not show strong similarities with one another. This means that the community structure was not cohesive enough and no meaningful communities were generated. The only exception was ABS-CBN News and ANC. This is expected as the two accounts are owned by the same organization and ABS-CBN News would often retweet from ANC. Outlets that exhibit very strong self-similarities include Rappler and Business World. In Rappler's case, they often repost their older tweets under "ICYMI" (in case you missed it). Similarly, Business World often has duplicate tweets as a result of them retweeting through some reporters.

It is note worthy that the Portuguese news accounts show lower levels self similarity with 8 out of 23 v.s. 2 out of 20. On the other hand the news dependence between outlets seems to be much less of an issue in the Philippines. Since the time constrained similarity shows overall slightly lower values.

4.2   What is the Relationship Between the Number of Followers and Follow Behavior with Other News Accounts

Using the follow to follow relationship graph we can understand who is following who and devise clusters of common following groups. Figure 3 shows us that directed graph, where the color is the community detected for each news outlet using following information and the size of the circle of each node is the in degree (or the number of other news outlets that follow that account). In this sense we can hypothesize that having a more news outlets following a certain account is a measure of importance and prestige in the news community. Firstly, we see that TVI Notícias has a very low in degree. In context, this may be the case because this is mainly known as a television channel and that may be their main platform. On the other hand, we see that Inimigo Público that is owned by Expresso, is still on the RTP, Púbilco cluster. Outlets with a main focus on online media, like: ionline, SAPO and Observador are in the same cluster. SIC Noticias and TVI that compete directly are in the same cluster. Jornal de Noticias, Diário de Noticias and TSF Rádio Noticias all belong to the Global media group [1], and are in the same cluster. We also see the pattern of following the competition since this group is in the cluster of SIC Noticias.

In the case of the Philippines, we see a less uniform distribution of in-degrees (seen by the sizes of the circles), suggesting a larger discrepancy in online influence. In the Philippines, there are a few big news outlets in the online space, with a lot of other news outlets with varying degrees of online focus. PTV has a very high in degree since it is the main broadcaster of the Philippines government [16], just like RTP in Portugal [2].

In Fig. 5, we see the relationship between the in degree and the Follower count. The node size is the out-degree. We can see a that in Portugal the relationship between the in-degree and the total number of followers is between linear and supra linear. We can also see a Log relationship between the number of followers and the in degree in the Philippines. With the exception of PTV, which may have a high relevance for other news outlets but not necessarily to the Philippines Twitter community. This is also the case for RTP 1 and 2 in Portugal. A noticeable outlier in the case of the Philippines is ABS-CBN, a large media broadcasting company that has the highest follower count, but has the smallest following among news organizations. A possible explanation for this is a controversial feud between the government and the network [9]; other news organizations may have tried to distance themselves from the organization to avoid complications (Fig. 4).

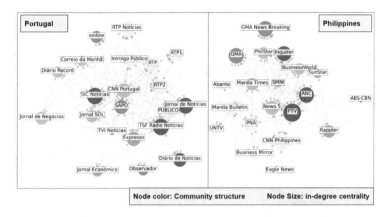

Fig. 3. News accounts follow in degree relationship. Bigger circles represent more follows. Portugal news on the left and Philippines on the right

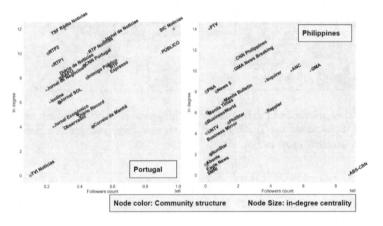

Fig. 4. Relationship between number of other news outlets following and the total number of followers. Portugal on the left, Philippines on the right.

4.3 Can We Identify Patterns Within News Organizations in Terms of Common retweeters?

Figure 5 show the retweet networks for Portugal and the Philippines. In these graphs, the red nodes represent the news outlets and the blue nodes represent the accounts that frequently retweet from these news outlets. The weight of the edges represent the amount of times that account has retweeted a popular tweet (tweets exceeding a threshold of retweets) from that news outlet.

There is a noticeable difference in terms of the polarization of the two networks. In the case of the Philippines, there is an apparent divide between two groups of news outlets. On the upper-right region, we see several accounts that often retweet from SMNI News as well as PTV and PNA. On the other side, we see other accounts that often retweet from the other news outlets. In the case of Portugal, the network is less polarized, with the retweet distribution more evenly spread out.

Interestingly, in the case of the Philippines, the polarization is consistent with the perceived alignments of the said news organizations. SMNI News is widely considered to be in support with the incumbent government, while PTV and PNA and state-owned news organizations. On the opposite end, the other news outlets are generally viewed to be critical of the government.

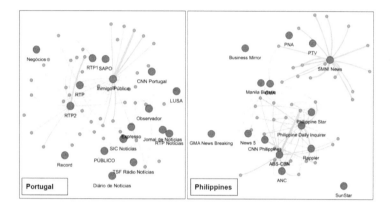

Fig. 5. Retweet network of Portuguese and Philippine news outlets. Red nodes are the news outlet, blue nodes are the frequent retweeters, edge weights are the frequencies of retweet from each news outlet.

To further understand the dynamics of polarization and communicability in this context a view into network robustness and other dynamic network properties may be analysed [4,5].

5 Conclusion

In this paper, we generated insights on the news organization landscape in Twitter for Portugal and the Philippines using network science techniques. The results showed similarities and differences between the two countries. In terms of the tweet content, there appears to be more similarity across news outlets in Portugal. However, in terms of retweeter behavior, there appears to be more polarization on the Philippine news outlets. For both countries, the organizational and user following is positively related, however, state ownership impacts organizational following. The results show the potential of social media platforms in extracting dynamics of communities like news organizations. We believe such insights are useful in understanding real-world communities better, which can equip decision and policy-making with knowledge to oppose information polarization by creating links in ideologically polarized communities. Future work on this area include the application of such analysis to other domains such as celebrities and niche communities, more detailed analysis of tweet content using more advanced natural language techniques.

Acknowledgements. We gratefully acknowledge financial support from FCT - Fundação para a Ciência e a Tecnologia (Portugal), national funding through research grant UIDB/04521/2020. This work is also supported by national funds through PhD grant (UI/BD/153587/2022) supported by FCT.

References

1. GMG quem somos. https://www.globalmediagroup.pt/o-grupo/quem-somos/
2. RTP relatório de contas 2021. https://cdn-images.rtp.pt/mcm/pdf/fd1/fd176c7e8e430880c96b4ab837af0dad1.pdf
3. Alzahrani, S., Gore, C., Salehi, A., Davulcu, H.: Finding organizational accounts based on structural and behavioral factors on twitter. In: Thomson, R., Dancy, C., Hyder, A., Bisgin, H. (eds.) SBP-BRiMS 2018. LNCS, vol. 10899, pp. 164–175. Springer, Cham (2018). https://doi.org/10.1007/978-3-319-93372-6_18
4. Aparicio, J.T., Arsenio, E., Henriques, R.: Assessing robustness in multimodal transportation systems: a case study in Lisbon. Eur. Transp. Res. Rev. **14**, 1–18 (2022)
5. Aparicio, J.T., Arsenio, E., Santos, F.C., Henriques, R.: LINES: multimodal transportation resilience analysis. Sustainability **14**, 7891 (2022)
6. Aparicio, J.T., de Sequeira, J.S., Costa, C.J.: Emotion analysis of Portuguese political parties communication over the COVID-19 pandemic. In: 2021 16th Iberian Conference on Information Systems and Technologies (CISTI), pp. 1–6. IEEE (2021)
7. Aparicio, S., Aparicio, J.T., Costa, C.J.: Data science and AI: trends analysis. In: 2019 14th Iberian Conference on Information Systems and Technologies (CISTI), pp. 1–6. IEEE (2019)
8. Broersma, M., Graham, T.: Twitter as a news source: how Dutch and British newspapers used tweets in their news coverage, 2007–2011. Journal. Pract. **7**, 446–464 (2013)
9. Cabato, R.: Philippines orders its largest broadcaster off the air as nation fights virus (2020)
10. Clauset, A., Newman, M.E., Moore, C.: Finding community structure in very large networks. Phys. Rev. E **70**, 066111 (2004)
11. Costa, C., Aparicio, M., Aparicio, J.: Sentiment analysis of Portuguese political parties communication. In: The 39th ACM International Conference on Design of Communication, pp. 63–69 (2021)
12. Costa, C.J., Aparicio, J.T.: Post-DS: a methodology to boost data science. In: 2020 15th Iberian Conference on Information Systems and Technologies (CISTI), pp. 1–6. IEEE (2020)
13. Dahlan, K., Terras, M.: A social network analysis of the oceanographic community: a fragmented digital community of practice, Preservation. Digit. Technol. Cult. (PDT&C) **49**, 159–181 (2020)
14. David, C.C., Tandoc, E.C., Katigbak, E.: Organizational adaptations to social media: how social media news workers in the Philippines are embedded in newsrooms and influences on editorial practices. Newsp. Res. J. **40**, 329–345 (2019)

15. Fócil-Arias, C., Ziiniga, J., Sidorov, G., Batyrshin, I., Gelbukh, A.: A tweets classifier based on cosine similarity. In: Working notes of CLEF 2017-Conference and Labs of the Evaluation Forum, Dublin, Ireland, pp. 11–14 (2017)
16. Gita-Carlos, R.A.: PCOO reorganized, renamed as office of the press secretary
17. Guarino, S., Trino, N., Celestini, A., Chessa, A., Riotta, G.: Characterizing networks of propaganda on twitter: a case study, Applied Network. Science **5**, 1–22 (2020)
18. Hayat, T., Samuel-Azran, T., Galily, Y.: Al-Jazeera sport's us twitter followers: sport-politics nexus?. Online Inf. Rev. (2016)
19. Ju, A., Jeong, S.H., Chyi, H.I.: Will social media save newspapers? Examining the effectiveness of Facebook and twitter as news platforms. Journal. Pract. **8**, 1–17 (2014)
20. Newman, N., Fletcher, R., Robertson, C., Eddy, K., Nielsen, R.K.: Reuters institute digital news report 2022. Reuters Inst. Study Journal. (2022)
21. Papadopoulou, O., Makedas, T., Apostolidis, L., Poldi, F., Papadopoulos, S., Kompatsiaris, I.: MeVer NetworkX: network analysis and visualization for tracing disinformation. Future Internet **14**, 147 (2022)
22. Samuel-Azran, T., Hayat, T.: Counter-hegemonic contra-flow and the Al Jazeera America fiasco: a social network analysis of Al Jazeera America's twitter users. Glob. Media Commun. **13**, 267–282 (2017)
23. Shearer, E.: Social media outpaces print newspapers in the U.S. as a news source (2020)
24. Soares, F.B., Recuero, R., Zago, G.: Asymmetric polarization on twitter and the 2018 Brazilian presidential elections. In: Proceedings of the 10th International Conference on Social Media and Society, pp. 67–76 (2019)
25. Tsuriel, K., Dvir Gvirsman, S., Ziv, L., Afriat-Aviv, H., Ivan, L.: Servant of two masters: how social media editors balance between mass media logic and social media logic. Journalism **22**, 1983–2000 (2021)

Multidimensional and Multilingual Emotional Analysis

Sofia Aparicio[1]([⊠]), Joao T. Aparicio[1]([⊠]), and Manuela Aparicio[2]([⊠])

[1] INESC-ID, Instituto Superior Técnico, Universidade de Lisboa, Rua Alves Redol 9, 1000-029 Lisbon, Portugal
{sofia.aparicio,joao.aparicio}@tecnico.ulisboa.pt
[2] NOVA Information Management School (NOVA IMS), Universidade Nova de Lisboa, Lisbon, Portugal
manuela.aparicio@novaims.unl.pt

Abstract. In order to monitor informal political online discussions and to lead a better understanding of hate speech on social media, we found that it was necessary to use sentiment quantification for languages with few training datasets. Previous studies mainly rely on languages with enough data to train a model. Several statistical and machine learning models were produced and compared in three languages (English, Portuguese and Polish). This work shows promising results when inferring sentimental dimensions, even in languages other than English.

Keywords: Emotional ratings of text · Affective norms · Long Short-Term Memory · Recurrent Neural Networks · Machine learning

1 Introduction

Sentiment analysis relies on the necessity to extract either negative and positive evaluation or estimate emotion. Two leading families of methods have been developed to represent human emotions [8]. One is categorical, based on six universal basic emotions (BE) [7]. The other is dimensional, advocating continuous numerical values that progress through multiple dimensions [23]. Since it takes significant human resources to annotate words and textual utterances regarding sentiment, it was necessary to produce automatic methods to infer sentiment. Several studies were conducted using convolutional neural networks (CNN) (i.e., that consider the spatial organization of a sentence) and recurrent neural networks (RNN) (i.e. that consider the sequential organization) [1,5,26]. These works were mainly applied to the English language. So, to the best of my knowledge, there is still a gap when using deep learning to quantify sentiment from languages with few or no training resources. The results show that three trained models performed better (Attention Concat, Attention Feature Bassed, and Attention Affine Transformation); however, the Average word-level prediction model also showed promising results. LSTM tends to perform slightly better than CNN models. The difference was more evident in the arousal dimension.

A. Rocha et al. (Eds.): WorldCIST 2023, LNNS 802, pp. 13–22, 2024.
https://doi.org/10.1007/978-3-031-45651-0_2

2 Related Work

2.1 Assigning Emotion to Textual Utterances

In emotion analysis, word-level prediction differs greatly from assigning emotion values to larger linguistic units, such as paragraphs and sentences. [3] recognized three different approaches for emotion detection: keyword-based, learning-based, and hybrid. However, all these methods resort to different linguistic analysis tools (e.g., semantic level, sentence segmentation, parts of speech recognition, token level). However, word-level problem solving cannot solve high-level linguistic prediction because of the way these words are combined [12]. The other use sequential input data, typical for recurrent neural networks (RNN), long short-term memory (LSTM), and general regression neural networks (GRNN). In Alswaidan et al. [1] work, three models were considered, gated recurrent unit followed by CuDNN concatenated with a CNN and a frequency-inverse document frequency (TF-IDF) to better label the text according to emotional categories.

2.2 Spatially and Sequential Architectures

In this study, the primary goals were to evaluate the two models of sentiment representation, namely the dimensional and the categorical models, and determine their applications and expected accuracy. Facebook posts were rated, firstly, considering the valence and arousal dimensions separately. In sum, 2895 messages were evaluated, and VA parameters were compared through the age and gender of the writer, with the authors concluding that female post-writers express more arousal and valence. Later, a two-linear regression model using a BoW representation, on 10-fold cross-validation with this data, reaches a high correlation to the annotated results, obtaining a Pearson correlation of 0.650 and 0.850 for valence and arousal, respectively. With the limited research on the use of sequential input data and the need for more emotionally rated data, [26] started a new investigation. Since CNN! (CNN!) are not ideal for processing sequences and RNN! (RNN!)s perform slowly, the authors induced four sequence-based convolution neural networks (SCNN).

3 Using Neural Word Embeddings for Extending Lexicons of Emotional Norms

We propose thirteen models based on models described in the related work section. All the studies were conducted in six different languages: English, Spanish, Portuguese, Italian, and Polish. In this section, we will start by describing the need for word embeddings, followed by an explanation of the models created in this study.

3.1 Word Embeddings

Word embeddings are vector representations for words, responsible for capturing their semantic or syntactic meaning. Our model used FastText word vectors pretrained on Common Crawl and Wikipedia, which are available in 157 languages.

3.2   Models Exploring Statistics

In this study, we assign sentiment to words considering three dimensions, and we compared four different methods to predict the emotion of each word. One of the models that had a good performance was a simple multi-layer perceptron (MLP) [18], a set of neurons fully connected. The MLP model was built through Keras[1], an open-source library integrated on top of TensorFlow to allow building deep learning models. On the output layer, we have a Dense layer with three neurons, one for each emotional dimension we consider. This MLP was trained through 200 epochs, with a batch size of 64 and an Adam [11] optimizer. We adapted the MLP from our previous work and trained it with datasets affective normas for words from six different languages: English [4,21,24], Spanish [17], Portuguese [22], Italian [15], German[19], and Polish [10].

3.3   Models Exploring Machine Learning

Yann LeCun, inspired by the human visual cortex, discovered by Hubel and Weisel [9], developed the Convolution and Polling architecture [13], also known as Convolutional Neural Networks (CNN). LeCun applied this technique to images, and it was years later that CNN was applied to NLP. The main goal of CNN is to detect patterns across space, by firing when a determined pattern of words compared to a determined filter.

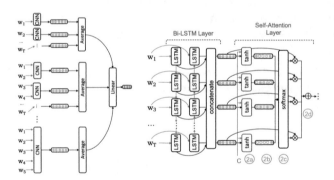

Fig. 1. Convolution and Polling operations applied to a sentence (Left) and BiLSTM and Attention (Right)

With the CNN model, we made four minor alterations. The first model is identical to the one shown in Fig. 1 on the left. In the next three, we applied the MLP model. First, passed the embeddings through the MLP model, and they were the input to the Convolution Layer. Second, we applied the MLP to the output of the Convolution Layer, and after applying the Average Pooling. In third place, we applied the MLP model in the end, after the linear operation.

---

[1] https://github.com/keras-team/keras.

Long Short-Term Memory

Even though CNN has fast performance, LSTM is more successful when working with natural language processing [25], such as sequences of words expressed as time series. In Eq. 1, we can observe all the operations that an LSTM cell requires.

$$s_t = R_{\text{LSTM}}\left(s_{t-1}, x_t\right) = [c_t; h_t]$$
$$c_t = f \odot c_{t-1} + i \odot z$$
$$h_t = o \odot \tanh\left(c_t\right)$$
$$i = \sigma\left(x_t W^{xi} + h_{t-1} W^{hi}\right)]$$
$$f = \sigma\left(x_t W^{xf} + h_{t-1} W^{hf}\right)$$
$$o = \sigma\left(x_t W^{xo} + h_{t-1} W^{ho}\right) g$$
$$= \tanh\left(x_t W^{xz} + h_{t-1} W^{hz}\right)$$
$$y_t = O_{\text{LSTM}}\left(s_t\right) = h_t$$

$$(1)$$

To enhance the position of each word in the sentence [20], we choose to use a Bidirectional LSTM (BiLSTM). The idea is to have two LSTMs traveling through the sentence simultaneously, one that encodes the sentence left to right and, separately, another that travels from the end to the beginning of the sentence. In the end, we concatenate these two representations. This is translated into the BiLSTM Layer of the Fig. 2. However, as Yin et al. [25] referred in their paper, tracing the whole sentence with an LSTM can disregard the keywords. So, align with LSTM, we also used a Self-Attention Layer.

Attention

In our models, we used a Keras SeqSelfAttention layer with a sigmoid attention activation. This layer can be translated into the Self-Attention Layer from Fig. 2 and the Eq. 2.

$$h_i, j = \tanh\left(x_i^\top W_1 + x_j^\top W_x + b_i\right) \tag{2a}$$
$$e_{i,j} = \sigma\left(W_a h_{i,j} + ba\right) \tag{2b}$$
$$a_i = \text{softmax}\left(e_i\right) \tag{2c}$$
$$\text{self}_a\text{ttention}_i = \sum_j a_{i,j} x_j \tag{2d}$$

In Self-Attention, it is first necessary to calculate $h_{i,j}$ (2a) by summing the values of the current position and the previous, all previously multiplied by a weight matrix. After multiplying the values by the alignment weights, we get the alignment scores (2b). On 2c, we apply softmax to the attention scores for the values to vary between 0 and 1 and determine the probability of each given word. At the end (2d), a_i, the amount of attention j^{th} should pay to i^{th} input, and summing all the results.

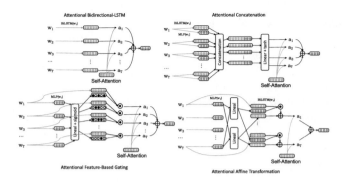

Fig. 2. Proposed models applying Self-Attention and BiLSTM Layers.

LSTM Models

First, we considered models with LSTM layer and a Self-Attention Layer, as shown in Fig. 2. We also analyzed an alteration to this model, instead of receiving the embeddings as the input, we applied the pre-trained MLP to all the embeddings and provided the operation results to the LSTM Layer.

We also produced three models inspired by the work developed by Margatina et all. [14]. These models were given the names they had in this paper.

Attentional Concatenation

The Attentional Concatenation model, Fig. 2 and Eq. 3, we calculate the BiLSTM of each embedding. In parallel, the MLP pre-trained model was applied for every word of the sentence. Then, we proceed to the concatenation of both operations and pass that concatenation through a Self-Attention Layer. In the end, calculate a Dense Layer with three dimensions to predict the three emotional dimensions.

$$x_1 = \tanh\left(W_c\left[\text{BiLSTM}\left(w_1\right) \| MLP\left(w_i\right)\right] + b_c\right) \tag{3a}$$

$$\text{operations 2a -2d}$$

$$d = l \cdot 3 + b \tag{3b}$$

Attentional Feature-Based Gating

The second method, described in Fig. 2 and Eq. 4, we apply the MLP pre-trained model to the word embeddings and later use linear plus sigmoid operations. Appling the gating mechanism, by applying the sigmoid function, we will have a mask-vector where each value varies between 0 and 1 that will later be applied to the embeddings of each word by an element-wise multiplication, $\odot$. Lastly, we used a Self-Attention Layer.

$$f_g\left(h_i, \text{MLP}\left(w_i\right)\right) = \sigma\left(W_g \text{MLP}\left(w_i\right) + b_g\right) \odot h \tag{4}$$

Attentional Affine Transformation

In the final model 2, the feature-wise affine transformation is applied; in other words, a normalization layer preserving collinearity and ratios of distances. Primarily, we apply the pretrained MLP model to the word embeddings, and enforce a scaling and shifting vector to the results of the MLP. This model, initially inspired by Perez et al. [16], allow to capture of dependencies between features by a simple multiplicative operation. The results of the linear operation γ over the MLP results are later multiplied element-wise with the results from the BiLSTM Layer over the embeddings. After, we add these values to β, and apply a Self-Attention Layer.

$$f_a\left(h_1, \mathrm{MLP}\left(w_i\right)\right) = \gamma\left(\mathrm{MLP}\left(w_i\right)\right) \odot h_i + \beta\left(\mathrm{MLP}\left(w_i\right)\right) \tag{5a}$$

$$\gamma(x) = W_\gamma x + b_\gamma \tag{5b}$$

$$\beta(x) = W_\beta x + b_\beta \tag{5c}$$

4   Experimental Evaluation

This section describes the experiments conducted to infer sentiment from textual utterances. In the third set of experiments, it was necessary to access the result of statistical models to predict the sentiment of textual utterances. One of the models that had a better performance was the MLP. An MLP was pre-trained with seven datasets in different languages. Figure 3 established a correlation between the dimensional distribution of the datasets (Table 1).

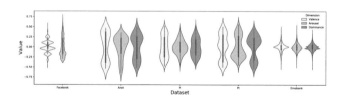

Fig. 3. Comparison of the dimensional distribution of the datasets in several languages.

Table 1. Results obtained for statistical sentiment prediction of textual utterances, in terms of Pearson's correlation coefficient and MAE.

		Pt		Pl		Emobank		ANET		Fb	
		Pearson	MAE	Pearson	MAE	Pearson	MAE	Pearson	MAE	Pearson	MAE
MLP Average	V	**0.686**	0.234	**0.499**	0.227	**0.359**	0.086	0.639	**0.301**	**0.384**	0.154
	A	0.511	0.216	0.222	**0.160**	**0.152**	0.101	**0.542**	0.319	**0.111**	0.234
	D	0.470	**0.238**	0.312	0.187	0.058	0.093	0.261	0.263	–	–
Average MLP	V	0.625	**0.232**	0.429	0.226	0.284	0.073	**0.697**	0.312	0.298	**0.132**
	A	0.342	0.218	0.109	0.187	0.122	0.089	0.433	0.355	0.790	0.237
	D	**0.579**	0.234	**0.436**	0.194	**0.123**	0.122	**0.622**	**0.258**	–	–
Pooling Average MLP	V	0.482	0.256	0.453	0.231	0.201	0.091	0.491	0.323	0.192	0.149
	A	0.187	0.231	0.166	**0.160**	0.110	0.122	0.420	**0.316**	0.79	0.250
	D	0.310	0.257	0.358	**0.183**	0.057	**0.092**	0.397	0.277	–	–
Pooling MLP Average	V	0.537	0.249	0.456	0.231	0.224	0.094	0.492	0.323	0.193	0.148
	A	0.266	0.230	0.168	**0.160**	0.098	0.130	0.420	**0.316**	0.82	0.244
	D	0.405	0.263	0.359	**0.183**	0.068	0.100	0.396	0.360	–	–
MLP Pooling Avg	V	0.339	0.317	0.402	**0.222**	0.083	**0.071**	0.605	0.312	0.137	0.161
	A	0.330	0.253	**0.335**	0.188	0.029	**0.088**	0.515	0.336	0.152	**0.208**
	D	0.219	0.342	0.256	**0.183**	0.039	0.182	0.327	0.268	–	–

The Affective Norms for English Words (Anew) comprised 1,034 unique words. The early work on sentiment analysis is annotated, considering the three dimensions of valence, arousal, and dominance. It was essential to consider other languages and provide richer data such as gender and education level.

4.1 Models Exploring Statistics

Despite the simplicity of the model Average (i.e., the MLP model is applied to each word of the text, and an average of all the outputs is calculated to deliver a final output), it was the model that showed a better performance of the word-level solutions in almost every dataset. All the models tested in these experiments are described in the section above.

4.2 Models Exploring Machine Learning

The set of experiments considering text-level sentiment prediction was conducted using cross-validation. This method allows for the validation of a model (e.g., by calculating its precision) by dividing a dataset into splits, usually between 2 and 5. A number of those splits are used to train the model, and the other is used to validate it. The table displays the results for each model through each dataset, considering Pearson's correlation, MAE and MSE.

The LSTM performs better with classification tasks than the CNN and the BiLSTM with regression tasks. It is possible to see a great improvement in more extensive datasets, such as Facebook. The dimension that was more difficult to tackle was arousal, especially in the Facebook dataset. This work shows better values for the dimension arousal than the work from other teams. The results were obtained using a model composed of Bi-LSTM+MP+Attention.

Table 2. The prediction of valence, arousal and dominance with several models. The training and testing data are textual utterances form datasets English, Polish and Portuguese.

		Pt			Pl			Emobank			ANET			Fb		
		Pearson	MAE	MSE	Pearson	MAE	MSE	Pearson	MAE	MSE	Pearson	MAE	MSE	Pearson	MAE	MSE
LSTM	V	0.641	0.184	0.059	**0.507**	**0.184**	0.055	**0.536**	0.070	0.009	**0.769**	0.207	0.059	0.547	0.100	0.018
	A	0.608	**0.164**	0.047	0.333	0.166	0.034	**0.333**	0.088	0.013	**0.617**	0.188	0.053	0.494	0.177	0.060
	D	0.576	**0.164**	0.056	0.445	0.149	0.042	0.092	0.120	0.065	0.439	0.231	0.082	–	–	–
MLP + LSTM	V	0.319	0.246	0.087	0.258	0.225	0.073	0.150	0.276	0.012	0.236	0.316	0.120	0.065	0.126	0.026
	A	0.232	0.241	0.071	0.108	0.146	0.034	0.016	0.297	0.013	0.254	0.282	0.097	0.126	0.235	0.081
	D	0.345	0.232	0.069	0.296	0.192	0.054	0.022	0.552	0.093	0.112	0.375	0.252	–	–	–
CNN	V	0.632	0.228	0.062	0.415	0.211	0.072	0.434	0.070	0.021	0.672	0.261	0.092	0.495	0.102	0.020
	A	0.312	0.241	0.050	0.241	0.148	0.059	0.170	.087	0.032	0.493	0.221	0.168	0.260	0.212	0.058
	D	0.427	0.234	0.063	0.247	0.235	0.109	0.040	0.258	0.075	0.261	0.329	0.091	–	–	–
MLP + CNN	V	0.584	0.236	0.076	0.397	0.212	0.067	0.466	**0.069**	0.009	0.657	0.249	0.087	0.501	0.109	0.019
	A	0.345	0.221	0.063	0.281	0.146	0.034	0.136	0.089	0.013	0.536	0.204	0.060	0.316	0.215	0.065
	D	0.419	0.227	0.078	0.282	0.202	0.067	0.040	0.251	0.085	0.167	0.410	0.302	–	–	–
CNN + MLP	V	0.552	0.223	0.066	0.395	0.215	0.066	0.449	0.071	0.009	0.523	**0.080**	0.066	0.485	0.107	0.019
	A	0.343	0.219	0.034	0.197	0.145	0.034	0.214	0.088	0.013	0.393	**0.114**	0.058	0.315	0.215	0.067
	D	0.342	0.227	0.061	0.243	**0.147**	0.061	0.066	0.182	0.076	0.408	0.291	0.149	–	–	–
Attention Concat	V	**0.691**	**0.177**	0.057	0.435	0.202	0.064	0.507	0.073	0.010	0.649	0.238	0.007	**0.561**	0.101	0.019
	A	**0.620**	0.165	0.046	0.297	0.144	0.035	0.302	0.089	0.014	0.481	0.209	0.051	**0.565**	0.176	0.052
	D	**0.663**	0.167	0.049	0.348	0.182	0.050	**0.363**	0.122	0.074	0.283	0.276	0.004	–	–	–
Attention Fracture Based	V	0.641	0.184	0.050	0.501	0.192	0.059	0.531	0.069	0.001	0.680	0.226	0.056	0.557	**0.098**	0.021
	A	0.608	0.164	0.042	**0.391**	**0.137**	0.031	0.320	**0.083**	0.014	0.538	0.198	0.051	0.545	**0.174**	0.057
	D	0.576	0.173	0.058	**0.470**	0.160	0.043	0.082	**0.116**	0.065	0.479	**0.217**	0.084	–	–	–
Attention Affine Transformation	V	0.569	0.206	0.072	0.434	0.206	0.065	0.501	0.074	0.010	0.728	0.225	0.010	0.523	0.108	0.057
	A	0.540	0.177	0.050	0.268	0.148	0.036	0.270	0.092	0.015	0.608	0.189	0.056	0.491	0.184	0.436
	D	0.473	0.218	0.075	0.338	0.180	0.051	0.075	0.143	0.067	**0.481**	0.266	0.119	–	–	–

Similar to my results using the Attention Feature Based model, which obtained results of 0.531 for Emobank, 0.557, and 0.545 for Facebook. Comparing the results and considering that my model performed lower, but being trained with several idioms, the lower performance can be justified. Possible applications of this model can be used in multilingual context such as tweets [2,6] (Table 2).

5   Conclusions and Future Work

This research provides three trained models and one word-level model that show promising results compared to the state-of-the-art. An MLP was pre-trained with lexicons from six different languages. Four models that do take into consideration the syntactic structure and do require training were created. LSTMs tend to perform slightly better than CNN models, and this difference was more evident in the arousal dimension.

This study provides, as theoretical implications, a comparison between statistical models and machine learning models. Possible practical applications to the findings in this study could be to monitor informal political online discussions and to lead to a better understanding of hate speech on social media. It could also be interesting to experiment with word embeddings trained on different types of corpora for future work.

Acknowledgement. We gratefully acknowledge financial support from FCT - Fundação para a Ciência e a Tecnologia (Portugal), national funding through research grant UIDB/04152/2020. This work is also supported by national funds through PhD grant (UI/BD/153587/2022) supported by FCT.

References

1. Alswaidan, N., Menai, M.E.B.: KSU at SemEval-2019 Task 3: hybrid features for emotion recognition in textual conversation. In: Proceedings of the 13th International Workshop on Semantic Evaluation, SemEval@NAACL-HLT 2019, Minneapolis, MN, USA, 6–7 June 2019, pp. 247–250 (2019)
2. Aparicio, J.T., de Sequeira, J.S., Costa, C.J.: Emotion analysis of portuguese political parties communication over the COVID-19 pandemic. In: 2021 16th Iberian Conference on Information Systems and Technologies (CISTI), pp. 1–6. IEEE (2021)
3. Binali, H., Wu, C., Potdar, V.: Computational approaches for emotion detection in text. In: 4th IEEE International Conference on Digital Ecosystems and Technologies, pp. 172–177. IEEE (2010)
4. Bradley, M.M., Lang, P.J.: Affective norms for English words (ANEW): instruction manual and affective ratings. Technical report, Technical report C-1, the center for research in psychophysiology ... (1999)
5. Buechel, S., Hahn, U.: Word emotion induction for multiple languages as a deep multi-task learning problem. In: Proceedings of the 2018 Conference of the North American Chapter of the Association for Computational Linguistics: Human Language Technologies, NAACL-HLT 2018, New Orleans, Louisiana, USA, 1–6 June 2018, Volume 1 (Long Papers), pp. 1907–1918 (2018)
6. Costa, C., Aparicio, M., Aparicio, J.: Sentiment analysis of Portuguese political parties communication. In: The 39th ACM International Conference on Design of Communication, pp. 63–69 (2021)
7. Ekman, P.: An argument for basic emotions. Cogn. Emot. **6**, 169–200 (1992)
8. Ekman, P., Friesen, W.V.: Constants across cultures in the face and emotion. J. Pers. Soc. Psychol. **17**, 124 (1971)
9. Hubel, D.H., Wiesel, T.N.: Receptive fields, binocular interaction and functional architecture in the cat's visual cortex. J. Physiol. **160**, 106–154 (1962)
10. Imbir, K.K.: Affective norms for 4900 polish words reload (ANPW_R): assessments for valence, arousal, dominance, origin, significance, concreteness, imageability and age of acquisition. Front. Psychol. **7**, 1081 (2016)
11. Kingma, D.P., Ba, J.: Adam: a method for stochastic optimization. arXiv preprint arXiv:1412.6980 (2014)
12. LaBrie, R.C., Louis, R.D.S.: Information retrieval from knowledge management systems: using knowledge hierarchies to overcome keyword limitations. In: 9th Americas Conference on Information Systems, AMCIS 2003, Tampa, FL, USA, 4–6 August 2003, p. 333 (2003)
13. LeCun, Y., Bengio, Y., et al.: Convolutional networks for images, speech, and time series. Handb. Brain Theory Neural Netw. **3361**, 1995 (1995)
14. Margatina, K., Baziotis, C., Potamianos, A.: Attention-based conditioning methods for external knowledge integration. arXiv preprint arXiv:1906.03674 (2019)
15. Montefinese, M., Ambrosini, E., Fairfield, B., Mammarella, N.: The adaptation of the affective norms for English words (ANEW) for Italian. Behav. Res. Methods **46**, 887–903 (2014)

16. Perez, E., Strub, F., De Vries, H., Dumoulin, V., Courville, A.: FiLM: visual reasoning with a general conditioning layer. In: Thirty-Second AAAI Conference on Artificial Intelligence (2018)
17. Redondo, J., Fraga, I., Padrón, I., Comesaña, M.: The Spanish adaptation of anew (affective norms for English words). Behav. Res. Methods **39**, 600–605 (2007)
18. Ruck, D.W., Rogers, S.K., Kabrisky, M., Oxley, M.E., Suter, B.W.: The multilayer perceptron as an approximation to a bayes optimal discriminant function. IEEE Trans. Neural Netw. **1**, 296–298 (1990)
19. Schmidtke, D.S., Schröder, T., Jacobs, A.M., Conrad, M.: ANGST: affective norms for German sentiment terms, derived from the affective norms for English words. Behav. Res. Methods **46**, 1108–1118 (2014)
20. Schuster, M., Paliwal, K.K.: Bidirectional recurrent neural networks. IEEE Trans. Signal Process. **45**, 2673–2681 (1997)
21. Scott, G.G., Keitel, A., Becirspahic, M., Yao, B., Sereno, S.C.: The glasgow norms: ratings of 5,500 words on nine scales. Behav. Res. Methods **51**, 1258–1270 (2019)
22. Soares, A.P., Comesaña, M., Pinheiro, A.P., Simões, A., Frade, C.S.: The adaptation of the affective norms for English words (ANEW) for European Portuguese. Behav. Res. Methods **44**, 256–269 (2012)
23. Wang, J., Yu, L.-C., Lai, K.R., Zhang, X.: Community-based weighted graph model for valence-arousal prediction of affective words. IEEE/ACM Trans. Audio Speech Lang. Process. **24**, 1957–1968 (2016)
24. Warriner, A.B., Kuperman, V., Brysbaert, M.: Norms of valence, arousal, and dominance for 13,915 English lemmas. Behav. Res. Methods **45**, 1191–1207 (2013)
25. Yin, W., Kann, K., Yu, M., Schütze, H.: Comparative study of CNN and RNN for natural language processing. arXiv preprint arXiv:1702.01923 (2017)
26. Zahiri, S.M., Choi, J.D.: Emotion detection on TV show transcripts with sequence-based convolutional neural networks. arXiv preprint arXiv:1708.04299 (2017)

Explaining Wikipedia Page Similarity Using Network Science

Joao T. Aparicio[1]($\boxtimes$), Valentina Timčenko[2], and Carlos J. Costa[3]($\boxtimes$)

[1] INESC-ID, Instituto Superior Técnico, Universidade de Lisboa,
Rua Alves Redol 9, 1000-029 Lisbon, Portugal
`joao.aparicio@tecnico.ulisboa.pt`
[2] Institute Mihailo Pupin, University of Belgrade, 11000 Belgrade, Serbia
`valentina.timcenko@pupin.rs`
[3] Advance/CSG, ISEG (Lisbon School of Economics and Management),
Universidade de Lisboa, Lisbon, Portugal
`cjcosta@iseg.ulisboa.pt`

Abstract. The relationship between sciences and scientific production is often established in a very organized way, with a top-down approach. But the link between sciences is much more dynamic and organic. To analyze this connection, we use similarity between Wikipedia science pages. We used network science and cluster analysis techniques. Results presented show us the boundary scientific fields between different areas and also suggests that this network is an ultra small world.

Keywords: Science · network science · small world · Wikipedia · similarity network

1 Introduction

In science we often find how new fields and sub-fields are related with one another via a set of broader definitions or applications. To understand the relationships between them, the study of ontologies and their instance via knowledge graphs has been the subject of some studies [9,11,14,15]. From which, we have seen several authors use Wikipedia as a large source of open data [18] to generate knowledge graphs and edit interaction graphs [14,15]. However, the study of meta-science through similarity has not been thoroughly analysed in a complex domain [19].

In this project we are analysing a network which shows the similarities among different branches of science based on Wikipedia pages in outline of natural, formal, social and applied science. We analysed the graph and collected our metrics using Python, powerlaw: A Python Package for Analysis of Heavy-Tailed Distributions[1] and NetworkX [10], the package for network analysis.

The relevance of this research extends to a number of different stakeholders. Publishing institutions may utilize the findings of the identification of emerging research subjects to help them decide whether or not to launch new journals on unique research topics. This makes the identification of emerging research

A. Rocha et al. (Eds.): WorldCIST 2023, LNNS 802, pp. 23–32, 2024.
https://doi.org/10.1007/978-3-031-45651-0_3

topics a very beneficial activity for publishing institutions. For the purposes of watching and monitoring the dynamics of science, policymakers and research-funding organizations will find that the identification of newly developing and transdisciplinary scientific fields is beneficial [16].

With this experiment we intend to answer a set of questions followed by methods (Table 1):

Table 1. Questions and measurements

Question	Measure
Which science has a page with the highest level of similarity?	Degree centrality
What are the sciences that boundary the high similarity sciences?	Eigenvector Centrality
Which science pages bridge different fields?	Betweenness centrality
Is similarity with a page related with similarity with its neighbours?	Cluster Coefficient
Is this network a candidate for a small world? and how would the network grow?	Average Path Length (APL) Power Law fitting of degree distribution

2   Methods

This study follows the POST-DS [7] methodology employing CRISP-DM [17]. The modelling is focused on network representation to understand higher level patterns. We measure different network metrics, analyse and interpret the meaning on this context. All the metrics used in this study are open source [10] and further introduced by [13]. These metrics have been used in on networks in other contexts [2,3]. This research is focused on a prescriptive data analytics task [4].

2.1   Data

The dataset is based on the one extracted by Alberto Calderone in 2020 [5]. As previously mentioned, this dataset depicts the different branches of science on Wikipedia. Nodes represent Wikipedia pages of a science field. An edge between two nodes is their page similarity above a 0.3 threshold according to cosine similarity. In other words, if the pages of node i and node j have a cosine similarity above 0.3, then there is an edge between node i and j. The graph is undirected and has a total of 687 nodes and 6523 edges with an average degree of 18.9898, which means that each field is somewhat similar to 18.9898 other fields on average. To model the information in each node we used the following data structure: `(0, {'name': 'Accounting', 'class':`

'Applied', 'url': 'https://...'}) Each page has a class that can be either Social, Formal, Natural and Applied. The percentage of each is about 33.33% (Pink), 28.09% (Green), 24.89% (Orange) and 13.68% (Blue) respectively. The graph is represented in Fig. 1 using the Fruchterman Reingold algorithm.

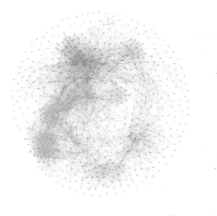

Fig. 1. Graph of the Wikipedia Map of Science, color-coded by Science Class.

By analysing the graph, we encountered five single connected components. Four of those were either pairs or triplets of nodes. These were sciences such as Coastal geography (As both natural and social sciences) and Hydrography; Oceanography and Marine Biology; And Food Science and Nutrition.

3   Results

3.1   Centrality Measurement

How do we classify a branch of science as the most similar to every other science? What are the bridges between different science fields? These are the type of questions we are trying to answer when computing the centrality of the nodes. By intuition, we may conjecture that the amount of links to other nodes would classify the page as important. Due to an impact on a higher number of fields, since they have more similarity and hence more influence on the concepts shared between one another, regardless of the importance of the concepts themselves.

Science with a Page with the highest Level of Similarity. To find the Hub in this network, we will start by comparing different centrality measures. Firstly we calculated the **degree centrality** to understand which science has a page with the highest degree of similarity to the remaining pages. In this case the hub is **School Psychology, as a social science** (https://en.wikipedia.org/wiki/School_psychology). This probably happens because this page has contents from many other sciences, for Education and Advocacy. This does not mean that this is the most influential branch of science. Instead, this may be an indicator of the ease of multidisciplinary integration of this particular sub-field.

Sciences that Boundary the High Similarity Sciences. To know the science that has a higher page similarity to the sciences with the highest degree of page similarity, we use the eigenvector centrality, which **yields the same result as degree centrality**, has similarity to the most prestigious sciences, or with the sciences with the highest similarity degree.

Science Pages Bridge Different Fields. To understand which of the science pages is a bridge between different parts of the network, we use the **betweenness centrality**. In this case this tells us the hub on this centrality measure is the node that connects the most different science pages to each other. In this network that science page is **Population Biology, as a natural science** https://en.wikipedia.org/wiki/Population_biology.

We might hypothesize that this happens because this is a field that derives from the intersection of formal sciences and natural science. The second highest betweenness centrality is **Algorithm, as a formal science** https://en.wikipedia.org/wiki/Algorithm. This is intuitive because they can be applied in the context of any other science. Another science with a high degree of betweenness centrality that can be easily guessed is **Economics** (https://en.wikipedia.org/wiki/Economics) (actually the third highest betweenness centrality). This happens because it is a strong link between the social sciences and the formal sciences.

3.2   Similarity with a Page Related With similarity with Its Neighbours

Usually science fields have similar aspects between one another. By using cluster coefficients we may compute how likely that two neighbor nodes have a connection with each other. In other words, if they create triangles. This metric allows for the depiction of the network by clusters, where nodes from a cluster have similar properties.

To analyze this we computed, using NetworkX, transitivity, clustering and average clustering. The first one computes the fraction of all possible triangles present in the graph (transitivity), and the ones that are in fact connected. The second computes the local cluster coefficient. And finally, the last one computes the global clustering coefficient, given and weighted adjacency matrix A (Fig. 2).

$$C = \frac{\sum_{i,j,k} A_{ij} A_{jk} A_{ki}}{\sum_i k_i (k_i - 1)}, \; where \; k_i = \sum_j A_{ij} \tag{1}$$

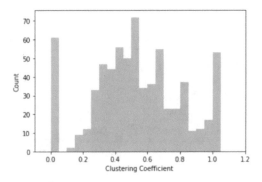

Fig. 2. Cluster coefficient per node.

The Global Clustering Coefficient is about 0.5302 and the transitivity is approximately 0.4692.

As we can observe, the majority of nodes have a cluster coefficient between 0.4 and 0.8, and others are really close to either 1 or 0. This will result in a global clustering coefficient and transitivity of approximately 0.53 and 0.47, respectively, which are both huge. These results imply that our network is very related, which makes sense since we are talking about branches of science, which have inter-class relations, whether it is math, physics, philosophy, etc. Nodes like "Automated reasoning", "Statistical theory", "Time series", have the highest values (approx. 0.984, 0.990, 0,994 respectively), which means that they can be correlated to every branch of science.

Looking at concrete examples and cross referencing with results of betweenness centrality (BC), we observe that pages that bridge different communities - i.e. have the highest BC value - such as "population biology", "algorithm", "economics" also have a very low local clustering coefficient (approx. 0.1328, 0.2387, 0.2664, respectively). This was expected, since these nodes are bridging different communities causing similarity to diminish.

Another interesting aspect is that nodes which are close to the bridges have also a high local clustering coefficient, for instance "population ecology" (cc approx. 0.9) which is close to "population biology" as seen above. This would suggest that these nodes are in fact bridging communities.

3.3 Degree Distribution

The **degree of a node** (or k) is the number of edges that the node has. The **Degree Distribution** is the probability distribution of these degrees over the whole network.

After collecting the degree of each node, we made an histogram with the results. We can observe that the majority of the nodes have only one connection. In practice, this says that each Science Wikipedia page is most likely to only have one other science page with high similarity.

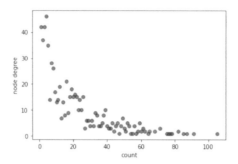

Fig. 3. Network degree distribution

We believe that this dataset also follows a power law distribution, since it is based on the number of Wikipedia pages and they usually have a rapid growth in terms of cardinality and reference. So we tested this hypothesis, to **understand how the network grows**.

Power Law. An important thing that we thought to analyse in our dataset was the question of, theoretically, how would our network grow?

A first intuitive interpretation of our problem: when a new Science arises, the tendency is to have new branches or derivations of that same science. Which makes a probable power growth in the number of Wikipedia Science pages. As they as well follow the creation of new science fields.

For this reason, we decided to analyze the dataset, using the power law package [1], and the follow up article on it, Statistical Analyses Support Power Law Distributions Found in Neuronal Avalanches [12]. The Fig. 3, shows us the distribution of degree probability in our network and in a power law. On the y axis the p(k) - degree distribution and on the x axis the k, denoting number of nodes in the network.

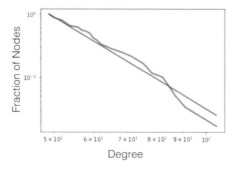

Fig. 4. Power law fit: target in blue, result in red.

The slope of our growth is $\gamma = 1.414412955754193$, following the power law formula:

$$p(x) \propto x^{-\gamma} \tag{2}$$

Typically the scaling parameter lies in the range $2 < \gamma < 3$, however there are occasional exceptions, which we believe to be in one of them. More often the power law applies only for values greater than some minimum $Xmin$. In such cases we say that the tail of the distribution follows a power law. [6] The γ fit lets us understand if this network is a small world, i.e. a sort of network where most nodes are not neighbours but can be accessed with few hops. In this case we find that the data described follows an ultra small world behaviour (i.e. since $\gamma < 2$). This is interesting because,

$$k_{\max} = k_{\min} N^{\frac{1}{\gamma-1}},$$

i.e. K_{max} (maximum degree) grows faster than N (number of nodes). This also means that the average $<k>$ and the variance $<k^2>$ will diverge. The implication of this is that the similarity of the science pages is related to the clear preferential attachment. As new science pages are added to the network, there is a tendency for them to be similar to the higher degree pages, and the preexisting pages also get more similar with those pages. This is interesting because it means that over time science pages get more and more similar.

Given the γ value smaller than 2, we may conclude that the tail of the distribution may follow a power law, and the network does not behave like a random network, see Fig. 4.

3.4 Average Path Length (APL)

In this case, the average shortest path length is the smallest amount of hops I have to take, on average, to be on another science field page, within Wikipedia. We could hypothesize that **the smaller the APL the more similar** (in terms of cosine similarity) **all science fields are**, since they are more interrelated with one another.

To calculate this we only used one of the single connected components and did not use the four pairs and triplets that were disconnected from the remaining massive (relatively speaking) component with 677 nodes. The average shortest path length, approximately is 3.433. Since each link represents a higher similarity between 2 pages, this means that on average the science pages are fairly indirectly similar with every other page, with the exception of the disconnected components. In other words, we can change subjects/classes with only 3.4 hops on average, meaning that pages are fairly similar.

The two of the small-world properties are obtained [8] by having a small APL and a high clustering coefficient, which is the case of our network. This implies that the growth of the APL scales in a logarithmic way with the growth of the network.

$$L \propto logN \tag{3}$$

In other words, by adding more pages to our network the average path length would not differ much from what it is now. In addition, these pages would also be highly correlated to the pages already in the network since the global clustering coefficient is also high.

3.5   Network Based Clusters of Scientific Pages

Using the network topology, a distance was defined based on number of edges, on the shortest path between two nodes, and respective heights summed. It was possible identifying the distance between different sciences within clusters. A dendogram allowed the visual identification of potential sciences groups, considering the linkage between pages.

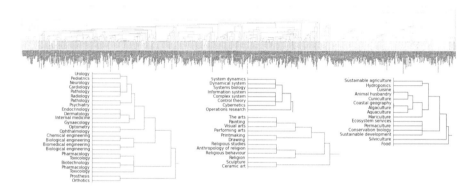

Fig. 5. Science clusters dendogram based on topological distance

In Fig. 5 we see the full dendogram on top and four samples of subclusters formed bellow: Medicine sub fields, complex systems and related sciences, art fields, and Sustainable development and related fields. This is useful representation for grouping sciences together based on how they are described.

4   Conclusions

Summarizing the main ideas on this analysis, we can conclude that it is probable that a new science page being added to the graph would have a high probability to have a similarity to only one other science page. This means that the degree distribution has a very high probability with k = 1. Given the results of the power law analysis, we can conclude that this network does not behave like a random network, and is a ultra small-word network. This result can be corroborated by the results from the cluster coefficient and average path length analysis. The analysis of the centrality of our network were interesting and surprising. The School Psychology page has the highest similarity and with most similar neighbors to other science Wikipedia pages. The nodes that have the highest betweenness centrality, i.e., are important bridges between the science pages, do not have a high degree centrality but they have key edges to those who have. Some science pages with this characteristic are Population Biology, Economics and Algorithms. Also, we got a high clustering coefficient (≈ 0.5), which means that sciences neighbors are very likely to be connected. Combining this with the relatively small APL, creates the small-world property that shows us the slow growth of the APL when the graph increases its size. This shows us that there is a tendency for an increasingly interdisciplinary scientific knowledge. In the future, the scientific field similarity can be studied to better understand gaps ease of collaboration in different fields.

Acknowledgements. We gratefully acknowledge financial support from FCT - Fundação para a Ciência e a Tecnologia (Portugal), national funding through research grant UIDB/04521/2020. This work is also supported by national funds through PhD grant (UI/BD/153587/2022) supported by FCT.

References

1. Alstott, J., Bullmore, E., Plenz, D.: PowerLaw: a python package for analysis of heavy-tailed distributions. PLoS One **9**(1), e85777 (2014)
2. Aparicio, J.T., Arsenio, E., Henriques, R.: Assessing robustness in multimodal transportation systems: a case study in Lisbon. Eur. Transp. Res. Rev. **14**(1), 1–18 (2022)
3. Aparicio, J.T., Arsenio, E., Santos, F.C., Henriques, R.: LINES: multimodal transportation resilience analysis. Sustainability **14**(13), 7891 (2022)
4. Aparicio, S., Aparicio, J.T., Costa, C.J.: Data science and AI: trends analysis. In: 2019 14th Iberian Conference on Information Systems and Technologies (CISTI), pp. 1–6. IEEE (2019)
5. Calderone, A.: A Wikipedia based map of science (2020). https://doi.org/10.6084/m9.figshare.11638932.v5
6. Clauset, A., Shalizi, C.R., Newman, M.E.: Power-law distributions in empirical data. SIAM Rev. **51**(4), 661–703 (2009)
7. Costa, C.J., Aparicio, J.T.: Post-DS: a methodology to boost data science. In: 2020 15th Iberian Conference on Information Systems and Technologies (CISTI), pp. 1–6. IEEE (2020)
8. Davidsen, J., Ebel, H., Bornholdt, S.: Emergence of a small world from local interactions: modeling acquaintance networks. Phys. Rev. Lett. **88**(12), 128701 (2002)
9. Gregorowicz, A., Kramer, M.A.: Mining a large-scale term-concept network from Wikipedia. Technical report, Mitre Corp Bedford, MA (2005)
10. Hagberg, A., Swart, P., S Chult, D.: Exploring network structure, dynamics, and function using NetworkX. Technical report, Los Alamos National Lab. (LANL), Los Alamos, NM (United States) (2008)
11. Hussain, M.J., Wasti, S.H., Huang, G., Wei, L., Jiang, Y., Tang, Y.: An approach for measuring semantic similarity between Wikipedia concepts using multiple inheritances. Inf. Process. Manage. **57**(3), 102188 (2020)
12. Klaus, A., Yu, S., Plenz, D.: Statistical analyses support power law distributions found in neuronal avalanches. PLoS One **6**(5), e19779 (2011)

13. Lewis, T.G.: Network Science: Theory and Applications. Wiley, Hoboken (2011)
14. Moro, A., Navigli, R.: WiseNet: building a Wikipedia-based semantic network with ontologized relations. In: Proceedings of the 21st ACM International Conference on Information and Knowledge Management, pp. 1672–1676 (2012)
15. Ponzetto, S.P., Strube, M.: Knowledge derived from Wikipedia for computing semantic relatedness. J. Artif. Intell. Res. **30**, 181–212 (2007)
16. Wang, Q.: Studies in the dynamics of science: exploring emergence, classification, and interdisciplinarity. Ph.D. thesis, KTH Royal institute of Technology (2016)
17. Wirth, R., Hipp, J.: CRISP-DM: Towards a standard process model for data mining. In: Proceedings of the 4th International Conference on the Practical Applications of Knowledge Discovery and Data Mining, Manchester, vol. 1, pp. 29–39 (2000)
18. Wu, F., Weld, D.S.: Open information extraction using Wikipedia. In: Proceedings of the 48th Annual Meeting of the Association for Computational Linguistics, pp. 118–127 (2010)
19. Yang, P., Colavizza, G.: A map of science in Wikipedia. In: Companion Proceedings of the Web Conference 2022, pp. 1289–1300 (2022)

Crosscheck Information—The Digital Resilience and Contribute of PMO

Helcio Mello[1]([⌂]) and Carlos J. Costa[2]

[1] ISEG, Lisbon School of Economics and Management, Universidade de Lisboa, Lisbon, Portugal
helcio@phd.iseg.ulisboa.pt
[2] Advance/CSG, ISEG (Lisbon School of Economics and Management), Universidade de Lisboa, Lisbon, Portugal
cjcosta@iseg.ulisboa.pt

Abstract. Through a comprehensive study of the literature, this paper examines the fundamentals of digital resilience and the contribution of the Project Management Office (PMO) and adaptive capacity within organizations. The main objective was to identify the main dimensions, proximity, and convergence between project management and information generating center (PMO), information systems, and digital resilience. In addition, a bibliometric study was performed to identify the co-occurrences, dimensions, and cluster analysis. The results are presented in this article, and where 4 clusters were identified and at the end a conceptual framework is presented.

Keywords: Project Management Office · Digital Resilience · IS/IT Management

1 Introduction

The decision to implement a PMO within an organization conforms to principles of maturity, experience in project management practices, and alignment of business objectives to the project portfolio, which must necessarily be reflected in the role and structure of the PMO. The PMO must possess advanced capabilities to manage multiple projects [5, 8]. Likewise, the PBO (project-based organizations) have been widely recognized as autonomous organizations with recognized value in knowledge generation and innovation [12, 22, 24]. Still, in [12], the difficulty of transferring this knowledge base and innovation for strategic use within the organization is observed. This expertise is restricted to the project limit. In the role of the PMO, the need for advanced capabilities configuration is perceived, as indicated by [8]. The applied methodology and dynamic and reliable systems for cross-referencing information are crucial for the expected results to be achieved in the success and management of the project portfolio [6, 16]. According to [10, 23], many organizations choose and perceive the PMO as a working structure to improve project management maturity and the chances of project success. Since the founding of PMI in 1987, it is evident that the evolution of theories and practices has been spread through the various organizations worldwide, which within the most different levels have

A. Rocha et al. (Eds.): WorldCIST 2023, LNNS 802, pp. 33–40, 2024.
https://doi.org/10.1007/978-3-031-45651-0_4

been converging to a constant interest in this topic and applicability in their organizations [15, 21]. However, perhaps not yet so well perceived by many organizations and project managers is the contribution that the adaptive capacity of the PMO to integrate the various areas, cross-referencing information for management reports during stable periods as well as in periods of instability and crisis, being a strategic entity and output for the governance of the organization and success of project portfolio management [20]. In this way, the primary goals are the following: (1) Finding how digital resilience and DT fit into the PMO context; (2) identifying the main components related to the PMO in bibliographic research; (3) identifying compact dimensions between PMO and digital resilience.

2   Perspectives PMO

2.1   Contribute

According to [11], companies have been increasingly investing in technologies that enable them to gain a competitive advantage and have improved capabilities to face a dynamic and highly competitive market. Similarly, the project portfolio's success suggests the PMO's contribution to better implementing the organization's strategy. However, [5], indicate that this contribution may fail in the long term. Furthermore, he adds that the absence of a theory of adaptation and change in PMO configuration increases the possibility of failure from a long-term perspective. He notes that more recent research should be better analyzed to understand and capture the relationship between organizational context and dynamics in a more advanced PMO typology.

Similarly, there is an understanding that the concept of project success derives from the multidimensional perceptions of the many stakeholders, and this characteristic is widely perceived within the industrial sectors [4, 9, 19]. Similarly, the importance of the PMO for the NPD is noted, where strategic areas of the economy are highlighted, such as the construction and IT industries [4]. Regardless of the configuration, structure, and methodology the company wants to adopt for the PMO, what is important is the alignment of the IT infrastructure according to the company's strategy and objectives and clearly communicating within the various levels of the company how projects will be managed and the goals to be achieved.Again, the critical thing to realize is that the adoption and absorption of DT by firms is what makes it possible to decide to employ an A, B, or C configuration according to the size of the project, resources, and organization. In [13], observed that DT enables this adaptive capacity and is strategically fundamental for resilience, growth, and response to the market, partners, and investors, adjusting with flexibility, methodology, and maturity in project management.

2.2   Digital Resilience

The resilience of a business is associated with the ability to respond to its customers and business partners in periods of crisis [18]. This response capacity is related to the projects contracted and developed by the company. Investing in digital technologies is a transformational factor that has become crucial for product development, service innovation, and firm survival [18]. Recent years, in particular the pandemic period, have shown how digital technologies and the ability to implement digital transformation are crucial to the organization's future, regardless of whether it is a company involved in technology products and services [25]. During this same period, organizations had to discontinue their workforce, and at the same time, workers and companies had to adapt to working remotely, managing their projects with information sharing and results [17]. This time gap created a circumstance of powerfully disruptive change. All functional areas of business organizations were impacted. All companies that were able to adapt quickly to the market and partners through the use of the technological infrastructure and digital platforms were the ones that had success and digital resilience.

3   Methodology

This paper was developed according to the following steps and methodology. First, the research was conducted through the Scopus database, where the published articles were searched with "Project Management Office" using the following query TITLE-ABS-KEY ("project management office"). In this step, 359 articles were found. Secondly, the articles already published in the final stage and in Conferences were selected. After this step, have been found 343 papers. Next, the papers were arranged with the highest number of citations in descending order. In addition, an exhaustive study was conducted within the literature with subsequent bibliometric analysis using the software program VOSviewer, which van Eck and Waltman developed to extract the results and data for this paper (Fig. 1).

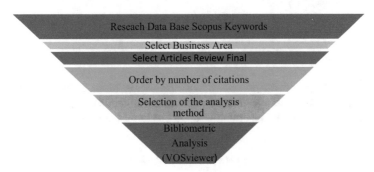

Fig. 1. Selection method and filter

Then the next step was to choose the type of analysis method to select the main authors, citations, and link strength. This method aims to understand better the strength and connection between the various authors and the proximities between the proposed themes. Next, a bibliographic data map was generated considering a co-occurrence analysis having as a unit of analysis all the words associated with a minimum number of 5 occurrences per keyword. To conclude, the main titles and words were identified using cluster analysis to associate terms and abstracts. In the end, we suggested titles for each block of the cluster, originated from the network map, and developed a conceptual work proposal. This paper follows the applied method observed by [13].

4   Results

A map based on the titles and abstracts was generated using VOSviewer software with 7426 terms, 189 met thresholds and a minimum of 10 co-occurrences. A total of 113 words were selected, showing the strength between them and their relationship of proximity (Fig. 2).

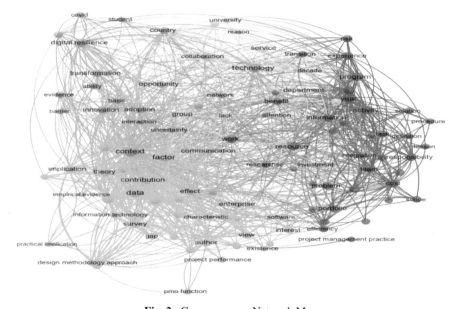

Fig. 2. Co-occurrence Network Map

The most cited papers are presented in the following Table. Those papers addressed various subjects, from change management to sustainability. One topic that is also especially relevant is the management of large and complex projects (Table 1).

Table 1. The most cited publication in "Project Management Office".

Cit.	Link	Docs	Description
999	25	23	This paper presents a theoretical contribution to the study of organisational project management and of the project management office (PMO) [1]
555	9	7	The paper presents an investigation of the creation and the reconfiguration of project management offices (PMOs) as an organisational innovation [14]
299	7	7	This paper presents empirical results from a research on Project Management Offices (PMO) in transition [2]

(continued)

Table 1. (*continued*)

Cit.	Link	Docs	Description
140	13	7	This paper aims at positioning organizational design as an important phenomenon in the field of project management with a high potential of contributing to organizational theory [1]
103	12	7	Project-based organisations have emerged as new forms of organisation in the last few decades [3]
9	17	6	Today's project management offices (PMOs) in the construction sector need to be equipped with breakthrough capabilities necessary for making a difference in multi-project management [8]
9	17	6	Environmental and ecological impacts associated with construction activities have become an ever-increasing concern, especially the considerable amount of waste generated on construction sites [7]
7	15	5	Purpose: The purpose of this study is twofold: first, to identify major project management (PM) complexities in principal construction contracting, and second, to study the contribution of project management offices (PMOs) to addressing such complexities [8]

We identified four clusters supported in the analysis performed and illustrated in Fig. 3. Those clusters allow grouping the papers into the following groups: PMO Advanced, Strategic Alignment, Digital Transformation, and External Factors.

This way, we can infer and identify association clusters and dimensions, which can be deepened in future research. To a conceptual proposal based on the findings of this study, we present below a suggestion that can serve as a basis for companies and new research on the PMO and its importance within business organizations. The dimensions represented in the conceptual model below are grounded according to Adaptive, Information Systems, and Strategic theories (Fig. 4).

Cluster1 PMO Advanced	Cluster2 Strategic Alignment	Cluster 3 Digital Transformation	Cluster 4 External Factors
- activity	- characteristic	- ability	- addition
- benefit	- communication	- adoption	- attention
- best practice	- contribuiton	- barrier	- community
- cost	- data	- collaboration	- demand
- definition	- design	- context	- department
- effectiviness	methodology	- country	- effort
- effiency	approach	- covid	- end
- execution	- effect	- digital resili-	- evolution
- experience	- empirical	ence	- government
- goal	evidence	- element	- group
- information	- enterprise	- evidence	- lack
- integration	- existence	- innovation	- person
- lesson	- focus	- interaction	- reason
- organizational	- form	- member	- service
structure	- gap	- network	- training
- place	- implication	- opportunity	- university
- portfólio man-	- importance	- resilience	
agement	- influence	- student	
- practioner	-information	- technology	
- problem	techonology	- transformation	
- procedure	- interest	- uncertainty	
- product	- interview		
- program	- investment		
- project man-	- organisation		
agement oficce	- perception		
- project man-	- pmo function		
agement profis-	- practical impli-		
sional	cation		
- project portfolio	- project		
management	performance		
- recomendation	- project sucess		
- research	- project team		
- responsibility	- questionaire		
- risk	- relantionship		
- scope	- software		
- solution	- survey		
- task	- theory		
- team	- view		
- technology			
- trasition			

Fig. 3. Cluster Analysis

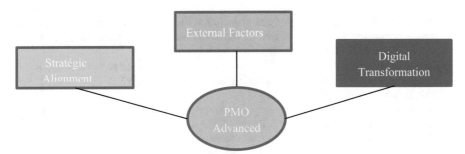

Fig. 4. Conceptual Framework

5 Conclusion

This paper analyzed the principles of digital resilience and the role of the Project Management Office (PMO) and adaptive capacity within businesses. The primary goal was to discover the critical dimensions, proximity, and convergence between project management and the information generation center (PMO), information systems, and digital resilience. To achieve this objective, we performed a literature review. Specifically, we conducted a bibliometric investigation to detect co-occurrences, dimensionality, and cluster analysis. It allowed identifying four clusters. The main output of this paper is a conceptual framework.

Acknowledgments. We gratefully acknowledge financial support from FCT - Fundacao para a Ciencia e a Tecnologia (Portugal), national funding through research grant UIDB/04521/2020.

References

1. Aubry, M., Hobbs, B., Thuillier, D.: A new framework for understanding organisational project management through the PMO. Int. J. Proj. Manage. **25**(4), 328–336 (2007)
2. Aubry, M., Müller, R., Hobbs, B., Blomquist, T.: Project management offices in transition. Int. J. Proj. Manage. **28**(8), 766–778 (2010)
3. Aubry, M., Richer, M.C., Lavoie-Tremblay, M.: Governance performance in complex environment: the case of a major transformation in a university hospital. Int. J. Proj. Manage. **32**(8), 1333–1345 (2014)
4. Barbalho, S.C.M., Silva, G.L.: Control of project data and team satisfaction as results of PMO effort in new product development projects. Int. J. Manag. Proj. Bus. **15**(1), 121–149 (2022)
5. Bredillet, C., Tywoniak, S., Tootoonchy, M.: Why and how do project management offices change? A structural analysis approach. Int. J. Proj. Manage. **36**(5), 744–761 (2018)
6. Duarte, R., Deschamps, F., de Lima, E.P., Pepino, A., Clavijo, R.M.G.: Performance management systems for project management offices: a case-based study. Procedia Manuf. **39**, 923–931 (2019)
7. Ershadi, M., Jefferies, M., Davis, P., Mojtahedi, M.: Achieving sustainable procurement in construction projects: the pivotal role of a project management office. Constr. Econ. Build. **21**(1), 45–64 (2021)

8. Ershadi, M., Jefferies, M., Davis, P., Mojtahedi, M.: Project management offices in the construction industry: a literature review and qualitative synthesis of success variables. Constr. Manag. Econ. **39**(6), 493–512 (2021)

9. Fernandes, G., Pinto, E.B., Araújo, M., Machado, R.J.: The roles of a programme and project management office to support collaborative university–industry R&D. Total Qual. Manag. Bus. Excell. **31**(5–6), 583–608 (2020)

10. Fesenko, T., Fesenko, G.: Developing gender maturity models of project and program management system. Eastern-Eur. J. Enterp. Technol. **1**(3–85), 46–55 (2017)

11. Gomes, J., Romão, M.: Improving project success: a case study using benefits and project management. Procedia Comput. Sci. **100**, 489–497 (2016)

12. Hadi, A.: Facilitating multidirectional knowledge flows in project-based organizations: the intermediary roles of project management office. Int. J. Syst. Innov. **7**(1), 66–86 (2022)

13. Hajishirzi, R., Costa, C.J., Aparicio, M., Romão, M.: Digital transformation framework: a bibliometric approach. In: Rocha, A., Adeli, H., Dzemyda, G., Moreira, F. (eds.) WorldCIST 2022. LNNS, vol. 470, pp. 427–437. Springer, Cham (2022). https://doi.org/10.1007/978-3-031-04829-6_38

14. Hobbs, B., Aubry, M., Thuillier, D.: The project management office as an organisational innovation. Int. J. Proj. Manage. **26**(5), 547–555 (2008)

15. Jugend, D., Barbalho, S.C.M., da Silva, S.L.: Contributions of the project management office to product portfolio management. Producao **26**(1), 190–202 (2015)

16. Ko, J.H., Kim, D.: The effects of maturity of project portfolio management and business alignment on PMO efficiency. Sustain. (Switz.) **11**(1) (2019)

17. Kohn, V.: How the coronavirus pandemic affects the digital resilience of employees. Association for Information Systems IS in the Workplace and the Future of Work How the Coronavirus Pandemic Affects the Digital Resilience of Employees. December 2020 (2022)

18. Mehedintu, A., Soava, G.: A structural framework for assessing the digital resilience of enterprises in the context of the technological revolution 4.0. Electron. (Switz.) **11**(15) (2022)

19. Monteiro, A., Santos, V., Varajão, J.: Project management office models - a review. Procedia Comput. Sci. **100**, 1085–1094 (2016)

20. Pastor-Sanz, L., et al.: Managerial framework for a large multi-centre clinical trial within an EU-funded collaborative project – the "PREVIEW" case study. J. Res. Adm. **52**(2), 15–50 (2021)

21. Pirotti, A., Rahim, F.A.M., Zakaria, N.: Implementation of project management standards and project success: the mediating role of the project management office. J. Eng. Proj. Prod. Manage. **12**(1), 39–46 (2022)

22. Sergeeva, N., Ali, S.: The role of the project management office (PMO) in stimulating innovation in projects initiated by owner and operator organizations. Proj. Manag. J. **51**(4), 440–451 (2020)

23. Sithambaram, J., Nasir, M.H.N.B.M., Ahmad, R.: A compilation of factors associated to the governance and management of agile projects: a systematic literature review. Malays. J. Comput. Sci. **34**(3), 266–307 (2021)

24. Teller, J., Unger, B.N., Kock, A., Gemünden, H.G.: Formalization of project portfolio management : the moderating role of project portfolio complexity. JPMA **30**(5), 596–607 (2012)

25. Zeng, X., Li, S., Yousaf, Z.: Artificial intelligence adoption and digital innovation: how does digital resilience act as a mediator and training protocols as a moderator? Sustain. (Switz.) **14**(14) (2022)

Digital Marketing and Communication, Technologies, and Applications

ESG in Advertising Narratives: Case Analysis of Golden Lion Winning Campaigns at Cannes 2022

Daniel Ladeira de Araújo[1] (iD), Jorge Esparteiro Garcia[2,3,4(✉)] (iD),
Manuel José Serra da Fonseca[2,5] (iD), and José Gabriel Andrade[6,7] (iD)

[1] Instituto Superior de Ciências Empresariais e do Turismo, Porto, Portugal
[2] Instituto Politécnico de Viana do Castelo, Viana do Castelo, Portugal
jorgegarcia@esce.ipvc.pt
[3] ADiT-LAB, Instituto Politécnico de Viana do Castelo, Viana do Castelo, Portugal
[4] INESC TEC, Porto, Portugal
[5] UNIAG - Unidade de Investigação Aplicada em Gestão, Viana do Castelo, Portugal
[6] Universidade do Minho, Campus de Gualtar, 4710-057 Braga, Portugal
[7] CECS - Centro de Estudos de Comunicação e Sociedade, Braga, Portugal

Abstract. The key question that will be addressed in this research is whether advertising narratives feature initiatives within the ESG - Environment, Social and Governance - framework. Investors increasingly value companies that promote sustainable impact actions, something that the promoting companies report in detail in their ESG reports directed at corporate stakeholders. To identify and understand how brands present such initiatives to the consumer market, we checked campaigns awarded with the Golden Lion of Advertising at Cannes 2022. The methodological cut involved three award categories, being analyzed 19 campaigns from 10 different countries. Besides the bibliographical survey on discourse analysis and content analysis, this exploratory qualitative and quantitative research had the inductive method to direct the individual analysis of each campaign. As a result, we identified that more than half of the campaigns address aspects related to ESG initiatives, being the social scope addressed by most of the campaigns. A result that provokes new hypotheses and possibilities of research continuity.

Keywords: ESG · advertising narrative · sustainability · social media · outdoor

1 Introduction

The main objective of this study is to investigate whether content related to the ESG criteria - Environment; Social; Governance - is present in advertising narratives. This paper is part of a developing study and is based on case analyses of advertising campaigns that won the Golden Lion at Cannes 2022. The questions answered in this preliminary study involve two points of analysis: i. Is there the presence of elements that dialogue with ESG criteria in the analyzed campaigns? ii. Which of the ESG aspects are addressed in the campaigns? Global events, such as the 27th session of the Conference of the

Parties on Climate Change - COP27, organized by the United Nations Organization (UNO), highlight in a large part of the world media the aspects related to the planet's sustainability. Nations, companies, and civil society mobilize themselves in the debate, promoting individual and collective goals of socially responsible impact. However, it is not only climate issues that are under debate. The business world is increasingly attentive to and values all aspects related to socially responsible business. For this reason, brands seek and should seek more and more concrete actions and the promotion/communication of these actions within ESG criteria. Following the example of COP27, the construction of a sustainable world involves several players. In addition to governments, society and brands, investors increasingly value companies with ESG strategies.

Thus, many brands have already realized the strategic importance of ESG actions and of communicating these actions to the stakeholders involved. Nassar and Pereira [1] point out that communication about how sustainability is part of an organization may involve specific reports. As an example, we can cite those that adopt the GRI (Global Reporting Initiative) standards; those of the SASB (Sustainability Accounting Standards Board), the System B - a methodology of metrics to measure the company's positive impact and the ISE B3 (Corporate Sustainability Index). It is worth noting the distinction between the formats and standards of reports involving ESG, produced to serve stakeholders such as investors and regulatory agencies, as stated by Puriwa and Tripopsakul [2] and other audiences such as end consumers who may have a different perception of the company's sustainability performance, since ESG reports do not necessarily reflect on marketing communication. The production of these reports is a reality in the financial environment and is increasingly necessary for doing business. For George and Schillebeeckx [3], since the Covid-19 pandemic, investors have realized more intensely that the global pandemic crisis that began in 2020 may be just the preamble to other systemic disruptions and therefore need transparency regarding corporate positioning in relation to environmental and, we may add, social and governance activism.

If financial market players are expressing increasing interest in these actions and receive communications in the report formats cited, what will be the level of interest of the consumer market? This study under development seeks to understand how in the relationship between brands and consumers ESG-related aspects emerge. If advertisements are tools that brands use to generate and/or fulfill consumer desires, would these consumers be interested in socially responsible issues? There are a few avenues to investigate this question, and in this study, we seek to punctually analyze advertising discourse and narratives to identify whether brands address initiatives involving ESG criteria in their advertising campaigns. For Puriwat and Tripopsakul [2], individuals develop perceptions about brands from external factors such as cultural context, inter-personal relationships, whether family, friendship or professional. This is how customers develop attitudes towards brands. They define whether they like, trust, and the degree of loyalty towards them. Thus, positive attitudes in customers' perception are built about companies that follow ethical, social, and ecologically responsible practices. Identifying the presence of these elements in advertising narratives can be an important indication of how much interest their consumers show in the subject. In this sense ESG actions properly communicated to the companies' stakeholders are important for building the reputation of corporations. For Pereira [4], communication plays a central role in this process of reputation building in a market in constant change and with huge challenges, where speeches, but especially practices, are today what allow companies to position themselves differently.

However, it is in the advertising relationship that brands seek to impact their audiences, whether in branding or sales processes. If themes such as the Environment, Social and/or Governance emerge in these narratives, this somehow dialogues with the receivers of the message in a positive way. This is not a study about the reception of the message, but an analysis of the emission of advertising content by brands to their stakeholders, both in off-line environments (Outdoor campaigns) and in digital environments (campaigns involving Social & Influencer and Mobile Experience). The remainder of this paper is organized as follows: next Section presents the Development of the proposed approach. Section 3 presents the results obtained. Conclusions, and future directions are presented in Sect. 4.

2   Development

The development of this study involves the following methodological assumptions:

   i. as to the purpose we approached the basic strategic research;
  ii. as to the objectives it involves the descriptive research;
 iii. the approach of this article is qualitative and quantitative;
  iv. the inductive method;
   v. in the procedures we resorted to the bibliographical research and the case study.

Regarding item i. the methodological choice of basic strategic research involves what Gil [5] conceptualizes as a research format that involves the acquisition of new knowledge aimed at solving practical problems. With this approach we worked with the descriptive research methodology which, according to Barros and Lehfeld [6], involves an observational study as well as descriptive recording and analysis of the characteristics and variables of the process or phenomenon. In this way, descriptive research enables the construction of scenarios from percentage analyses and is important for later qualitative analyses. We used the qualitative-quantitative approach for the preparation of this preliminary study. The quantitative study involved the analysis of advertising campaigns in three categories, which move between the digital marketing environment and offline.

The quantitative approach emerges from the qualitative analyses, namely using methodological resources of discourse analysis and narrative content analysis. In conjunction with this approach, we use the inductive method, to analyze the presence of ESG content in advertising narratives. We used literature research, to understand the ESG criteria, the methodologies of discourse analysis and narrative content analysis; as well as we used the case study procedure, specifically observing the case of the three cited categories of the campaigns awarded at Cannes 2022 with the Golden Lion.

2.1   Case Study

The case study involved the analysis of the categories: Outdoor; Social & Influencer and Mobile Experience. There were 19 campaigns analyzed, with those addressing ESG criteria subdivided according to their respective approach. Although in a first analysis we can include only the Social & Influencer and Mobile Experience categories, in the scope of digital marketing campaigns, the 10 campaigns analyzed in the Outdoor category are part of integrated digital marketing campaigns, involving impacts on the online environment from the offline experience. The process of choosing the categories and campaigns involved, for this ongoing study, two basic criteria. In the case of the curation of the campaigns, we analyzed content that had won awards at the Cannes Festival, an element that to some extent ensures the quality of the material chosen for analysis. In the case of the process of choosing the categories the choice involved their relationship with digital marketing, being in a direct way the categories of Social & Influencer and Mobile Experience, and in an indirect way the Outdoor category. It is worth mentioning that this study is still in progress, where it is intended in other publications to expand the basis of analysis for all categories.

Table 1. Analysis of the winning campaigns of the Golden Lion in Cannes 2022. Source: Self-elaboration.

Category	Sector	Title	Brand	ESG approach
Outdoor	Media/Entertainment	The Art of Stealing	Netflix	No
Outdoor	Food & Drink	Better with Pepsi	Pepsi	No
Outdoor	Promotial Items & Printed Media	Scratchboards	Activision	No
Outdoor	Special Build	Liquid Billboard	Adidas	Social
Outdoor	Special Build	The Refugee Jatoba	APIB	Environmental
Outdoor	Live Advertising and Events	The Art of Self Examination	MACMA	Social
Outdoor	Single-market Campaign	Nonartificial Mexico	Burger King	Social
Outdoor	Breaktrough on a Budget	Wilmore Funeral Home	Starmed Healthcare	Social
Outdoor	Corporate Purpose & Social Responsibiliy	Flags of Generosity	Cadbury	Social
Outdoor	Corporate Purpose & Social Responsibiliy	Plastic Fishing Tournament	AB INBEV/Corona	Environmental
Social & Influencer	Food & Drink	The Foamy Haircut	Brahma	No

(*continued*)

Table 1. (*continued*)

Category	Sector	Title	Brand	ESG approach
Social & Influencer	Automotive	BMW: Nothing But Sheer Joy	BMW	No
Social & Influencer	Not-for-profit/Charity/Government	SOS Children's Village India	Chatpat	Social
Social & Influencer	Influencer Marketing > Innovative Use of Influencers	Lu from Magalu	Magazine Luiza	No
Social & Influencer	Social Content Marketing > Social Film	The Lost Class	Change the Ref Inc	Social
Social & Influencer	Excellence in Social & Influencer > Multiplataform Social Campaign	The First Meta Sneaker	Under Armour	Social
Mobile Experience	mCommerce	A Chat away from Everything	Itaú Bank	No
Mobile Experience	mCommerce	The Eye Tracker	Supermax Supermarkets	Social
Mobile Experience	Content for User Engagement	Burger Glitch	Burger King	No

For this research paper it was necessary the theoretical contribution of several authors who address in their works issues related to the process of discourse analysis and narratives. For Charaudeau [7], the discourse analyst must observe from a distance the communicative phenomenon in question and, from this observation, seek to interpret how the meaning of what is said and not said, whether in images, silences, sounds and gestures, is constructed. In the case of the analyses in question, the distance occurs in a temporal way, since all the campaigns had already been aired before the analysis. There was also a geographic distance since campaigns from 10 different countries were analyzed. While the distance is important for the analysis, it can represent a difficulty, because it is necessary to understand the contexts. The context surrounding the advertising campaign may be contained in the narratives in a polyphonic way, as Bakhtin [8] bpoints out about the analysis of discourses. This element compels the analyst to a sensitive look at the plurality of voices present in the same ad. It is in this way that it is possible to understand the meaning and perform a proper analysis.

The use of the inductive method to analyze these campaigns requires some caveats. Authier-Revuz [9], states that within a discourse element such as humor and irony, for example, make analysis difficult. The focus of a discourse and content analysis process must take into consideration the reference to other texts, phrases, and words, as Orlandi [10] states. Since we analyzed campaigns in 10 different countries, it is possible that such specificities were not contemplated in the individual reading of each content. Thus, only more evident ESG aspects, regardless of the analyst's context, were contemplated for the analysis of the campaigns.

3   Results

Table 1 summarizes the analysis of the campaigns, with the last column highlighting whether there is the presence of elements we identified as ESG initiatives of the respective brands. Also, in this same column we identify in which ESG aspect the campaign fits. From the content analysis of the 19 campaigns described in the table above, we identified that 57.89%, or 11 campaigns in absolute numbers, have mention or content that demonstrates that the brand has some action under the ESG criteria. We emphasize that for this study there is only the identification of the ESG approach present in each of the campaigns analyzed. The analysis of the discursive layers presents in each of them, involving the image, the audio and the intertexts present, involves future studies as a continuation of this publication. Thus, the detailing of the ESG actions of each campaign will be contained together with the qualitative and in-depth analysis of each one since the purpose of this present study is to quantitatively identify the presence and initiative of brands in promoting their ESG actions to their respective stakeholders. Thus, future studies should deepen this theme, as well as the analysis processes.

It is worth highlighting 3 aspects that can be deepened in future studies: i. Expansion of the analyzed categories; ii. Qualitative analysis with the description of the discursive layers presents in a larger number of campaigns; iii. Comparative analysis with previous years' edition of the same festival and categories, to identify the growth curves of the ESG theme in brand communication.

Among these campaigns the social scope is preponderant, representing 81.82% of the ads. In second place, aspects related to the Environment represent 18.18%, with 2 campaigns in absolute numbers. In this case, in one of the campaigns the advertiser is an institution that represents the indigenous peoples of Brazil - APB/Articulation of the Indigenous Peoples of Brazil. Thus, it is not a commercial brand that promotes ESG-related aspects in the sense of building a relationship with investors and/or final consumers. This further reduces the number of commercial brand campaigns with the ESG approach to the environment. It is worth noting that of the 11 campaigns that contain content promoting ESG actions, none addressed the scope of Governance.

3.1   Discussion

By identifying that a little more than half of the analyzed campaigns address ESG aspects in their narratives, and that those that do address them do so largely in the social sphere, it is possible to trace the hypothesis that this theme in the content generates a connection with the audience receiving the message. For Owolabi [11], the observation of consumers' emotional aspects offers brands important elements in the process of creating marketing strategies. Thus, the context of the end of the Covid-19 pandemic, coupled with the international economic and geopolitical crises, may justify this investment of brands in including issues related to social aspects in their advertising narratives. It is possible to question whether the analyzed brands do not develop actions in the Environment and Governance fields, and therefore do not use this advertising narrative. We can also question if from a strategic point of view the marketing advertising narratives mostly emphasized social aspects to generate connection with the receptor audience.

The answer to these questions lacks an in-depth study, however it is possible to identify that some of the analyzed brands, which promoted social actions in their narratives, have a series of ESG initiatives in the Environmental and Governance spheres, in addition to Social. This is the case, for example, with the Under Armour brand, which makes clear in its 2021 sustainability and impact report, publicly available on its website, its goals, and actions, aligned with the Global Reporting Initiative (GRI) and the Sustainable Accounting Standard Board (SASB). Something similar also occurs with the brand Burger King Mexico, winner of the Gold Lion in the Outdoor category, where it highlights social aspects in its advertising narrative. However, this company discloses on its website, in a specific area for investors, the Restaurant Brands for Good report, where it discloses ESG actions in the three aspects: Social, Environmental and Governance.

4   Conclusions

Based on methodologies involving discourse analysis and content analysis, it was possible to map the ESG approach in advertising campaigns that won the Golden Lion in Cannes 2022, in three categories. The result proves the hypothesis about the presence of such narratives in advertising and shows that ESG is part of more than half of the analyzed campaigns but raises several possibilities for the continuity of this research.

We highlight three possibilities that require specific research: i. analyze the same award categories in previous years to see if there is an increase in the promotion of ESG initiatives in advertising campaigns; ii. Broaden the base of analysis to all award categories to obtain more precise data and indicators to corroborate the hypotheses raised; iii. Conduct research with the advertising agencies that created the campaigns to understand the reasons for the strategic choice of highlighting ESG initiatives in the social sphere and not in actions in the Environmental or Governance spheres. This study concludes that companies' ESG actions are also communicated to the public through advertising narratives, but the emphasis is mostly on social issues.

References

1. Nassar, P., Pereira, V.: ESG and its communication in organizations in Brazil. In: 25th International Public Relations Research Conference (2022)
2. Puriwat, W., Tripopsakul, S.: From ESG to DESG: the impact of DESG (digital environmental, social, and governance) on customer attitudes and brand equity. Sustainability (2022)
3. George, G., Schillebeeckx, S.J.D.: Digital transformation, sustainability, and purpose in the multinational enterprise. J. World Bus. **57**(3) (2022)
4. Pereira, M.: Comunicação: identidade e reputação. In: Comunicação nas PME: do conhecimento ao reconhecimento, Lisboa, Politécnico de Lisboa-ESCS, pp. 15–35 (2022)
5. Gil, A.C.: Como elaborar projetos de pesquisa – 5th edn., Atlas, São Paulo (2010)
6. Barros, A.J.P., Lehfeld, N.A.S.: Projeto de Pesquisa: propostas metodológicas. Vozes, Petrópolis (1990)
7. Charaudeau, P.: Discurso das mídias. Contexto, São Paulo (2013)

8. Bakhtin, M.: Estética da criação verbal, 6th edn. WMF Martins Fontes, São Paulo (2011)
9. Authier-Revuz, J.: Entre a transparência e a opacidade: um estudo enunciativo do sentido. EDIPUCRS, Porto Alegre (2004)
10. Orlandi, E.P.: Análise de discurso: Michel Pêcheux. Pontes Editores, Campinas (2017)
11. Owolabi, A.B.: Effect of consumers mood on advertising effectiveness. Eur. J. Psychol. **5**(4), 118–127 (2009)

The Digitalization of the Event Industry – Mobile and Internet Applications as a Tool to Improve Event Communication and Experiences: A Case Study of a French Event App Start-Up

António Cardoso[1] , Jorge Esparteiro Garcia[2,3,4]([✉]) , Manuel Sousa Pereira[2] ,
and Stécy Nasri[5]

[1] I3ID, Universidade Fernando Pessoa, Porto, Portugal
[2] Instituto Politécnico de Viana do Castelo, Viana do Castelo, Portugal
jorgegarcia@esce.ipvc.pt
[3] ADiT-LAB, Instituto Politécnico de Viana do Castelo, Viana do Castelo, Portugal
[4] INESC TEC, Porto, Portugal
[5] Universidade Fernando Pessoa, Porto, Portugal

Abstract. Following the lack of existing studies about the applications development and use in the event industry, this research aims to analyze the use of applications as a tool to improve event communication. The methodology used includes an exploratory method. This research methodology is based on a review of the literature, a semi-structured interview with the CEO of the company and a case study with participant observation of the French start-up Invent App which provides event applications. The study demonstrate that the event app use allows to ease the event organization, to improve the event communication but also the interactivity and the participant engagement. All of this, increasing the customer experience but also having a part in the customer digital experience improvement. Event applications could be thus considered as a marketing and event communication tool. The research has the limitation to its application within a single company in France. Furthermore, the point of view adopted is the one of the event organizers and not the event participant one. This research helped to better understand the role that event applications can have in the event industry and particularly in the communication around an event. In addition, this made it possible to understand the perception of market requirements and the characteristics expected of an event application.

Keywords: Event · Event Communication · Digitalization · Mobile Marketing · Event Application

1 Introduction

Today it is observable that globally, digital has a large part in everyday life and societies. Its use is predominant in most industries. The development of the Internet has changed the organization and working environment of many sectors. The development of digital

A. Rocha et al. (Eds.): WorldCIST 2023, LNNS 802, pp. 51–60, 2024.
https://doi.org/10.1007/978-3-031-45651-0_6

leads to the current fourth industrial revolution. The fourth industrial revolution refers to a hyperconnected world which induces fields such as artificial intelligence, robotics, the internet of objects or even nanotechnologies, a new benchmark which requires man to know how to reinvent himself [1]. The events industry is no exception to this phenomenon. The use of digital is gradually taking place in this industry. This seems at first glance paradoxical since an event tends to be considered as possessing a physical essence. According to Rapeaud [2], an event can refer to what happens and takes place in a specific place. The event industry is constantly evolving. Indeed, the event communication is a major issue for companies. It has a great place within the marketing strategy and internal and external communication. Furthermore, with the internet and mobile platforms development, it is possible to remark that business marketing is more and more digital. Besides, digital marketing refers to all marketing techniques used in digital media and channels, it covers marketing applications connected to the "traditional" Internet but also those connected to mobile phones and tablets [3]. According to Bathelot [3] key techniques of digital marketing can be distinguished such as email marketing, display advertising, web analytics but also mobile marketing. Mobile marketing, bringing together mobile websites and mobile apps, is playing an increasingly important role. According to Bathelot [3], mobile marketing brings together all marketing techniques based on the use of mobile devices, especially smartphones. Advertising on mobile websites and applications, the functionalities of locating points of sale, text messages represent mobile marketing techniques, among others. At the end of 2019, the Covid-19 pandemic appeared, upsetting economies and societies. This health crisis has generated great instability in the world. Here, digital could be perceive as a way for the events sector to reinvent itself [4]. Event applications are more and more used in the industry. They can now be used for physical events as well as for digital events. They therefore seem to bring something more to the organization of events and to event communication.

The main goal of this research paper is to research the role of applications in the event communication process, understanding the needs of an effective event communication and how event application can contribute to its improvement.

2   Literature Review

First of all, "Events were originally a celebration of ceremony and ritual – and were a reflection of a culture and a community" [5]. Besides, there is a great interest in generating "unique leisure" [6]. Thus, in order to communicate efficiently an event must be entertaining and show uniqueness. According to Pascal [7] "An event is a "live advertisement" for a defined audience at a given time". Indeed, the event interest is above all to communicate a message and make promotion, it must therefore be relevant, adapted, and effective. However, according to Rapeaud [8], the specificity of an event lies in the fact that it is ephemeral: it must therefore be dynamic, striking, and original in order to reach its targets. On the other hand, as stated by Getz [6], it is possible to define different typologies of planned events such as: cultural celebrations, business and trade, arts and entertainment, sport and recreation, political and state and private functions. This, according to their purpose, audience, and characteristics. According to

Pascal [7] "Men have always come together to celebrate or party" and thus, the event has always existed. It is the industrial revolution and the arrival of the consummation society who laid the foundations for today's event communication since the 19th century [7]. Today marketing has taken the major place and the budget is calculated to the nearest millimeter, everything is controlled and targeted. Pascal [7] emphasizes that "The events of the 1980s, which were above all festive, without any real framework or strategy, have given way to a very structured event". Rapeaud [8] distinguishes four types of audiences for events: consumers, it is about B to C or B2C communication, the employees, and collaborators of the company within the framework of internal communication, the professional partners when it is about B to B and finally the shareholders this is a so-called institutional communication. On the other hand, Rapeaud [8] recognizes different types of communication such as the corporate one which includes the institutional business communication and the internal and financial one, the "brand" communication promoting products and services and the brand values. Event agencies are the main players when it comes to event organization. According to Pascal [7] event communication market is part of the MICE industry (Meetings, Incentives, Congresses, and Event), organized between project sponsors (businesses, associations...) and organizers (event agencies, business tourism agencies, professionals of the hotel industry...). Pascal [7] states that the MICE industry represents nearly 8 billion euros in France and event communication revolve around 2.5 billion euros. Finance, cosmetic and pharmaceutical, computing and the trade industries are the main sectors who organize the most events [7]. According to Robert [9] 380,000 is the number of corporate events organized in France in 2018, representing 32 billion euros in economic benefits with 52% benefit for event production professionals and 48% for tourism stakeholders. According to Pascal [7] the events market "has evolved according to the various crises since 2008". With lower budget, agencies need to adapt enriching their offers, taking advantage of digital, opening up internationally. Pascal [7] describes the event market as a competitive market where reactivity is necessary.

2.1 The Digitization of the Event Industry

Rapeaud [8] stated that "digital fulfills the same objectives as events". According to Rapeaud [8] both of them allow to "reunite people, federate groups, develop information sharing, exchange". Thanks to digital participants can be even more involved [8]. At opposite of been passive during the event process, they are feeling actors of the event. Indeed, digital favorizes interactivity and involvement. As explained above, digital is everywhere and took slowly more and more space in the event industry. This occurred with the internet arrival and notably with the internet of social media, the web 2.0. Besides, the internet of social media drastically impacted communication. These social networks are unavoidable tools and implicate a real added value since it offers more visibility. According to Pascal [7] social media has become as important as a caterer, a place, or a setting when it comes to talking about an event planification. The combination of Web 2.0 and events has given rise to virtual events that take place only on the internet [8]. Digital can be perceived as an added value in the different steps of an event process. Before the event, it could inform and valorize, during the event it can give a resonance and is a way to make the event last after its end. Digital allows to extent the dialogue with the participants having a durable link with them. For example, it is still possible to share information after the event. Rapeaud [8] stated that before the events were one-off and fleeting, but thanks to digital technology and web 2.0, the event can be extended over time by creating a community. Pascal [7] also ensures that digital has not made the essence of the event disappear but enriched it before, during and after by allowing it to last over time. During an event, digital use allows to create a buzz and give it notoriety [8]. According to Rapeaud [8] and Pascal [7], after the event, it is possible to measure the impact it has thanks to social networks [8], Indeed, feedbacks, posts and reactions then allow to readjust the event strategy. Using a mobile app saves time for the end user [10]. According to Bresson [10], this helps participants to focus on what they are really looking for talking about a fair or exhibition app, visitors will be able to research in advance the exhibitors of interest. Having a mobile platform allows users to access information more easily than usual [10]. According to Martin and Lisboa [11], fully digital events are less expensive, representing a better cost-benefit ratio. Besides, event planification involves a lot of costs such as travel, food, accommodation, decoration, human resources and audiovisual for instance. Rapeaud [2] demonstrates that events represent a very polluting activity whether because of the transport used or the printed flyers. Rapeaud [2] underlines the importance of the environmental challenge. According to Rapeaud [8] with the digital use, there is a real difficulty to control the information shared. This represents one of the main disadvantages when talking about using digital.

3   Methodology

This research is linked to the possible benefits and contributions of the event applications use for event communication. In the literature it is possible to see the conceptual concern regarding the increasing use of digital in the event industry until the advent of completely digital events knowing that an event has fundamentally a physical essence. In addition, the literature demonstrates the requirements of effective event communication. In this sense, the present work aims to answer the following research problem: "How can event applications be used as a tool to improve event communication?".

The main goal is to develop a study to research the role of event applications in the event communication process, understanding the needs of an effective event communication and how event application can contribute to its improvement. To achieve the general objective mentioned above. In this sense, we define 5 research questions (RQ):

RQ 1. The use of event application eases the event organization?

RQ 2. Event apps facilitate the dialogue between and with the participants during the different steps of the event process?

RQ 3. Event app tool gives more interactivity to the event?

RQ 4. Event apps allow to develop participant engagement during the event?

RQ 5. Event apps allow to improve the customer experience but also the customer digital experience?

The impacts of the event applications use has not been studied, the aim of this project is to discover and determine ideas that can be used to further inquiry. Thus, the exploratory design is the more adapted for this investigation knowing that the subject is new and recent [12]. According to Yin [12] case study will use the same sources of evidence as history: direct observation of the studied event and interviews of the persons involved.

The present project uses a single case study research format [12] as it is only looking at as a case study and no other businesses. Furthermore, it is not an embedded case study as there are no subunits of analysis, the unique unit of analysis studied are event applications phenomenon. The semi-structured interview appears as the more coherent option in order to have a guided interview that able to obtain additional information about the event apps phenomenon. As a consequence, this qualitative data collection method is combined with the observation method to achieve the inquiry purposes. As addressed before, this study implicates a case study focusing on a French event application start-up. Moreover, thanks to the agreement of the event app start-up regarding the study, the investigator could have the opportunity to access customers cases, documents, participate to events and observe the employees and customers behaviors and thoughts regarding event applications use and development. Additionally, a semi-structured interview with the CEO of the event application company is applied as a complementary method of qualitative data collection to achieve this research.

4 Case Study of "Invent App"

"Invent App" is a French company specialized in event applications since 2018. In reality, the company already existed in Brazil under the name of Céu App but after 5 years of experience on the Brazilian market, Emmanuel Aires decided to expand its activity abroad and especially in France, his native country. Emmanuel Aires is then the CEO and owner of the French Invent App start-up. According to the Invent App website, the company is willing to support companies in the digitalization of their event, to facilitate the organization of events, to increase participant engagement and then dynamize events. Indeed, Invent App is working with many event professionals and provides them turnkey support solutions for professional event and corporate travel organizers. The main customers of Invent App are event agencies and great businesses organizing corporate events. Thus, the Invent App application (Fig. 1) is thought for corporate events such as meetings, congresses, symposiums, seminars, team buildings, or travel incentives for example. The app stands out from the other of its market thanks to its complete customization and its white label availability.

4.1 Findings of Content Analysis

In general, it is possible to identify several aspects resulting from the use of event applications through these customer cases analysis. First, the event app is a tool to digitize events. It allows to gather all the information in one single platform and to share information. The aim of this tool is usually to guide participants. Through these customer cases it is possible to observe that the customization of the app is important, and it has its part regarding the brand image, then, it takes fully part of the event communication process. In practically all cases, it is pointed out that the application made it possible to increase the interactivity, the attractiveness, and the interactions with and between the participants. The "ecological" aspect comes up through several customer cases, in fact the use of an application makes it possible to reduce the use of paper. Finally, in certain cases, the use of digital allowed to touch more people and to avoid the cancellation of

Fig. 1. Home page of the native App version. Source. Invent App documentation.

events because of health issues related to Covid-19. Then, it appears that digital use avoids the barriers of physical distance.

4.2 The Interview

The interview method that was chosen for this project was a semi-structured interview with the CEO of Invent App: Emmanuel Aires. Emmanuel Aires has been graduated of EMLyon, a French business school. Then, he created Invent App after more than 5 years of experience in Brazil as a business developer in an event applications company to help businesses modernize and digitize their events. Emmanuel Aires has long believed that mobile applications are one of the key elements for the future of organizing corporate events. The interview took place on October 6th, 2021, at the company premises. Table 1 establishes an analysis of the interview with the Invent App CEO by categories. To simplify the analysis, the categories are divided by themes which are the key themes aborded through the interview. The table is analyzing Emmanuel Aires responses according to the different themes of the interview.

4.3 Main Findings

The analysis of the interview with the Invent App CEO helps to bring out key ideas about the inquiry subject. Indeed, Emmanuel Aires highlighted some characteristics of the event applications. First, he underlines that it can be qualified as a marketing tool on its "ability to communicate" before, after and during the event but also to "promote the event". Thus, it is also important for the brand image and communication with the white

Table 1. Analysis of the interview with Emmanuel Aires. Source: Self-elaboration.

Categories by theme	Responses
Event App market	Event app market opportunity is mainly due to "the emergence of more and more smartphones" and growing apps creation. Plus, "events really need logistical information", the app can gather all of this Clients could be indirect (event agencies) or direct (companies). "Finance and pharmaceutical industry" are the main clients. Clients depends on companies' typologies, event budget and organization frequencies. Clients are willing to digitize and give interactivity to their events Invent App deliver a "turnkey application" with "full service", advice, support, and content integration
Events Organization	The App use allow to digitize the event, have a "digital guide", to deliver information, replace the paper and make interactivity The event app risk is to make "even more people dependent of their digital tools" and be "less attentive" to the event
Impact of use	- The event app use allows to "save money on printing costs, voting box costs, logistics" but also, to simplify the organization and to reach a remote audience". But it stays a fleeting tool, unless for fully digital events Event app tool completely has an impact on the brand image: application in white label "gives credibility to the event" and a "certain standing"
Event Communication	Event apps enhance event communication as it is used to "communicate before the event, to give information, during, to make interactivity as well as to also have the information and after the event to make the satisfaction survey this which allows the customer to have the Analytics data" It can be considered as a marketing tool for event in its "ability to communicate" by keeping "participants informed" and its ability to "promote the event"
Customer Experience	According to Emmanuel Aires, the "centralization of the information" improves the customer experience". Plus, it improves exchanges and interactivity and thus enhance the participant engagement It can be part of the customer digital experience strategies
The Future	According to Emmanuel Aires, event apps is a "growing market" with a "multiplication of players" and a product that has already "proven itself" and became "normalized" in events The Invent App CEO explains that the future of events will be "hybrid" with on-site and remote participants using an "event web application"

label for example. In addition, the Invent App CEO explains that event apps allow to digitize the event, make interactivity, interactions with the participants, create engagement, guide, and communicate, keep participants informed and centralize the information through a "digital guide" allowing to improve the "customer experience" and being part of the customer digital experience process. Event app tool permits to spend less money on logistical material. However, it is not necessary for an event, and it can make participants more dependent of their mobile devices during events. Regarding the event app market, it was developed thanks to the smartphones increasing using and applications increasing creation. Main customers of event apps are pharmaceuticals and finance industries and big companies making events. Event app market is a growing market and according to the Invent App CEO, the "future is on the hybrid". The Invent App CEO explained during the interview that the event apps emergence is due smartphones and apps explosion. This confirms Bresson [10] who qualifies smartphones as essential tools and demonstrates that it was linked with the app's creation explosion. Furthermore, Emmanuel Aires says during the interview that the main event apps clients are from pharmaceuticals and finance and big companies while Pascal [7] explained that finance, cosmetic and pharmaceuticals computing, and trade industry are the main sectors organizing events. Indeed, the Invent App CEO, the customers cases and the events analysis highlighted the importance of event app for the brand image and communication process, for example using customization and white label. This confirms the Bresson [10] hypothesis regarding the impact of event app use for a modern brand image. The customer cases and events analysis demonstrate that event apps serve as a guide for participants, centralizing and gathering information. The Invent App CEO talks about "digital guide" improving the customer experience. This matches Robert [9] who demonstrates that the experience proposed to the participant has a great importance. While Brassington and Pettitt [15] underline that promotion is part of the Marketing Mix, the Invent App CEO explains during the interview that event app could be considered as a marketing tool on its ability "to communicate before, during and after and to promote the event". Then, event app is a way of event promotion and takes place in marketing strategies. According to the Invent App CEO event apps enhance event communication as it is used to "communicate before the event, to give information, during, to make interactivity as well as to also have the information and after the event to make the satisfaction survey this which allows the customer to have the Analytics data". This statement confirms the Moon event website and Rapeaud [8] talking about the three stages of event communication: before (promotion), during (information provided, pushing for sharing and engagement) and after (feedback, prepare the next event). The data collected demonstrates that event apps allow to digitize the event, amplify interactions but above all, insists on the fact that it helps creating engagement from the participants and make interactivity. Pascal [7] exactly explained that to have an impact an event must be significant and draw the participant attention. Also, Calder, Isaac, and Malthouse [13] demonstrate that customer engagement is a major objective as it can increase advertising effectiveness. Jauréguiberry [14] states that thanks to digital consumer and advertiser into a real and virtual relationship and it is the case with event apps as it creates interactions. Furthermore, Rapeaud [8] stated that "digital fulfills the same objectives as events" which are "reunite people, federate groups, develop information sharing, exchange", stating that digital help people being more involved, which is demonstrated with the event app use that increase participants engagement and interactivity.

5   Conclusions and Future Research

As a result, different methods of data collection such as the interview, the events participant observation and the customer cases analysis has been used and thus allowed to highlight the same key findings which have been evocated. Indeed, the main results found demonstrate that event apps permit to facilitate the event logistic, ease the event organization and avoid the use of paper. It was possible to highlight the facts that event apps are a marketing and communication tool by allowing to communication before, during and after the event but also by promoting the event and having a great part regarding the brand image. Plus, the several data collection methods prove that event applications increase the participants engagement, the attractiveness of the event, the interactions, the event interactivity but also improve the customer experience by guiding the participants easily. As a conclusion, all the hypothesis which have been established regarding the inquiry can be validated. It has been proven that the use of event app eases the event organization by centralizing the information, the event logistic and guiding the participants. It also has been confirmed through the several methods of data collection that event apps facilitate the dialogue between the participants and this, during the different steps of the event process (before, during and after). In the same way, event apps allow to increase interactivity during the event thanks to several interactivity tools and thus, develop participant engagement, giving more attractiveness the event and the brand. It then represents a great marketing and communication tool. Event apps also allow to improve the customer experience by facilitating the event logistic, centralizing the information, and keeping the participants informed. Then, event apps have a part to enhance the customer digital experience. This inquiry enables to develop a recent phenomenon and to deeply study this phenomenon which is not very well-known and contains few data. Precisely, it helped to understand what the contributions are of using event applications and to understand the interests of a such tool in the event industry. It helped to demonstrate that event apps can have a role for the event communication and this all along the event communication processes. Besides through this study, it is possible to understand that event applications have a marketing and communication part for events as it can increase event interactivity, participants engagement, favorizes the interactions and enhance the customer experience but also the customer digital experience. Finally, the study helped to show the importance and issues of digital regarding the event industry and to know more about the event app development and market. Some limitations can be putted forward regarding this study. First, the study has been realized in France and within a French company. Even if the phenomenon is rather global and could be generalized, the study has been focused and realized in only one country and could have been extended for more observations and comparisons on other markets. Plus, here the study focuses on the organizer point of view but not on the participant one which could have allowed to have the participant feeling about the event app use. For further investigation, it could be possible to take the participant point of view to have the participant feeling and opinion about the subject but also to directly study and obtain the customer experience feedback. For example, a descriptive study could be made focusing on the event app customer digital experience or regarding the event app participant engagement. In addition, other methods of data collection can be used such as surveys with participants to obtain their feeling. Would also have been pertinent to do a comparative study with two events: one using event app and one without any event app use. This, in order to compare the event attractiveness.

References

1. Schwab, K.: La Quatrième Révolution Industrielle. Malakoff, Dunod. Translated by Jean-Louis Clauzier and Laurence Coutrot (2017)
2. Rapeaud, M.L.: La communication évènementielle, de la stratégie à la pratique (3ed). Vuibert (2019)
3. Bathelot, B.: Defining Digital Marketing. Définitions Marketing (2016). https://www.definitions-marketing.com. Consulted the [19/08/2022]
4. Cousin, C.: Lancer son événement digital. White paper, Win Win (2020)
5. Yeoman, I., Robertson, M., Ali-Knight, J., Drummond, S., McMahon-Beattie, U.: Festivals and Events Management an International Arts and Culture Perspective. Elsevier Butterworth-Heinemann, Oxford (2004)
6. Getz, D.: Event Studies Theory, Research and Policy for Planned Events (2ed). Routledge, New York (2012)
7. Pascal, C.: La communication événementielle. Dunod, Malakoff (2017)
8. Rapeaud, M.L.: La communication évènementielle, de la stratégie à la pratique (2ed). Vuibert (2016)
9. Robert, M.: L'événementiel: un secteur à 32 milliards d'euros. Echos, Paris (2019)
10. Bresson, A.: Mobile applications at the service of events. Tangram (2020)
11. Martin, V., Lisboa, R.: Eventos digitais: híbridos & virtuais. E-Book (2020)
12. Yin, R.K.: Case Study Research: Design and Methods, 4th edn. Sage Publications Inc., USA (2009)
13. Calder, B., Isaac, M., Malthouse, E.: How to capture consumer experiences: a context-specific approach to measuring engagement. J. Advertising Res. **56**(1), 39–52 (2016)
14. Jauréguiberry, F.: Les branchés du portable. Paris, Presses universitaires de France, coll, Sociologie des usages (2003)
15. Brassington, F., Pettitt, S.: Principles of Marketing, 4th edn. Pearson Education Ltd., Essex (2006)

Fundamentals of a Digital Marketing Plan for a Tourism Infrastructure in Alentejo

Mariia Popova[1], Manuel José Serra da Fonseca[2,3]($\boxtimes$) ,
Jorge Esparteiro Garcia[2,4,5] , and José Gabriel Andrade[6,7]

[1] Instituto Politécnico de Bragança, Bragança, Portugal
[2] Instituto Politécnico de Viana do Castelo, Viana do Castelo, Portugal
`manuelfonseca@esce.ipvc.pt`
[3] UNIAG - Unidade de Investigação Aplicada em Gestão, Viana do Castelo, Portugal
[4] ADiT-LAB, Instituto Politécnico de Viana do Castelo, Viana do Castelo, Portugal
[5] INESC TEC, Porto, Portugal
[6] Universidade do Minho, Campus de Gualtar, 4710-057 Braga, Portugal
[7] CECS - Centro de Estudos de Comunicação e Sociedade, Braga, Portugal

Abstract. There is no doubt that tourism benefits Portugal's economy. The area should be able to better target the market for its travel and tourist products and increase its performance with the use of a marketing strategy, especially in a field where global competition is always increasing. This digital marketing plan seeks to analyze the strengths and limitations of the Portuguese region in its target markets, as well as the most current advancements in the global travel and tourism sector. In this work, it is conducted an analysis of both micro and macro environments of Casa Pereirinha. In addition, it was created a marketing mix based on the 7P's. Furthermore, the work provides the prospective avenues for product growth and marketing following the markets' stated growth objectives. This project comprises creating a comprehensive digital marketing strategy for Casa Pereirinha as a significant example of the region's tourism business, and the primary purpose is to conserve its past, while developing a new strategy.

Keywords: Tourism · Digital Marketing Plan · Marketing Strategy

1 Introduction

Tourism is one of the most important sectors in the World. According to the World Tourism Organization [1] it accounts for 9% of the global Gross Domestic Product (GDP), generates 1 in 11 jobs, and represents 6% of total world exports [2]. The same organization noted a record 1.087 million international tourists in 2013 and predicts that in 2030 this figure will reach 1.8 billion. For Portugal specifically, is one of the most important economic activities developed in the country. A bright future for Portugal's tourism and hospitality industries can be predicted given the successful results of 2019 and the current pipeline of ongoing projects. This optimistic outlook has been supported by reputable industry experts. The World Economic Forum's placement of Portugal as

A. Rocha et al. (Eds.): WorldCIST 2023, LNNS 802, pp. 61–70, 2024.
https://doi.org/10.1007/978-3-031-45651-0_7

the 12th best country (out of 140) in the 2019 Travel and Tourism Competitiveness Index reflects its solid position at the beginning of this year. The rating has moved up three notches since 2015. According to WTTC predictions, travel and tourism will contribute 16.5% of Portugal's GDP in 2019, up 4.2% from 2018. Foreign visitors made up 24.6 million, an increase of 7.6% from 2018, and they generated €18.431 M in revenue, or 8.7% of Portugal's GDP (if local tourism is included, this percentage rises to 15% of GDP). Tourism accounted for 19.7% of all exports and 52.3% of service exports, according to the Instituto Nacional de Estatística [3]. During the last four years in a row, Portugal won the prize of Europe's best tourism destination. In the last edition of the World Travel Awards, Portugal received 24 prizes, including Europe's leading beach destination, leading adventure tourism destination, leading cruise destination, and leading city break destination [4]. In addition to this, the Executive Digest ranks Portugal first with the best tourism promotion in Europe, and the third best in the world [5]. Alentejo is a high potential area for tourism development, but only now it begins to be implemented. Tourism in Alentejo is one of key priorities of the Portuguese Government these days [6].

In this work we develop a digital marketing strategy for a pateo house in Vidigueira – Casa Pereirinha, that is a great example of the local traditional tourism scape, and our key priority is to create a modern approach while keeping all the values and the heritage of this place. The remainder of this paper is organized as follows: next Section presents a brief Literature Review. Section 3 presents the Methodology used in the is work. Section 4 gives an overview of the developed Digital Marketing Plan. Conclusions, Limitations, and future directions are presented in Sect. 5.

2   Literature Review

Businesses in the tourism sector must always be aware of their products' market positions, as well as the competition they face, and the most current industry innovations. They must also be able to analyze if they are on track to meet their previously established goals and, if not, what tactical changes are required. To conduct a proper marketing research and establish a marketing strategy it is essential to review the marketing definitions that will help to do so: marketing concept, marketing objectives, marketing situation diagnosis, marketing strategy, segmentation, targeting, positioning, marketing- mix, marketing plan, tourism marketing. Defining the research objective is the most complex, time-consuming, and valuable stage of the market research process. It requires a multi-faceted analytical effort to select and expertly assess a variety of factors in the macro and micro marketing business environment [7]. It allows studying the company's place in the market, identifying alternative ways of performing management tasks, defining marketing research objectives, and obtaining information for making informed decisions. The issue of defining the purpose of marketing research is mostly treated as a general method, without a specific methodology detailing the tools that could be immediately used in practical work. It is important to distinguish between the method of setting a marketing research objective, as a theoretically based approach, and the technique, which is a specific logical set of steps and tools to implement the method to achieve a predetermined result [8]. The technique for setting marketing research objectives should consider some general principles: feasibility, repeatability, relevance to the goals and objectives of the planned activity, validity, and effectiveness [9]. One of the weakest points of both theory of marketing and theory of strategic management is interpretation of relationships between overall (corporate) strategy and marketing strategy. The most popular approach is the idea of the so-called hierarchy of strategies [10]. In creating a marketing strategy, we use an STP approach which refers to Segmentation, Targeting and Positioning. Marketing research conducted to develop a company's marketing strategy is not only very difficult, as it is carried out in a highly competitive environment, but also very responsible. In this regard, one cannot underestimate the importance of carrying out activities to develop a company's marketing strategy [11]. Another approach to strategic marketing was pointed by McDonald who identified it as a series of logical steps that must be worked through to arrive at a marketing plan [12]. Nowadays, tourism marketing is a field of a wide demand due to rising customer expectations and growing competition between destinations [13]. Tourism marketing is raising into an important field of marketing because most tourism destinations must compete on a global level as the world-wide travelling is becoming more common. Gössling [14] pointed out that COVID-19 provided exceptional lessons for the tourism industry, policy makers and tourism researchers. The challenge is to collectively learn from this global tragedy to accelerate the transformation of sustainable tourism. Ozili and Arun [15] in their empirical study examined the pandemic situation throughout the economic environment as increased closure days, monetary policy decisions and international travel restrictions have seriously affected the level of economic activity.

3 Methodology

This research work is a practical roadmap and an applicable to implementation work. For this reason, the objective of this study is to develop an extended marketing strategy and tactics considering all the specifications of the Alentejo region and the regional tourism. To achieve this objective, we used both quantitative and qualitative methodological approaches. Since Casa Pereirinha is a new to the market business it is not possible to use its data base or ant data collected from the customers, however, we use the booking services – Booking.com and Airbnb to collect the feedback and create an extended SWOT analysis, as well as sum up the data to create a consumer portfolio for the analysis. Thus, we can conduct the following objectives of the research: Analyze the market of hotel tourism in the Alentejo Region and analyze the internal characteristics of the project; Develop a Marketing Mix considering the specifications of the region and the business project; Create a practically applicable operational plan that will cover the needs of the business and will help to achieve the marketing objectives. The quantitative method is broadly used in the digital marketing analytics and play the core role in optimizing the strategy based on the data provided by analytical services [16]. The data was collected in 2 stages. In the first one we gave a brief-questionnaire to the owner of the guest house to obtain the most basic and essential information for the research. The questionnaire consisted of 28 questions and 3 topics: product, competitive environment, and marketing. The responded was able to give an extended answer to the questions in case of a non-disclosure agreement or other reasons. The main goal of the questionnaire was to develop and overall image of the macro and microenvironment of the business from the owner's point of view to be able to proceed to a deeper and detailed research of the topic. Moreover, the analysis contains the data of the website traffic in "www.casapereirinha. com" with the geography, browsers, operating system, traffic, look-through and sources. The information obtained from a questionnaire provided and overall information and helped to create a roadmap of the future research. In addition to this we developed an extended marketing mix based on the empirical research method. In this research one of the main sources of sampling was a convenience sampling that is a type of non-probabilistic sampling method where the sample is taken from a group of people easy to contact or to reach. In the case of the present research, it was the sampling of the business owners and some of the customers. The questionnaire for the business owners consists of the marketing and business-related questions to understand their business awareness and marketing inclusion, as well as see their vision of the product and perspectives. The second target audience was the current customers, and the main objective of the research was to analyze their satisfaction with a product. As we did not have a wide range of the respondents due to the absence of CRM system, it was chosen to use an open-question questionnaire to get a broader picture and let the respondents justify their answers. The third questionnaire was created for the potential target audience, mainly the younger generation, to collect their view and preferences in terms on local tourism and see if your business has the potential in them as in our target audience.

3.1 Interim Analysis

The survey process allowed us to come to the certain intermediate conclusions about the target audience, as well and strong and weal points of the product. From the position of the business owner the main conclusions were stated as follows: Casa Pereirinha have a high potential due to the small competition in the area; the perspectives on the region mainly rely on the airport's accessibility; the main problem of Casa Pereirinha is the old maintenance, not big enough pool, small number of rooms, absence of own restaurant and the city itself as it does not provide any sort of entertainment to the tourists. The survey of the current customers brought up the following conclusions: The staff in the guest house is very polite and helpful; the atmosphere of Casa Pereirinha is charming. However, some furniture is better to be renewed; The animal-friendly policy is very important; they would not recommend to the families with children, because they did not see the necessary equipment for nursery; the territory is very green and it's very close to the wineries. And the third group of respondents highlighted the points of the product that can help to attract more audience and broaden the targeting: The most important thing about the place they go to is activities; they prefer the modern facilities but do not mind traditional design; They like wine-tourism in the "all inclusive" bindles where they do not have to think about where to eat; They prefer beautiful views to the calmness. The summary of the conclusions gives the opportunity to develop the product and adapt it to different generations, as well as work on the weak points of the service.

4   Digital Marketing Plan

The marketing plan consist of four main parts: 1 - Diagnosis of the situations; 2 - Marketing objectives that we will state using STP marketing model; 3 - Marketing Mix. In this research there will be used a 7P's approach: product, place, price, promotions, people, processes, physical evidence; 4 - Operational plan.

4.1   STP Model

The company must be aware and understand their customers; their needs and demands.

Segmentation and Targeting. Casa Pereirinha was opened to tourists for less than a year and because of this we do not have a CRM system of clients which would allow us to make an appropriate analysis of the consumers. Due to this reason, we will base our segmentation analysis on the knowledge of the market and the interviewing the management if the Guest House. As it was mentioned above, we will start from 2 main segments – foreigners and Portuguese tourists. This factor will depend on the type and language of content and communication, as well as the advertisements we will use for targeting. The geographical segmentation for tourists will be: Northern Europe, United Kingdom, USA, Eastern Europe. For the Portuguese tourists it is mainly Lisbon area and central Portugal. Based on statistics Portuguese tourists from the north chose the winery tourism in their own region and prefer ocean vacation on the south of Portugal. Thus, the main target audience in Portugal will be located in Lisbon, Faro, Leiria, Coimbra, Aveiro, Portalegre, Santarem, Setubal and Catelo Branco. From which the key cities will be Lisbon, Setubal, Grandola, Santarém, Peniche, Leira, Castelo Branco, Portimão, Faro and Albufeira. Secondly, we must elaborate the age range. The most appropriate age segments are 28–35, 36–48, 48–56, 57+.

The winery tourism implies the consumers with the average or above the average income, mainly the couples because Alentejo is a quiet and peaceful region that cannot offer much on an entertainment such as clubs, festival, and other events. However, we must never exclude the percentage of tourists who travel in companies. It can both freelancers, people with stable job and entrepreneurs. What is more, it is believed that besides wine tourism Alentejo attracts people with remote work for a temporarily or permanent resediment in there. Vidigueira mainly attracts people who like calm and peaceful pastime, as this region cannot offer many sports or hiking activities. Casa Pereirinha can offer high level of security and privacy as well as the atmosphere of the desired by tourists' Portuguese culture.

Positioning. Casa Pereirinha's main objective is to be for its consumers the "island of peace" where they can always escape from their daily routine and worries. Casa Pereirinha is about "Dolce far Niente" or the joy of doing nothing – the main theme of Alentejo region. This guest house is for people who value relationship, family, and emotional intimacy. The place that teaches you slowing down, spending time with yourself and enjoying the moments of simple life. For our target audience it is important to have the fast and direct communication with the owners, with whom you can agree on all the personal matters and find the best solution. Casa Pereirinha's very strong characteristic is the stuff – very polite, helpful. The questioning the second category of respondents they mentioned that this place make you feel like "childhood and grandmother's house". Most of the moderns megapolis-citizens do not have the luxury of the countryside house, and Casa Pereirinha can become one for them. In addition to this, Casa Pererinha is located only 2,5 h from Lisbon which makes it a perfect escape place and a wonderful destination for a "family trip". It is located in the heart of Alentejo, but it is cheaper than the guest houses of the same level in Evora or Beja. So, when the consumer is planning the trip to Alentejo he chooses the most advantageous in terms if the price-location-service offer. Thus, the positioning defined presents an identification principle (how it wants to be seen by consumers) and a differentiation principle (how it distinguishes itself from the competitors).

4.2   Marketing Objectives

Touristic sector is fast growing sphere and due to this fact, we will present short-term and long-term objectives. Based on our values and the analysis we present the short-term objectives for 1 year as following: O1 – To increase the reach and the engagement on the Instagram account by 1000%; O2 – Increase the number of followers to 2000 users; O3 – Reach a monthly conversion rate of 17% for the featured Ads; O4 – Prolong the LTV of the customers; O5 – Optimize the search engine on the website and increase the position in Google Search Results; O6 – Create minimum 2 substantiable collaborations with local businesses; O7 – Increase the number of new visitors by 35% and organize minimum 3 events.

In long-term objectives, 5 years planning, we suggest the following objectives: O1 – Open 12 new rooms (right now we have 5); O2 – Build collaborations with restaurants and develop a special voucher system; O3 – Reach 10.000 followers on Instagram; O4 – Create a blogger community around our business; O5 – Take part in collaborative events from the Portugal Ministry of Tourism to spread brand awareness.

4.3   Marketing Mix

The objective of the marketing mix used by the Casa Pereirinha in Vidigueira is to draw clients or visitors to the place, thing, or service that the tourism firm plans to sell them. The seven components of the travel and tourist industry's marketing mix are the following: product, place, price, promotion, people, processes, and evidence.

Product. Hotels sell goods such as room service, banquet halls, restaurants, parking lots, and the labor. The service must meet the wishes of the guests. The offer of hotels is based on the definition of a potential category of guests and on their wishes to receive one or another set of services. Therefore, when we talk about a product as a constituent element of the marketing mix for the hotel industry, we must distinguish between a product that takes a physical form and services that are aimed at meeting the needs of customers associated with their stay in a hotel. The set of products provided can be identified as: Hotel service for guests; SPA service; Event venue; Weddings and Resto-bar service.

Place. The following service can only be obtained in Casa Pereirinha itself in Vidigueira. It is in the historical center of the city by address: Doutor António Carlos Da Costa

Street, 26. The location is easily reached by car or by the buses, presented by FlixBus and Rede Nacional Expressos transport services. As it was, the bright perspectives of the developing of Beja and Evora airports would play a significant role in the boosting the tourist traffic to the region. Casa Pereirinha presented on such booking platforms as Booking.com and Airbnb. In addition to this, it has a website that allows the direct messages to the manager with the booking option. As well, the Guest House is presented on the Instagram Platform "@casa_pereirinha" where customers can leave feedback, ask questions and book through direct messages or by call.

Price. The pricing policy for the tourism sector plays a significant role in the decision-making process. In Alentejo tourism industry, pricing is a key component of the marketing mix. Pricing must be set up to be competitive with any rivals offering the same service or a comparable alternative service. Casa Pereirinha has a competitive price among the other hotels and guest houses in Vidigueira and offers us the best result in the price-quality ratio. Casa Pereirinha has special offers for the event planners and have private approach to clients. For this reason, it is more relevant to contact directly before booking the room online. This may not be very convenient, but this is one of the aspects of traditional businesses in Portugal where real communication dominates over online connections. The pricing policy practiced implies several forms of payment, all of them secure, as well as the possibility of cancellation.

Promotion. Promotion plays a key role in the marketing mix for Casa Pereirinha. It encompasses all strategies used by the business to promote and sell its products and services. In the marketing of consumer goods, the main goal is to ensure that goods are available to the consumer, to know when, where, in what quantity and in what range they are needed. Nevertheless, distribution channels in the hospitality industry can be divided into direct and indirect. Direct channels include mail, phone calls and fax, while indirect channels are represented by intermediaries, namely travel agents, tour operators and independent hotel agents. The main channels of Casa Pereirinha's promotion would be Contextual Advertising, Paid Search in Booking.com, Airbnb, and Instagram. Facebook could be an irreplaceable instrument for building the brand awareness among the Portuguese tourism sectors. Touristic sector is widely promoted on the Instagram which even created a definition "instagrammable" – the way to describe the place that looks attractive in the pictures. This social media is the main tool to spread the brand awareness among potential customers.

People. Spend money on hiring staff that has the traits of a quality tour guide. They have a significant impact on how satisfied your customers are and how well they will speak of your company. Hiring enthusiastic and energetic tour guides will help you expand your customer base and improve your visitors' overall experience. Casa Pereirinha – is a family business which became the business itself just recently. The owner of the Maison is currently the head of the family - Diogo Pulido Pereira Freire de Andrade. His son – Tiago Freire de Andrade is the one responsible for handling communication right now, and in the future will take over the business and will be responsible for its growth. There are people who permanently work in Casa Pereirinha and take care of the guests' comfort and experience. These are: hosts, cleaning service, cooking team, pool service, massage specialist, etc.

Processes. How well the clients are served will depend on your processes. Convenient procedures will benefit your clients as well as your staff. It ensures that all corporate procedures and operations go off without a hitch. The process point is especially important in the hotel business because it covers many points of contact with a customer: Booking process; Cancellation policy; Check-in process; Check-out process; Cleaning service among others. To make the booking management process easier many sources recommend an online booking service like Rezdy booking service to automate the management of your processes. Currently Casa Pereirinha is in the process of integration the Smoobu as a vacation rental app this easily integrated with other services. An Hotel business is interested in direct booking of their services from their own website not to pay a commission to the services (Booking, Airbnb). Nowadays Portuguese residents prefer to pay by MBway, which is very fast and secure.

Physical Evidence. Physical evidence is an important part of the product. However, as we mentioned before, the product of Casa Pereirinha – is the mix of service and facilities. Currently Casa Pereirinha does not have branded product or the tangible branded items. Right now, the business is in the process of creating a collaboration with a local winery to offer the clients the welcome gift – branded wine from the local vineyards.

4.4   Operation Plan

To achieve the objectives defined before in 4.2, we shall start by stating a TOV of the project and elaborate the design and the mood-board for the social media, specifically for O1 – To increase the reach and the engagement on the Instagram account by 1000% and O2 – Increase the number of followers to 2000 users. The stated TOV: chilled and relaxed conversation with a friend, happiness is peace, and the real feeling of life can be obtained in "*dolce far niente*". We want the customers to get out of the rush and stress and remind them about what is important. After the mood-board we proceed to content-plan, Instagram planning and copywriting. After the account packaging, we can proceed to the other marketing tools such as: User-Generated Content, collaborations with Influencers, Reels and TikTok Production. A very important aspect of Social Media marketing in 2022 is content creation as the audience becomes more demanding and judgmental to the businesses Social Media accounts, and for this reason we must create the technical task for each photoshoot. To achieve "O3 – Reach a monthly conversion rate of 17% for the featured Ads", it is required to hire a targeting specialist and provide a monthly budget for testing campaign and for targeting promotions. The objective of "O4 - Prolong the LTV of the customers" has a very prolonging effect, but approximately we can achieve this goal in a 6-month long period by working with the CRM system and rely on the client orienteering techniques and instruments. To achieve the "O5 - Optimize the search engine on the website and increase the position in Google Search Results", may take around 3–4 months and will depend on the medium CPC and CPV in the region. The objective of "O6 - Create minimum 2 substantiable collaborations with local businesses", "has the "season" element which means the local businesses have the tendency to be more active in the warm time of the year which will postpone the implementation of this actions till May 2023. However, the period before that can be effectively used for the preparation for the hot tourist season. The goal of "O7 - Increase the number of new visitors by 35% and organize minimum 3 events", requires a sustainable and respected brand identity to increase the trustworthy image in the eyes of the counterparties.

Specific Objectives and Long-Term Actions. The specification of the tourist business has a high frequency of changing environment, and according to this it is challenging to predict the situation in 5 years for a small business. However, any business ought to have a plan a long-term action strategy to have an overall vision and the direction of growth. The long-term objectives of Casa Pereirinha were stated as follows:

5 Conclusions, Limitations, and Future Work

The goal of the research, which was developing a practically applicable marketing strategy, was successfully reached. It is a very important point at this case because the main goal was to make the strategy useful for the Casa Pereirinha business owners and help them to expand their business and show both weak and strong features of it. Touristic business is a rapid changing economic sector that and a high volatility as it is not the first or even second priority need for people. In this research we conducted a deep and detailed analysis of the tourism in Alentejo and Vidigueira which was essential to understand the macro and microenvironment of Casa Pereirinha guest house. Most of the actions are centered alongside the Digital Marketing domain due to the dynamic of this marketing field and the priority of our consumer's presence in the digital field. In addition to this, digital marketing requires less personal involvement from the business owners and can be operate by the independent team which can help to reduce the cost whole improving the quality and the results. However, the research has certain limitations: i. Casa Pereirinha does not have a big experience in the market which limited the data for the analysis, such as LTV of the consumers, marketing KPI and analytics of advertisements; ii. Some of the data is (such as financial documents) confidential, and for this reason the marketing budget is not adapted to the actual marketing budget ability of Casa Pereirinha; iii. Seasonality in Alentejo region has a serious impact on the marketing strategy which as not covered in the study due to the lack of information on the Guest house's occupation and activity in the period from December to March; iv. Vidigueira does not have a wide range of the competitors with a higher quality of the service, what I, as a marketing specialist find very discomforting as it reduces the level of competition which is not good for the growing market.

This research requires future continuation and expanding together with the new data and the changing environment. Marketing strategy is more of a tool, rather than a study, and it needs regular adaptation to the rapid modifications and the hypothesizes testing.

Acknowledgment. This work is supported by the Fundação para a Ciência e Tecnologia (FCT) under the project number UIDB/04752/2020.

References

1. World Tourism Organization. UNWTO Annual Report 2012 (2013)
2. UNWTO. International Tourism Highlights 2019 Edition. World Tourism Org. (2019)
3. Instituto Nacional de Estatística. Turismo em Portugal 2020. INE (2020)
4. World Travel Awards. Europe Winners 2020. WTA (2020)
5. Executive Digest. Country Brand Awards: Portugal é eleita a melhor marca turística da Europa e a terceira a nível mundial. (2020). https://executivedigest.sapo.pt/country-brand-awards-por tugal-e-eleita-a-melhor-marca-turistica-da-europa-e-a-te
6. Serdoura, F., Moreira, M.D.G., de Almeida, H.: Tourism development in alentejo region: a vehicle for cultural and territorial cohesion. Sustain. Archit. Urban Dev. **2**, 619–633
7. Starostina, A., Kravchenko, V., Petrovsky, M.: An innovative technique to define marketing research objective. TEM J. **11**, 955–963 (2022)
8. Mirskij, E.M.: Methodology. New Philosophical Encyclopedia (2001)
9. Senkina, G.E., Emelchenkov, E.P., Kiseleva, O.: Methods of mathematical modeling in teaching. Smolensk State University (2007)
10. Marek, P.: A critical analysis of the concept of marketing strategies for small and mid-sized companies. Econ. Manag. Financ. Markets **9**, 254 (2014)
11. Malhotra, N.: Marketing Research: An Applied Orientation (6th ed.), Prentice Hall (2010)
12. McDonald, M., Christopher, M., Bass, M.: Market segmentation. Marketing, pp. 41–65 (2003)
13. Middleton, V., Fyall, A., Morgan, M., Ranchhod, A.: Marketing in Travel and Tourism (9th ed.), Routledge (2009)
14. Gössling, S., Scott, D., Hall, C.M.: Pandemics, tourism and global change: a rapid assessment of COVID-19. J. Sustain. Tourism **29**(1), 1–20 (2021)
15. Ozili e, P.K., Arun, T.: Spillover of COVID-19: Impact on the Global Economy. SSRN Electronic Journal (2020)
16. Saheb, T., Amini, B., Kiaei, F.: Quantitative analysis of the development of digital marketing field bibliometric analysis and network mapping. In: International Journal of Information Management Data Insights, vol. 1 (2021)

Integrating Online and Offline Distribution Strategies – A Portuguese Case Study

Ana Santos[1], Jorge Esparteiro Garcia[2,4,5](✉) 🆔, Lia Coelho Oliveira[2,4,5] 🆔,
Daniel Ladeira de Araújo[6] 🆔, and Manuel José Serra da Fonseca[2,3] 🆔

[1] Instituto Politécnico do Porto, ISCAP, São Mamede de Infesta, Portugal
[2] Instituto Politécnico de Viana do Castelo, Viana do Castelo, Portugal
`jorgegarcia@esce.ipvc.pt`
[3] UNIAG - Unidade de Investigação Aplicada em Gestão, Viana do Castelo, Portugal
[4] ADiT-LAB, Instituto Politécnico de Viana do Castelo, Viana do Castelo, Portugal
[5] INESC TEC, Porto, Portugal
[6] Instituto Superior de Ciências Empresariais e do Turismo, Porto, Portugal

Abstract. The online channel, particularly in the food retail area, has been evolving positively and exponentially in the world, including Portugal. Currently, this type of purchase is increasingly part of people's daily lives, even more so with the emergence of the Covid-19 pandemic. Consequently, in Portugal, most companies adopt a multichannel strategy, where the physical store and the online store operate independently from each other. However, it is necessary to rethink this channel integration model, which may go through an omnichannel strategy, where the physical store and the online store operate as a single store, and where several advantages are already recognized in terms of the consumer's shopping experience. The main objective of this study is to understand the strategy implemented by the company studied, Pingo Doce, through an analysis and description of its channels. To better understand the strategy of the company under study, a semi-structured exploratory interview was carried out with one of the people in charge of Pingo Doce's digital channels, to understand the strategy used by the company and thus complement the data obtained through direct observation and bibliographic research. At the end of the work developed it was possible to understand the positioning of Pingo Doce in the online food retail area and their online and offline distribution strategies.

Keywords: Omnichannel · Online Food Retail · Pingo Doce · Multichannel

1 Introduction

The retail sector has undergone major changes since the online channel and the constant development of technologies, and the Internet started to be part of people's daily lives [1]. The digital world has been showing a relevant growth and predominance and, as a result, retailers started to implement multichannel strategies, i.e., through more than one channel, where their consumers could do their shopping through several channels, whenever and wherever they wanted [2]. Still, the consecutive arrival of new shopping

A. Rocha et al. (Eds.): WorldCIST 2023, LNNS 802, pp. 71–80, 2024.
https://doi.org/10.1007/978-3-031-45651-0_8

channels, such as the mobile channel, and the constant evolution of technologies and information systems, influenced retailers to operate through several channels, recognizing the advantages of each one, and, as a result, changed the habits and expectations of consumers [3].

This growth in all fields of work, especially in retail, accompanied by the change in consumer behavior in the purchase process [1], has led entrepreneurs in various industries to rethink their strategic models, wondering about possible operational difficulties with the integration of channels and the costs associated with multichannel strategy [4].

The multichannel strategy, already implemented by several retailers, quickly led to an omnichannel strategy, a synergy between channels that Mosquera, A. et al. [5] believe has revolutionized several areas of work, such as communication, marketing, and retail. The integration between physical and digital channels has demonstrated the many advantages it brings by allowing consumers to navigate between channels of the same brand creating a complete and efficient consumer experience [6]. However, Alves et al. [7] refer that this commercial strategy is still not fully diffused since most retailers are not aware of its approaches and possible benefits.

In Portugal, the emergence of the Covid-19 pandemic in the first quarter of 2020 became one of the main influencers of the growth of online shopping. According to Observador [8], the Covid-19 pandemic contributed to a significant increase in the number of new users in digital channels. Before the pandemic, the growth of e-commerce was not as high as expected (31%), compared to the proliferation of the Internet (71%). After the outbreak of the pandemic, e-commerce in the food retail sector grew between 40% and 60%, due to the increase in purchases by regular online consumers and the entry of new consumers in digital channels. Influenced not only by digital transformation and changes in consumer behavior, but also by the most recent global events, most of the major food retailers in Portugal already have an online store.

Pingo Doce, the company addressed throughout this study, is one of the companies that already has an online platform and, more recently, a mobile application, which is still in its initial stage. However, unlike the other competitors, this company's digital approach is carried out in a different way, since they collaborate with a partner instead of owning their own online channel.

Thus, the main goal of this study is to answer the following research questions:

- What is the strategy used by Pingo Doce in distribution channels and the reasons for their choice?
- What are the strategies used by Pingo Doce's main competitors in their distribution channels?

2 Literature Review

Neslin et al. [9] define the channel as a customer touch point or a way through which the customer interacts with the company. Initially, the only existing channel of purchase was the offline channel, or also called physical, with stores being the main responsible for the process of acquiring goods and services. With the rapid growth of technology, the concept of the online channel was born, and quickly became part of people's daily lives.

Currently, there are 3 types of shopping channels: the offline channel, the online channel, and the mobile channel, more recent and in the process of expansion, but equally relevant today. Understanding the characteristics of each channel and what they can offer allows the consumer to prefer or not a specific channel. For example, the value of a channel is assessed by the ability to communicate information to consumers and help them make decisions in that channel and not only varies between channels but also varies within a single channel [10].

The offline channel, also called the physical channel, allows the customer to touch and feel the products and give immediate gratification [11]. Sarkar & Das [12] characterize the offline channel in five stages of a consumer decision model, comparing them also with the online channel: recognition, information search, alternatives evaluation, purchase, and post-purchase evaluation. In the authors' opinion, the offline channel is the consumers' preferred channel when talking about the tangibility of the products and the possibility to try them on site, giving the consumer more confidence when purchasing them. Also, the offline channel is more advantageous since the product or service can be consumed immediately.

Chatterjee [13] characterizes the online channel by the extended temporal and geographical accessibility as well as a greater assortment, information, and novelty of products. However, this remote characteristic of an online channel makes the delivery of the product temporally separated from its entire order and payment process. Thus, regardless of whether the pre-purchase research has been done online or not, if the customer opts for an online product delivery, it implies that he must bear the transport and waiting costs inherent to his purchase.

The mobile channel is the most recent channel associated with the channel typologies of a purchase process. With the evolution of technologies, including mobile devices, more and more Internet is incorporated into more portable devices than computers, such as cell phones and tablets. According to Brynjolfsson et al. [11], the mobile channel can help both channels (physical and online) by reaching new customers and expanding their markets. This channel is also referred to as the channel that provides the customer with the Internet 24/7, anytime, at their disposal, wherever they are.

Multichannel shopping has been growing rapidly with companies investing in more and more channels and customers using more devices anywhere, anytime. Channel integration has been a major challenge for retailers, as some services that include integration between remote and physical channels - such as online shopping and in-store pickup - depend on the customer's device or location [2]. The multichannel strategy, also known as "bricks-and-clicks," supplements conventional stores with online services [14]. It is also characterized, according to Kourimsky & Berk [15], by involving several channels of the shopping process that work individually, meaning that they exist but do not interact with each other. According to Verhoef et al. [1], multichannel strategy is driven by the growth of online channels that affect businesses and consumers using traditional channels. These channels are developed and managed independently or with rather limited integration within companies.

Over the years, consumers have created expectations for consistency in their shopping experience across all channels [16] and the lack of integration by the multichannel strategy resulted in inefficiency and consumer confusion [4]. Retailers with multiple channels had two options: operate the various channels as independent or integrate them allowing cross-channel movements between products, money, and information [13]. As a result, cross-channel strategy is considered the sequel to multichannel strategy, where the first steps are taken in the integration between physical stores and online channels [5]. Cross-channel allows the exchange between some channels and touchpoints and the partial sharing of data between channels. Consequently, cross-channel retailers must maintain standardization of pricing, positioning, and merchandising strategies [13].

Omnichannel is defined by Kourimsky & Berk [15] as a strategy where all channels are integrated. The consumer can purchase their goods or services through any channel and information about what they want is available in any channel and in real time. [1] define omnichannel management as a synergistic management of the various available channels and customer touch points, so that the customer experience across channels and their performance are optimized. According to Piotrowicz & Cuthbertson [17], the concept of omnichannel results from an evolution of multichannel, where consumers, unlike in multi-channel, navigate freely between the online channel, mobile devices, and the physical store, all in a single transaction. The omnichannel strategy must therefore provide a unified and integrated experience for the consumer, regardless of which channel they are using. Compared to other multichannel strategies, in this typology consumers gain other benefits such as the availability of new in-store information technologies (virtual screens, virtual fitting rooms, vending machines or smart self-service kiosks), cost savings and convenience [17]. As mentioned by Gao & Su [18], one of the big challenges in the omnichannel environment is efficient information delivery, as consumers are becoming increasingly sophisticated and demanding enough to optimize their shopping experience by considering all possibilities across all channels.

3   Methodology

According to Menezes et al. [19], the case study is a type of research that focuses on a specific case and whose objective is to know its causes comprehensively and completely. The case study is considered a research method used to understand, in a deep and multifaceted way, a complex problem in its real context [20].

In the present study the single case variant is used where the aim is to investigate a single company in the food retail sector in Portugal, Pingo Doce. Thus, the problem addressed in this case study presents an analysis of the company's distribution channels, with the objective of concluding the channel integration strategy implemented. Thus, in addition to using the single case study methodology, descriptive research is also carried out as a methodological approach, where qualitative data about the company is used.

Initially, bibliographic research was carried out to obtain a better understanding of the theme addressed. Next, a qualitative methodological approach is used through an exploratory semi-structured with 8 questions, held on 11th February 2022, at the shop on Estrada Dom Miguel in São Pedro da Cova, Gondomar, and direct observation. In this study, the search engine Google Scholar, as well as the scientific repositories Science Direct, Research Gate and RCAAP were used to obtain secondary data. Additionally, websites of magazines in the field under study, such as Distribuição Hoje and Jornal de Negócios were visited. In this study, direct observation was used to obtain information about the channels used by the company under analysis.

Visits were also made to some stores in order to confirm the information shared by the company, namely the Marquês and Marechal Gomes da Costa stores, in Porto, and the Pingo Doce & Go Nova SBE store, in Lisbon. In the context of the research, it was opted for a semi-structured exploratory interview with discussion topics, to give greater freedom of response to the interviewee and greater flexibility and spontaneity to the whole interview. Even so, a script was made with some key questions necessary to understand the choice of channels used by the company and to know some imperceptible points during direct observation.

4 Case Study Analysis

4.1 Pingo Doce

The company's main distribution channel is offline, with the online one being operated by an external company and independently, Mercadão. This option *"has a lot to do with the return and the investment... And the time in which you get that money back"*, as mentioned by PC. Another advantage that Mercadão brings is its agility, as it only works with online business and can offer a delivery time of 2 h. However, there are some limitations in the products that can be ordered on the platform as not all products are available in all shops. *"(...) the assortment that is presented to the customer is always the assortment of the shop, of the area... That serves that customer area (...) In the case of the online catalogue, what happens is that when there are products that only exist in a certain week, they often do not make it into Mercadão's assortment"*.

Recently, the company decided to invest in the mobile channel by developing the Pingo Doce app, as this is the most personal and easiest way to communicate with the customer. *"(...) the app is Pingo Doce in the customer's pocket. It is the most intimate and personal way that we have of communicating with the customer and it is the easiest way to get there. Sending a notification is at the very least seen, even if it's not clicked."*

According to the interviewee, the Pingo Doce app has made the customer more involved with the company, increasing their purchases, compared to other customers who use the Poupa Mais card (loyalty card made available since 2013), but do not have the app. To encourage its use, it has invested in offering a different own-brand product every week on the first purchase of €20 or more. *"...If I don't give the customer an incentive to use the Poupa Mais card for small purchases, because until now I didn't (...). With this, we are giving a reason for the customer to use the card for purchases over 20€. (...) what we also see is that the app can have the effect of making the customer more involved with us and increase their purchases"*.

When questioned about the possibility of creating an exclusively mobile shop, PC stated that *"What we are thinking of doing is to follow the same path as we have online in terms of access to products through catalogues and receipts. We will certainly want to make this happen. Then we will see if we take an extra step, that is, if the conclusion of the purchase is made within the app instead of going to Mercadão"*.

4.2 Strategies Used by the Main Retailers Operating in the Portuguese Market

According to a study conducted by Lapa [21], the food retailer of the Sonae group, Continente, operates with a well consolidated omnichannel strategy. The brand allows customers to return purchases made online at any Continente, Continente Modelo or Continente Bom Dia physical store. Customers of this retailer are also able to collect their orders, with no minimum purchase amount and only by going to the service counter and picking up a ticket for the "Click and Collect" service. According to the author, all Continente's channels share customer information with each other, facilitating all its processes.

Regarding technological solutions in physical stores, Continente allows customers to create and use their Continente card through the app and also has the Continente Siga app, where the customer can carry out the entire shopping process via cell phone while in the physical store. In this app, customers can scan all the items they need using their smartphone, have the possibility of obtaining virtual passwords for all the service counters through the "Tira-vez" service, and can also pay for their purchases using the Continente Pay service, an e-wallet developed exclusively by Continente for its customers. Continente also has an electronic vehicle charging service, Continente Plug & Charge, while customers are shopping in the physical store and a parking feature in the Continente card app that allows customers to enter and exit paid parking lots by scanning a QR code. In summary, Continente is a retailer with a very complete omnichannel strategy and a solid integration between the physical, online, and mobile channels.

After researching through its channels, we were able to conclude that Lidl currently operates with a multichannel strategy. In Portugal, the company does not yet have an online channel and its mobile platform does not have a shopping functionality either. Through the Lidl Plus platform, the customer can access the store's weekly flyers, digital coupons, and discount coupons. The only cross-channel aspect of this retailer is that customers can activate their coupons using their smartphone and use them when shopping in the physical store. Regarding in-store technological solutions, the retailer does not have any functionality that allows customers to integrate digital channels with Lidl's physical stores.

The retailer Intermarché also operates with a multichannel strategy, where its channels do not show evidence of integration and consistency of information and products made available. Currently Intermarché has an online channel exclusive to the company, with pick-up delivery options in store, on the drive or at home. The online platform is updated according to the store selected by the customer, so not all items may be available in all stores and the minimum purchase amount may vary. Also depending on the store choice, delivery options may not all be available. The online platform does not provide payment methods through the site, so the customer can only pay for his purchases in the store when he picks up the order, or by bank card when he receives the order at home. Regarding the mobile channel, Intermarché only presents functionalities for consulting leaflets, shopping lists and recipes, as well as the possibility of using the savings card in the store through the app or checking the balance. The customer can't make any purchase or order through the Intermarché mobile app. Intermarché physical stores do not have any technological solution that allows customers to integrate Intermarché physical stores with digital channels.

[21] reveals in his study that the retailer Auchan works, similarly to Continente, with an omnichannel strategy. Although it does not provide as many technological solutions

in the physical store as Continente, at Auchan customers can also scan all the desired items and add them to their digital shopping cart and then make their payment at the Express Scan boxes located in the store. Also, in this retailer the customer can make the return to the physical store any item purchased through the digital channels as well as pick up their orders online. The author also mentions that the strategy implemented by the company results in an integrated, homogeneous, and synchronized experience between the channels, sharing all the necessary customer information.

4.3 Data Analysis

After an interview and analysis of the data described above, and in response to the first research question about the strategy used by the company under study, we can see that currently the Pingo Doce brand operates with a multi-channel strategy, since the channels used by the brand are not integrated in any way.

On the online platform operated by Mercadão, there is no sharing of information or data that could provide Pingo Doce with complete knowledge of the consumer profile and a uniform and integrated purchasing experience. The only omnichannel aspect existing at Mercadão is that the customer has the option of picking up his purchases in the physical store. Although there is information from the source contacted from Pingo Doce on the positive impact that the partnership between the brands brings to the company, since they do not incur costs or investments, it is not perceptible in the data shared by the reports Jerónimo Martins' financial reports that there is a significant contribution of online sales in the remaining sales of the brand.

As far as the Pingo Doce mobile application is concerned, it is still at a very early stage, with only permission to consult products, prices, and shopping lists. Here, the client also has access to his card balance and to the weekly promotions in force. However, the whole online purchasing environment is non-existent in this channel.

Regarding the second research question defined in this study, concerning the strategies used by the main Portuguese food retailers, we were able to show that only two works with a well consolidated omnichannel strategy, Continente and Auchan. Both have a complete and advanced online and mobile platform and plenty of technological solutions in the physical store, which allows their customers to purchase products from any of the channels, anytime and anywhere.

Regarding the other retailers, Lidl and Intermarché, both operate with a multichannel strategy with some limitations. Lidl with only a limited mobile app and no online store, and Intermarché, unlike Lidl, with a limited online platform and no mobile app.

5 Conclusions, Limitations, and Future Work

In terms of channel integration strategies, Pingo Doce still shows little evolution when compared to its main competitor and the demands of its target public, with its channels working independently and disintegrated. The financial results obtained by the brand, mainly from the second quarter of 2020, make us believe that this business strategy is not the most appropriate for the company. As mentioned before, Pingo Doce, which adapted to the same Covid-19 pandemic measures that were imposed on all food retailers in Portugal, had negative results when compared to its main competitor.

Through the semi-structured exploratory interview to the person responsible for Pingo Doce's digital channels, used as a qualitative methodological approach, it was possible to perceive that there is no intention on Pingo Doce's part to replace its current strategy by one that includes the creation of a website from within the brand itself. Although there are still some weaknesses, there are also measures that can be taken to improve the customer's shopping experience. These measures include the weekly exclusive discounts from the "Poupa Mais" card on the Mercadão platform, which to date does not happen. The same applies to the discount coupons, which are only available in paper format and not digitally. These opportunities may result in greater migration to online platforms.

As a result of this migration, another measure to be taken would be to eliminate the minimum purchase amount at Mercadão, which is set at €50, and to make it possible to return items purchased in the online store to the physical store, which currently constitutes a barrier in the post-purchase phase. For greater integration of the physical and online channels, it would be important for Pingo Doce to opt for more technological solutions in the physical store, such as the possibility for customers to purchase and pay for their purchases using their smartphones, even when they are in the physical store, which is currently only available in a single store in Lisbon, the Pingo Doce & Go Nova SBE.

Another solution also involves the creation of an integrated information system, between Mercadão and Pingo Doce, where transactional and non-transactional information is shared, enabling the company to have a complete and detailed knowledge of all its customers so that it can later issue personalized discounts or anticipate its customers' purchases through a history of orders and products most consumed by each of its consumers.

About the mobile channel, it would be essential that the brand develop a mobile application, even in partnership with Mercadão, where customers can also place orders using their cell phones. Also, the integrated information system could be extended to this modality so that they can integrate as many channels and information as possible. Although the company intends to move forward with the mobile channel in the short term, always in partnership with Mercadão, it currently proves to be an obstacle to the digital growth of its business model, since it does not have any purchasing process functionality. The application is still quite simple, with the sole function of providing its customers with information on its products and current campaigns.

Regarding retail competitors, only two are advanced in terms of channel strategies, Continente and Auchan, with an online and mobile channel already consolidated and with a very complete integration between channels. Intermarché and Lidl assume, just like Pingo Doce, a multichannel strategy. Both can also improve through the creation of online and mobile platforms integrated with the physical stores, which allow the consumer to have a uniform and efficient shopping experience.

Concerning the main limitations of the study, we highlight the limited information obtained from the company under study, since it was only possible to speak with one of the people responsible for digital marketing and it was not possible to obtain other types of information from other departments at Pingo Doce, namely at the strategic level. There was also a lack of communication with Pingo Doce's partner company, Mercadão, meaning that it was not possible to obtain data or information on the long-term intentions of this partner with the company under study. Another limitation was the lack of financial data regarding some Portuguese retailers, which did not allow for a more comprehensive analysis among the main retail companies in Portugal.

For future research it is suggested that some aspects that were not explored in this study should be addressed, namely a similar and detailed analysis of other Portuguese food retailers and their market position, as well as the definition of their strategic positioning. Within the same theme, it may be relevant to carry out new interviews with different departments of the Pingo Doce company, in order to understand what strategy to follow in the long term and to analyze the distribution processes inherent to the company.

References

1. Verhoef, P.C., Kannan, P., Inman, J.J.: From multi-channel retailing to omni-channel retailing - introduction to the special issue on multi-channel retailing **91**(2), 174–181 (2015)
2. Beck, N., Rygl, D.: Categorization of multiple channel retailing in multi, cross, and omni-channel retailing for retailers and retailing. J. Retail. Consum. Serv. **27**, 170–178 (2015)
3. Kazancoglu, I., Aydin, H.: An investigation of consumers' purchase intentions towards omni-channel shopping - a qualitative exploratory study. Int. J. Retail Distrib. Manag. **46**(10), 959–976 (2018)
4. Zhang, J., Farris, P.W., Irvin, J.W., Kushwaha, T., Steenburgh, T.J., Weitz, B.A.: Crafting integrated multichannel retailing strategies. J. Interact. Mark. **24**, 168–180 (2010)
5. Mosquera, A., Pascual, C.O., Ayensa, E.J.: Understanding the customer experience in the age of omni-channel shopping. In: ICONO 14, vol. 15, no. (2) (2017)
6. Quach, S., Barari, M., Moudrý, D.V., Quach, K.: Service integration in omnichannel retailing and its impact on customer experience. J. Retail. Consum. Serv. (2020)
7. Alves, S., Serra da Fonseca, M.J., Esparteiro Garcia, J., Oliveira, L., Teixeira, A.: The omnichannel strategy in portuguese companies: an overview. In: 16th Iberian Conference on Information Systems and Technologies (CISTI), pp. 1–6, December 2021. https://doi.org/10.23919/CISTI52073.2021.9476612
8. Observador: O impacto Covid-19: a segunda vida do e-commerce em Portugal? Obtido de Observador, 05 de junho de 2020. https://observador.pt/opiniao/o-impacto-covid-19-a-segunda-vida-do-e-commerce-em-portugal/
9. Neslin, S.A., et al.: Challenges and opportunities in multichannel customer management. J. Serv. Res. **9**(2), 95–112 (2006)
10. Maity, M., Dass, M.: Consumer decision-making across modern and traditional channels: E-commerce, m-commerce, in-store. Decis. Support. Syst. **61**, 34–46 (2014)
11. Brynjolfsson, E., Hu, Y.J., Rahman, M.S.: Competing in the Age of Omnichannel Retailing. MIT Sloan Manag. Rev. (2013)
12. Sarkar, R., Das, D.S.: Online shopping vs offline shopping: a comparative study. IJSRST **3**, 424–431 (2017)
13. Chatterjee, P.: Multiple-channel and cross-channel shopping behavior - role of consumer shopping orientations. Mark. Intell. Plan. **28**(1), 9–24 (2010)
14. Xing, Y., Grant, D.B., McKinnon, A.C., Fernie, J.: Physical distribution service quality in online retailing. Int. J. Phys. Distrib. Logist. Manag. **40**(5), 415–432 (2010)
15. Kourimsky, H., Berk, M.V.: The Impact of Omni-Channel Commerce on Supply Chains. ITelligence (2014)

16. Dholakia, R.R., Zhao, M., Dholakia, N.: Multichannel retailing: a case study of early experiences. J. Interact. Market. **19**(2) (2005)
17. Piotrowicz, W., Cuthbertson, R.: Introduction to the special issue information technology in retail: toward omnichannel retailing. Int. J. Electron. Commer. **18**(4), 5–15 (2014)
18. Gao, F., Su, X.: Online and offline information for omnichannel retailing. Manuf. Serv. Oper. Manag., 1–15 (2016)
19. Menezes, A.H., Duarte, F.R., Carvalho, L.O., Souza, T.E.: Metodologia Científica - Teoria e Aplicação na Educação à Distância. Petrolina (2019)
20. Crowe, S., Cresswell, K., Robertson, A., Huby, G., Avery, A., Sheikh, A.: The case study approach. BMC Med. Res. Methodol. (2011)
21. Lapa, T.G.: O omnicanal- Um estudo de caso de plataformas de comércio eletrónico dos retalhistas alimentares em Portugal (Master thesis). Instituto Superior de Contabilidade e Administração do Porto, Politécnico do Porto, Porto, Portugal (2021)

The City Makes Its Mark in a Review on Digital Communication and Citizenship

José Gabriel Andrade[1,2]([envelope]) [ORCID], Adriano Sampaio[3,4] [ORCID], Jorge Esparteiro Garcia[5,6,7] [ORCID], and Manuel José Fonseca[5,8] [ORCID]

[1] Universidade do Minho, Campus de Gualtar, 4710-057 Braga, Portugal
jgandrade@ics.uminho.pt
[2] CECS - Centro de Estudos de Comunicação e Sociedade, Braga, Portugal
[3] Universidade Federal da Bahia, Barão de Jeremoabo, 1649-023 Salvador, Brasil
[4] Grupo de Pesquisa Logos – Comunicação Estratégica, Marca e Cultura, Salvador, Brasil
[5] Instituto Politécnico de Viana do Castelo, Viana do Castelo, Portugal
jorgegarcia@esce.ipvc.pt
[6] ADiT-LAB - Instituto Politécnico de Viana do Castelo, Viana do Castelo, Portugal
[7] INESC TEC, Porto, Portugal
[8] UNIAG - Unidade de Investigação Aplicada em Gestão, Viana do Castelo, Portugal

Abstract. This article delves into the intersections of place branding, digital strategic communication, citizenship, and tourism. It explores the dynamic relationship between these concepts, particularly within the context of Brazilian city governments. With an emphasis on reflexivity, the study investigates how governments manage their public image and engage citizens through digital channels. Simultaneously, it examines how these governments strategically position their cities as attractive tourist destinations. By analyzing these tensions and synergies, the article provides insights into the complex landscape of communication strategies employed by Brazilian city governments, which aim to balance citizen engagement and tourism promotion.

Keywords: Place branding · Digital Strategic Communication · Citizenship · Tourism

1 Local Branding, Tourism and Citizenship

In the late 1970s, a groundbreaking communication campaign in the United States captured global attention and set a precedent for how cities, states, and countries would be managed worldwide. The campaign, famously known as "I Love NYC" or "I Love New York," propelled the city of New York into the spotlight and revolutionized place branding. By strategically leveraging the domains of communication, culture, and tourism, New York City successfully crafted a comprehensive place brand that garnered widespread visibility. Place branding refers to the communication and marketing efforts undertaken by territories, such as cities, states, and countries, to position themselves as desirable destinations. It involves devising strategies that enhance visibility in the

A. Rocha et al. (Eds.): WorldCIST 2023, LNNS 802, pp. 81–90, 2024.
https://doi.org/10.1007/978-3-031-45651-0_9

tourism market while effectively promoting products and services in international markets. The success of New York's place branding campaign serves as a notable example of how these efforts can shape the perception and reputation of a location on a global scale [19].

The emergence of local branding as a significant phenomenon on a global scale occurred primarily in the 1990s, coinciding with the rise of globalization processes. This shift was driven by the increased need for regions and cities to engage in competitive practices. According to Milton Santos [30], in a globalized world, the imperative for regions and cities to compete becomes essential due to the prevailing rules of production and consumption. The concept of competitiveness extends beyond economic factors and encompasses the overall coexistence and interactions between people. Therefore, local branding plays a crucial role in positioning regions and cities within the global landscape, allowing them to compete effectively and thrive in a world driven by globalized processes.

During the 1990s, governments at various levels began recognizing the significance of tourism as a crucial industry and a valuable source of income for their respective territories. In this period, tourism was increasingly acknowledged as a substantial economic driver. For instance, in 1990, the global tourism industry generated a staggering $268.9 billion in revenue. In the case of Brazil, the country collected $1,492.3 million, representing approximately 0.55% of the global market share for that year [26]. These figures underscored the growing importance of tourism as a lucrative sector and prompted governments to invest in its development, thereby capitalizing on the economic benefits it could bring to their territories.

Data from the World Travel & Tourism Council (WTTC) reveals the significant impact of the Travel & Tourism sector on Portugal's economy. The country's economic recovery has been largely attributed to the robustness of its tourism industry. Taking into account the WTTC, Travel & Tourism accounted for a remarkable 16.5% of Portugal's total economy. In 2019, the sector's GDP experienced impressive growth of 4.2%, surpassing the global GDP growth rate nearly threefold. Portugal's success in revitalizing its tourism sector can be attributed to the effective implementation of a comprehensive tourism strategy by the government. This strategy facilitated strong growth in visitor numbers from non-traditional markets, including the United States, Canada, China, and Brazil [38]. By diversifying its target markets, Portugal was able to attract a broader range of tourists and capitalize on the resulting economic benefits. This sustained growth in the Travel & Tourism sector underscores the positive outcomes achieved through a well-executed tourism strategy and its contribution to Portugal's overall economic performance.

Brazil has faced a severe health, social, and political crisis resulting in a significant loss of lives, with over 600,000 deaths. The country's economy has been heavily impacted and is at risk of entering a recession. However, despite these challenges, a report by the Brazilian Association of Tourism Operators [6] indicates that the Northeast region has accounted for 70% of tourism revenues in Brazil in 2021. The cities of Maceio, Natal, and Salvador have emerged as the primary tourist destinations in the country. It is worth noting that the states where these tourist hubs are located, namely Alagoas (ranked 26th), Rio Grande do Norte (ranked 16th), and Bahia (ranked 22nd) according to the Human

Development Index (HDI) [2], are among the lowest-ranking states in terms of development indicators in Brazil. This presents a striking contradiction between the choices made by public administrations to prioritize the construction of tourist destinations and the implementation of public policies aimed at improving the well-being of citizens. The contradiction between these approaches forms the central focus of this research's political analysis. Citizenship, as understood in this context, refers to a societal principle that applies to all individuals without distinction and empowers each person with the ability to have their rights respected in the face of force, regardless of the circumstances [29]. This notion of citizenship provides a framework for examining the tensions and conflicts arising from the prioritization of tourism development over policies that address the needs and rights of citizens.

Indeed, the cities included in the research corpus have been successful in attracting a significant portion of tourism resources, accounting for approximately 70% of tourism revenues [6]. However, it is evident that public policies aimed at improving the well-being of citizens have not been effectively implemented in the country. As exemplified by the low Human Development Index (HDI) rankings of the Northeastern states, there is a stark disparity between the focus on tourism promotion and the implementation of policies that address the needs of the local population. This situation leads us to hypothesize that the strategic communication choices made by city administrations, which prioritize promoting cities as commodities [15], are not aligned with the communication efforts and discursive practices aimed at public communication. In other words, there appears to be a discrepancy between the messaging and priorities conveyed in the promotion of cities for tourism and the actual implementation of comprehensive public communication strategies that address the needs and concerns of the citizens. This hypothesis highlights the need for a critical examination of the communication strategies employed by city administrations, particularly in relation to how they balance the promotion of cities as tourist destinations with the fulfillment of public communication objectives that prioritize citizen well-being and the overall development of the territory. Understanding and addressing this discrepancy is crucial for achieving a more balanced approach to strategic communication that aligns the interests of tourism promotion with the needs and aspirations of the citizens.

The present circumstances we find ourselves in, characterized by a severe health and political crisis, along with the ineffective implementation of inclusive public policies in our nation, have resulted in significant alterations to our daily lives. Moreover, our interactions with brands, businesses, and public administration at the local level have undergone profound transformations. Undoubtedly, we will emerge from this global crisis, which has had a particularly devastating impact on Brazil, fundamentally changed. Consequently, we must consider how our relationship with tourism and citizenship will be shaped in the aftermath of the pandemic and its subsequent repercussions.

2 Exploring the Intersection of Place Branding and Citizenship in Brazilian Cities

Recent studies conducted on place branding and public communication in Brazil have revealed a prevailing issue: the construction of place brands and communication guidelines for citizens is predominantly unilateral. During the COVID-19 pandemic, many authors have observed how the federative model and the proactive role of states and municipalities, particularly those in the northeast region of Brazil, have successfully overcome the inertia of the federal government when it comes to managing the pandemic, particularly in the cultural domain. Typically, communication models are developed by government entities in collaboration with specialists in strategic communication, including communication and marketing advisors, public relations professionals, and advertisers. These models are then disseminated to society and the market. However, this approach poses certain obstacles, namely the absence of public discussion and the potential for perpetuating representations based on stereotypes and social stigmas. This perspective brings forth a number of concerns. Firstly, the lack of public discussion deprives citizens of the opportunity to actively participate in shaping the image and communication strategies of their respective places. By excluding diverse perspectives and failing to engage in open dialogue, the resulting communication initiatives may not accurately reflect the aspirations, values, and identities of the local population.

In contrast to existing approaches, our research proposes a comparative investigation in two distinct phases: qualitative and quantitative. The qualitative phase aims to critically analyze the discursive frameworks surrounding the branding of cities, particularly in relation to tourism promotion. We will examine the brand campaigns targeted at tourists, as well as the subsequent public communication campaigns targeted at citizens. Our analysis will focus on the plans, campaigns, and communication materials that serve as manifestations of the brand project in these cities, as identified by previous scholars [33–36].

In a subsequent phase, drawing inspiration from the quantitative approach, we aim to conduct a comparative analysis of two key aspects: a) public communication and b) strategic communication for tourism and place branding. To accomplish this, we will employ document analysis [23] as our primary methodological framework. The corpus for this phase of the research will consist of official documents and data, including communication and tourism plans, budget allocations by communication, culture, and tourism departments, and the resources dedicated to a) public communication campaigns addressing the pandemic's impact on citizens, as well as b) strategic communication campaigns focused on self-promotion by these cities. This proposal is rooted in the well-established analytical framework of discourse analysis studies [8, 12, 22, 37]. It emphasizes the interplay between textual and contextual analysis, as recently redefined by [8] as textual, discursive, and social practices. In summary, our research proposal is structured as follows:

1. **Discourse Analysis:** Examining the "Possible World" [10, 11, 18, 31, 34–37] of a) Public Communication in the Observed Cities and b) Tourist Communication Campaigns for Place Branding in Brazilian Cities.
2. **Documentary Analysis** (Moreira, 2009) and Contextual Examination [8]: Analyzing Data and Official Documents in the form of communication and tourism plans, budget allocations, and other material and symbolic resources provided by the communication and tourism secretariats.

3 Unraveling the Brand: Analytical Grid, Discursive Positioning, and Place Brand Analysis

The concept of "place brand" is closely tied to the construction of universes of meaning through discursive positioning [1, 31] in the context of tourist destinations. It encompasses the self-promotion and public communication campaigns undertaken by these destinations. Discursive positioning refers to the strategies and discursive structures that shape the production of meaning for brands and their products, making it a key focus of discourse analysis studies. In this research, we aim to comparatively analyze the brand identity projects of northeastern cities. These cities, acting as competitors in the same symbolic arena, strategically communicate their unique attributes to attract tourists. To delve into this field, we draw upon relevant theoretical proposals that shed light on the intricacies of place branding and discursive positioning.

According to scholars such as Andrea Semprini and Jean-Marie Floch, three fundamental properties—credibility, legitimacy, and seduction—play a pivotal role in brand analysis and discourse studies [36]. These properties hold significant influence over the public's adhesion to a brand, as they evoke a strong sense of appeal and ease of identification. They are not only prevalent in various renowned global brands but also manifest in strategic communication discourses [20]. Guided by these concepts, each brand endeavors to construct a coherent universe of meaning, a "possible world" [9, 32], through its self-promotion strategies. For a brand project to succeed, it must establish a consistent and harmonious possible world that resonates with the public. This coherence allows the audience to align their expectations with the brand's promises, as exemplified in its campaigns and overall brand positioning. Ultimately, audience loyalty is nurtured through the careful cultivation of these expectations and the creation of a discursive unity that embodies the brand's essence [9–11, 18, 31, 32].

Within this framework, the elements that shape this context are closely linked to both the mediums through which brand manifestations are disseminated, such as online social networks, billboards, television commercials, and magazine ads, as well as the specific "ways of saying" [4]. These ways of saying encompass the strategic discourse of marketing communication [20], which aims to captivate and persuade the audience through seductive means. By utilizing these persuasive strategies, brands seek to create a compelling narrative that entices and engages their target audience.

As Aristotle suggested, the principles of seduction and persuasion are integral to the art of rhetoric. These principles are derived from three key aspects of communication: Ethos, Pathos, and Logos. Ethos refers to the construction of the enunciative subject, shaping the credibility and authority of the speaker. Pathos involves employing seductive strategies that appeal to the emotions and feelings of the audience. Lastly, Logos focuses on the argumentative foundation of the discourse, utilizing reasoning and logical appeals to support the statements being made. By skillfully combining these three horizons of expectations, communicators can effectively engage and persuade their audience.

This triad of Ethos, Logos, and Pathos finds its roots in rhetoric, which is considered by Paul Ricouer [27] as the oldest discipline in the study of language. Aristotle defines rhetoric as "the art of human discourse, more human" [27]. According to Aristotle, rhetoric follows three essential criteria: it is audience-oriented, argument-oriented, and aims at persuasion. The rhetorical approach is not limited to specific forms of speech but extends to any form of enunciation, a term closer to discourse analysis. It is governed by three perspectives: managing faces (Ethos), treating information in relation to a practical goal (Logos), and addressing emotions (Pathos). It is evident that Semprini's [32] analytical proposal, centered on the examination of the discursive construction of credibility, legitimacy, and seduction, draws inspiration from Aristotelian Rhetoric. Specifically, it aligns with Aristotle's three types of arguments or proofs that aim to persuade: Ethos, Logos, and Pathos. These three instances are also explored in discourse analysis, which seeks to understand the production of perlocutory effects: to please, inform/convince, and move [7].

It is important to highlight that these three types of effects production primarily aim to persuade through the discursive construction of evidence within the discourse. We posit as a working hypothesis that it is through these three variables that the possible world of the referenced brands is developed. In turn, this world seeks to engage the public, whether tourists or citizens, by constructing a discursive subject that pleases through Ethos, informs and convinces through Logos, and evokes emotions through Pathos. This process of symbolic construction [35] gives rise to the possible world, allowing audiences to identify with it through discursive strategies employed by contemporary brands. With this proposal, our intention is to conduct a critical analysis of this symbolic construction strategy, assessing how it either contributes to or challenges the reproduction of social stigmas [13] associated with the analyzed cities. Furthermore, we aim to evaluate the relevance of this strategy in relation to the cultural context in which it is embedded. For the discourse analysis of the campaigns, we will adopt the "project/manifestation model" developed by Andrea Semprini [35, 36]. This approach entails analyzing campaigns through three key aspects: the textual dimension (examining strategic communication materials), the discursive dimension (identifying the positioning employed), and the values dimension (considering social and discursive practices).

In summary, this analytical grid is built upon the examination of the textual, possible, and "real" worlds. Each brand, rooted in this premise, constructs a unique possible world in every brand manifestation, employing specific strategies to engage its audience. The first aspect involves the construction of a distinctive mode of enunciation (Ethos) aimed at pleasing the audience through the discursive construction of credibility. The second aspect encompasses the particular construction of the brand's perspective, which can be observed through discursive strategies of legitimation (persuasion) implemented at the level of the utterance (Logos). Lastly, discursive positioning is developed through a specific approach to establishing an emotional connection with the audience, employing seduction (Pathos). It is important to note that the separation of these three facets of discursive positioning is purely didactic, as they are interdependent and complementary. We have observed that the operations involved in constructing the meaning and impact of the utterances from enunciators to co-enunciators aim to legitimize the possible world constructed by the brand. In essence, the real world, the possible world, and the textual world intertwine in a process of social semiosis [37].

In discourse analysis, the textual world must be observed within the framework of a dialogical relationship [3]. The enunciator, who speaks through their discourse, anticipates a co-enunciator. As Benveniste [4] explains, "As a form of discourse, enunciation involves two equally necessary 'figures,' one being the origin and the other the end-point of the enunciation. It is the structure of dialogue" [4]. It is presupposed that every enunciation aims to establish a bond of feeling between the listener and the speaker (enunciator/co-enunciator) [4]. Authors such as Maingueneau [22] use the term "guar-antor" to describe this co-responsibility relationship between the speaker and the one addressed by the discourse. When place brands propose their possible world through strategic communication campaigns, they strive to establish a relationship of trust with their audience, whether tourists or citizens, by constructing meaning through Ethos (credibility), organizing their discourse through Logos (reasoning), and employing per-suasion and seduction strategies through Pathos (emotions). Through these discursive strategies, the enunciator becomes involved in the discourse by making a promise [18] to their co-enunciator. In making a promise, the enunciator becomes obligated to fulfill it in the co-enunciator's discourse. Once the public becomes engaged in the possible world suggested by the brand, they must be convinced of the belief [25] that this system represents the interpretive frameworks surrounding them.

4 Conclusion

In addition to the increased exposure facilitated by online social networks, there is also an opportunity to foster participation in public forums and debates through communication technologies. Since the 1990s, studies have emerged in this field, aiming to explore cit-izen engagement in various digital experiences such as e-government and e-citizenship [14]. This context of reflexivity has motivated governments to face scrutiny regarding their management practices, while the population has increasingly utilized informa-tion and communication technologies to engage in public discourse and participate in meaningful debates.

It is important to acknowledge that the interpretation and understanding of Public Communication (PC) can vary across different countries and contexts. As Brandão [5] emphasizes, in many countries, PC is commonly associated with organizational com-munication, which involves analyzing communication within organizations and between organizations and their audiences, aiming to develop strategies and solutions. However, in the context of this study, the concept of public communication that we adopt as a reference is characterized by its ascending and bilateral nature. Its primary objective is to facilitate public participation, aligning with the principles outlined in the 1989 Federal Constitution, which advocates for participatory democracy. This includes the institution-alization of communication, culture, and tourism councils, among other mechanisms that promote public involvement and engagement.

The development of these forms of management is institutionally facilitated by the Federal Constitution of Brazil, which adopts a federative model granting autonomy to the country's states. This framework encourages the establishment of municipal and state plans, supported by specific councils and legislation. For instance, in Brazil, the organic laws of culture and the establishment of funds are instrumental in promoting the development of public policies. Notably, this approach was exemplified during the administration of Gilberto Gil and Juca Ferreira (2003–2011) at the Ministry of Culture, where initiatives such as the Culture Conferences [28] were employed to engage the public in shaping cultural policies.

In this model, the CPF framework (Council, Plan, and Fund) plays a central role. It operates through a participatory and collegial approach, allowing citizens to elect representatives from organized civil society. Together with the government, these representatives collaborate in designing sector-specific plans (e.g., culture, tourism, communication) and proposing resource allocations from development funds for each segment. The utilization of communication and new information technologies can enhance the involvement of third-sector organizations in these participatory processes. E-governance designs enable people to follow their leaders' profiles and pages on social media platforms like Facebook, Twitter, and Instagram. User engagement through comments, likes, memes, emojis, and other interactions significantly impacts the reputation of leaders and their governments. Simultaneously, it necessitates the involvement of communication professionals from various fields, such as journalists, public relations experts, and advertisers, to effectively manage these communication channels. This heightened reflexivity poses considerable challenges and new dilemmas for governance and management.

While there have been notable advancements in the utilization of new technologies in urban areas [16, 17], it is crucial to consider the situation in rural regions within Brazilian states. According to the data from the aforementioned source, only 41% of the rural population in the country is connected to the internet, which amounts to less than half of the entire rural population. Moreover, in urban centers, telecommunication and high-speed internet service providers still face challenges in delivering reliable signal coverage to all Brazilians. This is particularly evident in locations that are not prioritized within public policy agendas, such as favelas and marginalized areas. As a result, there remains a significant digital divide, with unequal access to the internet and communication technologies across different regions and populations in Brazil.

Based on our working hypothesis, we assert that when citizens are actively involved in the decision-making processes related to management plans, councils, and funds, it paves the way for effective Public Communication (PC) within territories. In a context of reflexivity, it becomes crucial to prioritize the implementation of CP, particularly through the lens of Public Relations. Transparency and mutual understanding between different publics are key elements to be achieved [20, 21]. This will be a significant differentiating factor in the future management of territories, especially for public administrations, in terms of their public communication policies.

References

1. Andrade, J.G., Dias, P.: A Phygital approach to cultural heritage: augmented reality at Regaleira. Virtual Archaeol. Rev. **11**(22), 15–25 (2020). https://doi.org/10.4995/var.2020.11663
2. Atlas Brasil IDHM: Atlas Brasil [Atlas Brazil] (2021). http://www.atlasbrasil.org.br/ranking
3. Bakhtin, M.L.: Marxismo e Filosofia da Linguagem [Marxism and Philosophy of Language] Trad. de Michel Lahud e Yara Frateschi Vieira, 10 ed., 196 p. Hucitec, São Paulo (2002)
4. Benveniste, É.: Problemas de linguística geral II. [Problems of general linguistics II]. Pontes, Campinas (1989)
5. Brandão, E.: Conceito de Comunicação Pública [Concept of Public Communication]. In: Duarte, J. (ed.) Comunicação Pública. Estado, mercado, sociedade e interesse público. Atlas, São Paulo (2012)

6. Braztoa, Associação Brasileira das Operadoras de Turismo (2021). Anuário. [Yearbook]. https://drive.google.com/file/d/15dqqAnlX1kROS3QVFG-BZxhAIBBa8qwW/view

7. Charaudeau, P., Maingueneau, D.: Dicionário de análise do discurso [Dictionary of discourse analysis]. Contexto, São Paulo (2004)

8. Fairclough, N.: Discurso e mudança social [Discourse and social change]. Editora Universidade de Brasília, Brasília (2001)

9. Floch, J.: Identités visuelles. [Visual identities]. 221 p. Presses Universitaires de France, Paris (1995)

10. Floch, J.: Sémiotique, marketing et communication [Semiotics, marketing and communication]. Sous les signes, les stratégies. Presses Universitaires de France, Paris (1990)

11. Floch, J.: Le changement de formule d'un quotidien approche d'une double exigence: la modernité du discours et la fidélité du lectorat. [The change of formula of a daily newspaper approaches a double requirement: the modernity of the discourse and the loyalty of the readership] Les Medias - Expériences recherches actuelles applications. IREP, Paris (1985)

12. Foucault, M.: Arqueologia do Saber [Archaeology of Knowledge]. Forense Universitária, Rio de Janeiro (2012)

13. Goffman, E.: Estigma. Notas sobre a manipulação da identidade deteriorada [Stigma. Notes on the manipulation of deteriorated identity], 4 edn. LTC, São Paulo (1988)

14. Gomes, W.: 20 anos de política, Estado e democracia digitais: uma "cartografia" do campo [20 years of digital politics, state and democracy: a "cartography" of the field]. In: Silva, S.P., Bragatto, R.C., Sampaio, R.C. (eds.) Democracia digital, comunicação política e redes. Teoria e prática. Folio Digital: Letra e Imagem, Rio de Janeiro (2016)

15. Harvey, D.: From managerialism to entrepreneurialism. The transformation in urban governance in late capitalism. Geogr. Ann. Ser. B Hum. Geogr. **71**(1) (1989). Roots Geographical Change, 3–17 (1989)

16. IBGE: O IBGE apoiando o combate à Covid-19 [IBGE supporting the fight against Covid-19] (2021). https://covid19.ibge.gov.br/pnad-covid/trabalho.php

17. IBGE: População estimada [Estimated population] (2021). https://www.ibge.gov.br/pt/inicio.html

18. Jost, F.: La promesse des genres [The promise of genres]. Réseaux n. 81. CENT, Kavaratzis, Michalis, Paris. From city marketing to city branding: towards a theoretical framework for developing city brands. Place Branding, vol. 1. Henry Stewart Publications, 2004, pp. 58–73 (1997)

19. Kotler, P., Keller, K.: Marketing de lugares [Seat Marketing]. Como conquistar crescimento de longo prazo na América Latina e no Caribe. Prentice Hall, São Paulo (2006)

20. Kunsch, M.: A comunicação estratégica nas organizações contemporâneas [Strategic communication in contemporary organisations]. Media & Jornalismo, Lisboa **33**, 13–24 (2018). https://doi.org/10.14195/2183-5462_33_1

21. Kunsch, M.; Krohling, M.: Comunicação organizacional [Organisational communication], vol. 2. Linguagem, gestão e perspectivas. Editora Saraiva, São Paulo (2009)

22. Maingueneau, D.: Análise de textos de comunicação [Analysis of communication texts]. Cortez, São Paulo (2001)

23. Moreira, S.: Análise documental como método e como técnica [Documentary analysis as method and as technique]. In: Duarte, J., Barros, A. (eds.) Métodos e técnicas de pesquisa em comunicação. Atlas, São Paulo (2009)

24. Peruzzo, C.: Comunicação e terceiro setor [Communication and the third sector]. In: Duarte, J. (ed.) Comunicação Pública. Estado, mercado, sociedade e interesse público. Atlas, São Paulo (2012)

25. Quéré, L.: «dispositifs de confiance» dans l'espace public [Confidence-building devices in the public space]. Reseaux, n. 132, FTR&D Lavoisier, Paris (2005)

26. Rabahy, W.: Aspectos do turismo mundial, situação e perspectivas desta atividade no Brasil [Aspects of world tourism, situation and perspectives of this activity in Brazil]. Observatório de Inovação do turismo – Revista Acadêmica, vol. 1, no. 1, ago (2006)
27. Ricouer, P.: Les références du langage [The language references]. In: Ricouer, Paul. Anthologie. Seuil, Paris (2007)
28. Rubim, A., Barbalho, A., Calabre, L. (eds.): Políticas culturais no governo Dilma [Cultural policies in Dilma's government]. Edufba, Salvador (2015)
29. Santos, M.: O espaço do cidadão [The citizen's space]. Edusp, São Paulo (2014)
30. Santos, M.: Metrópole corporativa fragmentada: o caso de São Paulo [Fragmented corporate metropolis: the case of São Paulo]. Edusp, São Paulo (2009)
31. Sampaio, A.: A marca em produtos midiáticos: o estudo do posicionamento discursivo aplicado ao telejornalismo [The brand in media products: the study of discursive positioning applied to telejournalism]. In: Ferreira, G., Sampaio, A., Fausto Neto, A. (eds.) Mídia, discurso e sentido, pp. 217–236. Edufba, Salvador, BA (2012)
32. Semprini, A.: A marca pós-moderna. Poder e fragilidade da marca na sociedade contemporânea [The postmodern brand. Power and fragility of the brand in contemporary society]. Estação das Letras, São Paulo (2006)
33. Semprini, A.: La marque [The mark]. Une puissance fragile. Vuibert, Paris (2005)
34. Semprini, A.: CNN et la mondialisation de l'imaginaire [CNN and the globalisation of the imagination]. CNRS, Paris (2000)
35. Semprini, A.: Analyser la communication [Analysing communication]. Comment analyser les images, les médias, la publicité. L'Harmattan, Paris (1996)
36. Semprini, A.: El marketing de la marca [The marketing of a brand]. Una aproximación semiótica. Paidós, Barcelona (1992)
37. Verón, E.: La semiosis social, 2. Ideas, momentos, interpretantes [Social Semiosis, 2. Ideas, Moments, Interpreters]. Paidós Planeta, Buenos Aires (2013)
38. WTTC: Economic impact reports (2021). https://wttc.org/Research/Economic-Impact

The Influence of In-Store Music on the Consumer's Shopping Experience

Manuel Sousa Pereira[1]([✉]) [ID], António Cardoso[2,3] [ID], Manuel José Fonseca[1,4] [ID],
Joana Borges[1], Bruno Pereira[1], and Tatiana Carvalho[1]

[1] Polytechnic Institute of Viana do Castelo, Viana do Castelo, Portugal
{pereiramanuel,manuelfonseca}@esce.ipvc.pt, {jborges,bfilipea,
tatianacarvalho}@ipvc.pt
[2] University Fernando Pessoa, Porto, Portugal
ajcaro@ufp.edu.pt
[3] FP-I3ID – Instituto de Investigação, Inovação e Desenvolvimento, Fundação Fernando Pessoa,
Porto, Portugal
[4] UNIAG - Applied Management Research Unit, Bragança, Portugal

Abstract. The objective of this study is to understand the sensory experience, regarding the use of sound inside the ZARA store, to determine the genre of music, as well as the influence on the consumer's purchase decision. As methodology, a quantitative methodological approach was developed, through an online survey, in a random sample of customers and consumers of Zara. A total of 221 valid answers were collected. The results obtained were relevant to achieve a greater awareness of the impact of sound on the consumer's sensory experience, as well as to identify good practices to improve the atmosphere of commercial spaces.

Keywords: Consumer Sensory Experience · Zara Store · Type of Music · Sound · Retail

1 Introduction

Today, more than ever, consumers are exposed to a set of commercial dynamics, seeking to improve the attractiveness, communication, and involvement of brands in the relationship with shoppers. Thus, commercial spaces aim to contribute to a relevant sensory experience, seeking to meet customers' expectations and increase, in a sustained way, their loyalty and increase in sales. As a result, brands tend to choose to create an environment conducive to shopping with sensory stimuli in line with their own interests, values, and desires.

This study aims to analyze and understand the sensory experience, regarding the use of sound and musical genre that may have the most influence on the purchase decision, at the ZARA clothing store. The musical genres used in this study are electronics, rock, pop, funk, chill out, classical and Rhythm and blues (R&B). In this perspective, this investigation seeks answers and evidence about the effect of the use of sound in the ZARA store, and its influence on the consumer's purchase decision.

A. Rocha et al. (Eds.): WorldCIST 2023, LNNS 802, pp. 91–100, 2024.
https://doi.org/10.1007/978-3-031-45651-0_10

Regarding the specific objectives, we sought to understand whether or not the sound sensory variable is valued by the consumer in the physical environment of the store; determine what genre of music influences more the purchase decision (Electronic, Rock, Pop, Funk, Chill Out, Classical and R&B); understand what type of consumer this variable is more involved; understand the activation of this stimulus from the consumer's point of view.

To support this study, the collection of primary data occurred through an online questionnaire survey, with 221 valid answers), which are believed to provide essential insights regarding the analyzed topic.

This study is structured as follows: abstract, keywords, introduction, literature review, methodology, results analysis, conclusions, and bibliography.

2 Literature Review

Organizations are in constant interaction with their stakeholders, as their success depends on the ability to interact and add value to their surroundings and, fundamentally, to their customers. In this context, the relationship with the consumer and the degree of loyalty are fundamental factors for a sustainable maintenance in the market, both in its ability to sell its products and in the ability to maintain a commercial, but fundamentally emotional, relationship with the brand, through of its meaning and attributes, as mentioned [10]. In addition, the in-store experience is also a distinctive factor in the creation of emotions that mark the consumer's memory, accentuating its strategy of emotionally involving the consumer in the brand's values.

According to the understanding of Rieunier [11], sensory marketing should be a tool to improve the atmosphere in the store, with the objective directed to the personalization of the surrounding space, attributing to it the ability to stimulate certain emotions and positive ties in the customer's mind. Adding that, the use of activation of the five senses (smell, taste, vision, hearing, and touch) fundamental for the relationship of the human with the environment must be applied.

For Chen [3] atmosphere has been defined as the conscious design of spaces to create certain shopper effects, particularly the design of shopping environments to create specific emotional effects among shoppers that increase the likelihood of purchase. Thus, these authors also mention that music has been recognized as an important environmental component that influences consumer behaviour.

Authors such as Esfidani, Rafiei Samani and Khanlari [9], reviewed previous studies on the relationship between music and consumer behavior, and described that the most relevant variables were music genre, song duration and applied volume.

Regarding the store environment, it should provide an atmosphere of physical and emotional comfort, focused on the people who visit it, seeking a positive and experiential stimulus relevant to the customer's mind. As stated by Bonfanti and Yfantidou [2] in some studies that focused on the sports store environment both in physical and digital terms [7], and store atmospheres, especially music [6]. Complementarily some authors [4] states that, over the years, variables have been developed to study the impact of elements on retail experiences including atmosphere [1], visual, auditory, tactile, olfactory, and gustatory aspects [14] and merchandise with assortment and interaction with the team [15].

According to some authors [13], a large proportion of customers list the environment as the main reason to enjoy a retail store, above friendliness of staff or customer service. These environmental elements can invoke visual (e.g., lighting, brightness, contrast, zoom capabilities), auditory [5] (e.g., background music, in-store noises), olfactory (e.g., aromas from the bakery to some aisles) and the senses of touch.

On the impact of music inside commercial spaces Witek et al. [16] mention that music is a powerful driver of experiences related to shopping, affection, and movement. Listening to music often creates feelings of pleasure and induces spontaneous movements, such as tapping the foot, shaking, or shaking the head, contributing to the creation of positive and differentiating experiences that can improve relationships between customers and the commercial store.

3 Research Methodology

3.1 Data Collection

As a methodology, we developed a quantitative study, with the objective of obtaining ideas, attitudes and behaviors from consumers who know, visit, and usually buy in ZARA stores. We built a questionnaire with closed questions and only one open, to obtain user opinions. In the construction of the questionnaire, the works identified in the literature were taken into account, namely the investigations of Ballantine, Parsons, and Comeskey (2015), Bonfanti and Yfantidou (2021), Chen et al. (2022), Kapoor (2016), and Esfidani, Rafiei Samani, and Khanlari (2022). The objective is to understand this theme in theoretical and practical terms, with a non-probabilistic random sample.

The application of this online questionnaire considered the option of a period that coincides with a greater flow of customers to the ZARA store (promotional season - January 10th to 17th), seeking to obtain results that are more consistent with the consumers' shopping experience. Thus, to understand the effect of sound, in a commercial context and with the application of different musical genres (Electronic; Rock; Pop; Funk; Chill Out; Classical and R&B), using the scale adapted from the authors [5] that corresponds to the general objective and the specific objectives of studying.

This study also aims to answer the research objectives described above, using statistical approaches. The respondents to this research were consumers of legal age, who frequent the ZARA store in Portugal. 235 surveys were collected and 221 were validated (94.04%). The data was analyzed using SPSS 25.0 software.

To assess the internal consistency of the research instrument, Cronbach's Alpha was calculated ($\alpha = 0,812$), resulting in a score considered good [10]. Since this is an exploratory study with a descriptive design [8], univariate data analysis was performed, based on descriptive statistics (absolute and relative frequencies, means, and standard deviation).

3.2 Sample Characterization

As can be seen in the Table 1, the respondents were mostly female (72,3%), with 61 male respondents (27,7%). In terms of age, the respondents were between 35 and 44 years old" (28,6%), followed by the "18–24 years" age group with 52 respondents (23,6%), then the "25 to 34 years" age group (22,7%), and "45 – 54 years" age group (20%). Only 11 respondents (5%) were more than 55 years old. Regarding income, most respondents earn a monthly income between "500 to 999 euros" (43.6%), followed by the category "1000 to 1499 euros (20.5%). That 30 respondents (17.3%) have an income of less than 500 euros and 6 (4.5%) have no income, only 16 respondents (7.3%) have an income of more than 2000 euros.

Table 1. Sample Characterization

		F	%
Gender	Male	61	27,7
	Female	159	72,3
	18–24 years	52	23,6
	25–34 years	50	22,7
Age groups	35–44 years	60	28,6
	45–54 years	44	20,0
	55–64 years	9	4,1
	>65 years	2	0,9
	<500 euro	38	17,3
	500–999 euro	96	43,6
Income	1000–1499 euro	45	20,5
	1500–1999 euro	15	6,8
	<2000 euros	16	7,3
	No income	6	4,5

Source: Created by the authors

4 Results

4.1 Evaluation of Musical Stimuli in the Store Environment

Regarding consumer preference in relation to music genres, it is important to note that respondents were given the possibility to choose several options, as well as to add an option that is not present (Note: Respondents could choose more than one alternative).

Table 2. Music genres preferences

Genres							
	Pop	Rock	Electronic	Funk	Chill Out	Classic	R&B
F	146	71	34	37	61	62	53
(%)	(66,4)	(32,3)	(15,5)	(16,8)	(27,7)	(28,2)	(24,1)

Source: Created by the authors.

Based on Table 2, there is a clear preference for the "Pop" option, with the choice of 146 people (66.4%). Furthermore, the choice of musical genres, such as: "Rock" by 71 people (32.3%), "Classical" by 62 people (28.2%) and "Chill Out" by 61 people (27.7%).

As shown in the Table 3, when asked what respondents value in a store environment, most consumers prefer "organization", a choice of 199 respondents (90.5%). This is followed by "lighting" by 119 respondents (54.1%), "store layout" by 86 respondents (39.1%) and "music" and "temperature" by 73 respondents (33.2%). This demonstrates the prioritization of other elements instead of music (respondents could choose more than one alternative). Despite sharing the third most chosen option with "temperature", it represents just over a third of the choices, indicating a certain importance, however reduced, compared to the most chosen ones.

Table 3. Elements most valued in the store environment

	Lighting	Music	Organization	Temperature	Olfactory Stimuli	Store Layout
F	119	73	199	73	28	86
(%)	(54,1)	(33,2)	(90,5)	(33,2)	(12,7)	(39,1)

Source: Created by the authors.

About the Evaluation of musical stimuli in the store environment, more than half of the respondents feel that it is imperative that stores have a musical environment, with 166 people (75.5%) responding that they strongly agree, or agree, and only 9 (4.1%) fully disagreed. However, 33 individuals neither agree nor disagree that it is essential for a store to have a musical environment (M = 4,06; ST = 1,069).

Regarding the position on ambient music (level of indifference) in stores, 42 respondents (19.2%) fully disagreed about feeling indifferent to the music present in stores, 48 (21.8%) disagreed, which makes us realize that about 90 respondents (41%) do not feel indifferent to music in the store. Already 31.8% of respondents (70) responded that they agree or that they fully agree that they are indifferent to this factor. At the however, the largest share of responses, 60 respondents (27.3%) show that they neither agree nor disagree. It is concluded that we cannot effectively deduce that indifference in this case marks a decision in the existence of music in stores, since the answers are distributed, almost unanimously, for each option (M = 2,94; SD = 1,010).

Most respondents, 175 people (79.5%) strongly agree and agree that they feel more comfortable in a store with background music. Ten people disagree or partially disagree, a tiny part of a sample that is almost entirely dependent on the fact that music is one of the factors influencing the comfort of customers in the store. However, here are still some answers from people who neither agree nor disagree, corresponding to 15.9% of the sample (35). After this analysis, it is stated that the vast majority of people feel more comfortable with in-store music when they go to the ZARA store (M = 4,21; SD = 0,892).

Eighty-nine respondents (40,5%) strongly disagree that they have stopped making a purchase due to the store's ambient music, while forty (18,2%) strongly agree that they have stopped buying from this brand due to this factor. The remaining feedbacks are, however, more balanced by the other options. However, this graph refers to the importance of in-store music, when it claims that 62 people were highly influenced not to complete the purchase because the sound stimulus in the store was not pleasant (M = 2,51; SD: 1,533).

We can also see that the store's ambient music interferes with its customers' decision to stay in the store or leave, having a direct impact on the result of its sales (M = 3,19; SD = 1,315). Thus, 42 respondents (19.1%) fully agree that music influences their decision to stay or not in the space and 56 (25.5%) agree. Fifty-nine respondents (26.8%) neither agree nor disagree with this direct influence, while 28 respondents (12.7%) disagree with this direct impact and 35 individuals (15.9%) totally disagreed with being directly influenced by this factor in their decision to stay or not in the store. In the case of the influence that music has on the purchase decision, a total of 51.3% of respondents disagree that music influences their decision, and among these, the majority, 31, 8% strongly disagree, while only 21.8% agree that there is influence. On the other hand, 26.8% do not have a formed opinion on the subject. This result demonstrates that in most cases, music does not influence the purchase of respondents in this study, this may be linked to the fact that the influence is more related to the unconscious, as can be seen in the graph below (M = 2,48; SD = 1,300).

The time that a consumer spends in a store influences his purchase, because it can make him buy more, even if unconsciously. The data show that the type of music influences the time consumers spend on it (M = 3,31; SD = 1,383). Most respondents, 52.3% of respondents, say that music influences the time they spend on it. While only 30.4% disagreed on the matter. For this question, 17.3% of respondents neither agree nor disagree on the matter.

The answer to the question "Music in stores makes the shopping experience more pleasant" was almost unanimous, as 69.1% of respondents agreed with it, while only 10.9% disagreed with the matter. On the other hand, 20% did not show a formed opinion. These results show that music creates an impact on consumers and that, perhaps, if it were non-existent, they would notice a difference, and perhaps show displeasure (M = 3,89; SD = 1,083).

Predictably, as can be seen from the Table 4, only 17.8% of respondents recognize choosing a store based on its musical style, with only 7.3% fully agreeing with the statement. On the other hand, 56.3% say they do not have music as a basis for choosing the stores they decide to visit. Even so, 25.9% of the people questioned for the present study do not have an opinion on the subject (M = 2,35; SD = 1,237).

The rhythm of the music doesn't seem to have much influence on the shopping experience (M = 2,87; SD = 1,314). Most respondents (30%) have a neutral answer to this question. Whereas, 13.6% fully recognize that there is influence in it, and 21.8% fully believe that there is no influence whatsoever.

On the other hand, the data show that the volume of music has a strong impact on the consumer experience (M = 3,73; SD = 1,208). In this question, only 22.3% of the respondents (49) were indifferent, while 61.3% (135) said that the volume has, in fact, strongly agree and agree that music volume changes consumer shopping experience.

The genre of music that most respondents (115) associate with the Zara store (Table 5) is pop music (53.7%). This is followed by the musical genre "Chill Out" (30.4%) and "Classical" (23.8%).

Regarding the type of emotions that ZARA's music conveys to costumer (Table 4), most of them are positive emotions, which even allude to consumer happiness. The main ones are: comfort (with 53.6% of responses, the majority); indifference (23.2%) and enthusiasm and happiness (23.2% and 20.9%, successively). From the set of possible answers, these were the main ones selected, however there were other equally interesting answers, but with less focus. These are nostalgia (4.5%), melancholy (4.5%) and anxiety (3.2%). From these results we can see that, in general, ZARA consumers feel comfortable with the music that it broadcasts in their physical spaces.

Table 4. Kind of emotions does ZARA's music convey to the costumer

Emotions						
Nostalgia	Happiness	Melancholy	Comfort	Anxiety	Enthusiasm	Indifference
F 10	46	10	118	7	47	51
(%) (4,5)	(20,9)	(4,5)	(53,6)	(3,2)	(21,4)	(23,2)

Source: Created by the authors.

Regarding the assessment of background music in Zara stores (Table 5), the results show that there is high satisfaction with the musical environment of the store (M = 3,71; DP = 1,018)), but the genre of music has no influence on the choice of store (M = 2,01; DP = 1,188) or drives shoppers away from the store based on the music it plays (M = 3,95; DP = 1,019). In fact, the majority of respondents (127) are satisfied with the music played in the store (57.7%), they will continue to shop at Zara no matter what kind of music is going on (68,6%), but they don't choose Zara because of the genre of music that it plays compared to other stores (70,9%).

As can be seen from Table 6, about 49.1% of respondents (108) say that music has no impact on their purchase decision and about 25.9% (57 people) say it has little impact. On the other hand, about 3.6% of respondents (21 people) believe that music has a great or huhe impact on the purchase decision. However, 28 respondents (7.7%) consider that background music "has neither little nor much impact" on the purchase decision process.

5   Conclusions, Limitations and Future Research Lines

As the most valued elements in the store environment, we could verify that more than half of the respondents (75.5%) consider imperative that stores have a musical environment, and the majority of respondents (79.5%) agree that they feel more comfortable in a store with background music. However, 40.5% strongly disagree that they stopped making a purchase due to the store's background music. We can also observe that the ambient music in the store interferes with the decision of customers to stay in the store, as authors such as Esfidani, Rafiei Samani and Khanlari [9] mentioned in previous studies

Table 5. Zara store background music review

	Strongly disagree 1	Disagree 2	Undecided 3	Agree 4	Strongly agree 5	M	SD
	F (%)	F (%)	F (%)	F (%)	F (%)		
I am satisfied with the genre of music on ZARA store	7 (3,2)	13 (5,9)	73 (33,2)	70 (31,8)	57 (25,9)	3,71	1,018
I choose ZARA because of the genre of music that plays, compared to other stores	104 (47,3)	52 (23,6)	37 (16,8)	15 (6,8)	12 (5,5)	2,01	1,188
I still shopped at ZARA no matter what kind of music	7 (3,2)	8 (3,6)	54 (24,5)	72 (32,7)	79 (35,9)	3,95	1,019

Source: Created by the author

Table 6. Music impact on purchase decision

Items	No Impact 1	Little impact 2	Undecided 3	Lot impact 4	Huge impact 5	M	SD
	F (%)	F (%)	F (%)	F (%)	F (%)		
What impact does music have on your purchase decision?	108 (49,1)	57 (25,9)	28 (7,7)	17 (1,8)	4 (1,8)	1,84	1,057

Source: Created by the authors.

on the relationship between music, musical genres and duration. Can directly affect the sales result. As for the influence that music has on the purchase decision, 51.3% of the respondents disagreed that music influences their decision. About the time the consumer spends in a store, music can contribute to increase sales volume, even if not assumed. Thus, when asked about the impact of music on the purchase decision 49.1% of respondents say it has no impact and 25.9% say it has little impact. However, about the appreciation of the buying experience and its friendliness, 69.1% answered that music influences positively. As for the type of emotions that ZARA music conveys to people, most are positive emotions, which even evoke the consumer's happiness.

Regarding the limitations of this study, we can mention that a qualitative analysis should be carried out with the managers of the store, as well as a comparative analysis with other commercial stores, seeking to have a broader and more complete view of the subject. As future work, we can list the implementation of a more representative sample and the application and validation of the questionnaire in other commercial spaces.

Acknowledgment. UNIAG, R&D unit funded by the FCT – Portuguese Foundation for the Development of Science and Technology, Ministry of Science, Technology and Higher Education. Project no. UIDB/04752/2020.

References

1. Ballantine, P.W., Parsons, A., Comeskey, K.: A conceptual model of the holistic effects of atmospheric cues in fashion retailing. Int. J. Retail Distrib. Manag. **43**(6), 503–517 (2015)
2. Bonfanti, A., Yfantidou, G.: Designing a memorable in-store customer shopping experience: practical evidence from sports equipment retailers. Int. J. Retail Distrib. Manag. **49**(9), 1295–1311 (2021). https://doi.org/10.1108/IJRDM-09-2020-0361
3. Chen, D., et al.: How background music of shopping sites affects consumers during festival season. Cogn. Comput. Syst. **4**(2), 165–176 (2022). https://doi.org/10.1049/ccs2.12044
4. Dalmoro, M., Isabella, G., de Almeida, S.O., dos Santos Fleck, J.P.: Developing a holistic understanding of consumers' experiences: an integrative analysis of objective and subjective elements in physical retail purchases. Eur. J. Mark. **53**(10), 2054–2079 (2019)
5. Duman, D., Neto, P., Mavrolampados, A., Toiviainen, P., Luck, G.: Music we move to: audio features and reasons for listening (2022). https://psyarxiv.com/nye58
6. Kapoor, R.R.: The effects of in store music on shopping behaviour in a retail setting. Manag. Dyn. **16**(1), 96–108 (2016)
7. Koontz, M.L., Gibson, I.E.: Mixed reality merchandising: bricks, clicks – and mix. J. Fash. Mark. Manag. **6**(4), 381–395 (2002). https://doi.org/10.1108/13612020210448664
8. Malhotra, N.: Marketing Research: An Applied Orientation. Person Education, UK (2019)
9. Esfidani, M.R., Rafiei Samani, S., Khanlari, A.: Music and consumer behavior in chain stores: theoretical explanation and empirical evidence. Int. Rev. Retail Distrib. Consum. Res. **32**(3), 331–348 (2022)
10. Pestana, M., Gageiro, J.: Data Analysis for Social Sciences. The Complementarity of SPSS. Sílabo Editions, Lisbon (2014)
11. Ramos, M.G.: Branding sensorial: a relação marca x consumidor criada pela ambi- entação das lojas. In: Universitas: Arquitetura e Comunicação Social, vol. 8, no. (2) (2012). https://doi.org/10.5102/uc.v8i2.1333
12. Rieunier, S., Dion, D.: Le marketing sensoriel du point de vente-4e éd.: Créer et gérer l'ambiance des lieux (2013). https://doi.org/10.1177/076737010201700408
13. Roggeveen, A.L., Grewal, D., Schweiger, E.B.: The DAST framework for retail atmospherics: the impact of in- and out-of-store retail journey touchpoints on the customer experience. J. Retail. **96**(1), 128–137 (2020). https://doi.org/10.1016/j.jre-tai.2019.11.002
14. Spence, C., Puccinelli, N.M., Grewal, D., Roggeveen, A.L.: Store atmospherics: a multisensory perspective. Psychol. Mark. **31**(7), 472–488 (2014)

15. Terblanche, N.S.: Revisiting the supermarket in-store customer shopping experience. J. Retail. Consum. Serv. **40**, 48–59 (2018)
16. Witek, M.A., Clarke, E.F., Wallentin, M., Kringelbach, M.L., Vuust, P.: Syncopation, body-movement and pleasure in groove music. PLoS ONE **9**(4), e94446 (2014)

Marketing Plan for an Online Business: A Case Study

Yuri Souza[1], Manuel José Fonseca[2]([email]) [ID], and Sofia Cardim[1] [ID]

[1] Instituto Politécnico de Bragança, Braganza, Portugal
[2] Instituto Politécnico de Viana do Castelo, Viana do Castelo, Portugal
manuelfonseca@esce.ipvc.pt

Abstract. The aim of this study is to define the most appropriate marketing strategy, as well as develop a marketing plan to increase brand recognition, for a newly created Brazilian company called Probate for Dummies, which operates in the online resale of inherited assets. The data that provide the basis for this study was generated through semi-structured interviews applied to a sample formed by the owner of the company, followed by specialists in different fields of knowledge which may directly influence the business. As for the main outcomes, this project unfolds the key factors of both micro and macro environments of the company, in the form of a SWOT and PESTEL analyses, as well as the analysis of the Porter's five competitive forces. In the sequence, a marketing central strategy was delineated through the market segmentation, target selection and positioning definition, which guided the marketing objectives to be pursued. Thereafter, based on those marketing objectives, a marketing mix was composed based on the 7P's (Product, Price, Placement, Promotions, People, Process, Physical Evidence). Furthermore, an operational plan composed by specific actions addressed to each marketing goal was settled, along with the budget and schedule of implementation. Regarding the conclusions, a market penetration strategy shows to be the most suitable for this business, with development of new services for the costumers of the niche it acts. Yet, the strategy is intensively based in terms of promotion, especially in segmented digital marketing.

Keywords: Digital Marketing · Marketing Plan · Service Marketing · Probate

1 Introduction

The Brazilian online real estate resale and the online second-hand goods trading segment are greatly explored markets and with well stablished players, with decades of experience in these segments, however, from the point of view of a new online business called Probate for Dummies, there is the still room for exploring a new market segment, which is implicitly embedded within those two previously mentioned, namely this would be the market of online trading of inherited assets in Brazil. In this context, the new start-up considers that with the appropriate marketing strategy, it is possible to develop and explore a promising market, and that it has potential growth and profitability, especially for the company that takes the first action and recognizes the hidden potential of this market. Despite of having low resources and being in its early of development, the company believes that with a noticeable propose and vision for the future, the internet era makes it possible for an entrepreneur to start developing a new digital business, and attempt to conquer, even in a slowly pace, a position among the competitors.

Thus, it is under such circumstances that this research presents its relevance, and seeks to answer questions such as "What is the efficient way to practically apply the marketing theories for a new and small business and with limited resources?" or even "How to define a marketing plan for a company at is in an early stage of its existence, but with a very broad target audience and in a country as big as Brazil?".

Therefore, to try to answer these questions, and considering the context already presented, the main objective defined for this study is to define the most appropriate marketing strategy, as well as develop a marketing plan to increase brand recognition, of a newly created Brazilian company called Probate for Dummies, which operates in the online resale of inherited assets.

To support this study, the collection of primary data occurred through four semi-structured interviews, which were applied to the owner of the company and to specialists from three different areas of knowledge (a Lawyer, a Psychologist, and a Real Estate Agent), which are believed to provide essential insights regarding the analysed topic. The content of these interviews was recorded, transcribed, and processed for further qualitative analysis. In addition, secondary data was collected from the review of public documents, the company's website, and bibliographic research.

2   Literature Review

The concept of marketing took 70 years to evolve from the initial product-driven orientation to the human centric perception, however in the last decade as the world started to move towards the digital media and social networks, it was inevitable for the marketer to pivot, to follow the course of technological innovations, thus adapting its frameworks in order to serve customers in the hybrid of physical and digital stages of the new buying journey [3]. Nowadays, with technologies such as artificial intelligence (AI), big data analysis, internet of things (IoT) and augmented reality becoming mainstream, companies start to unleash the full potential of the use of tech-driven marketing in its strategies and operations [4]. By speeding up the decision making based on big data and creating value for the customer in an individual and customised level, marketing has become now more predictable and cost-effect than ever, but still with the fundamental concept of the human as its essential base [4].

Similarly, to the marketing of products, when thinking about services marketing we can say that the main goals pursued are to identify and fulfil the desires of the customer, thus creating value for all parts involved. However, when there is no tangible product, things become subjective and abstract, and that is the essential challenge of service marketing [2].

The difficulty when talking about service marketing starts while trying to define what is a service, since there are many interpretations regarding the concept, and although within the academic literature, multiple designations have been proposed, there is still no consensual definition of the concept [8].

As reported by Westwood [12], marketing strategy means the allocation of resources dedicated to marketing in order to meet companies' overall goals. The writer interprets marketing strategy as the process of establishing the segments, targets and positioning, the company should follow to face and fulfil costumers' demands. Naim [9] points out that top-down models like the STP became more appealing over the years as companies pivot to delivering content to their target audiences by digital channels and social media.

In addition Naim [9] claims that marketers should divide the market and adopt the fundamental scheme of strategic marketing, that initiates with segmentation, stage where companies figure out the essential characteristics of group of customers within the market, followed by the definition of the target, when the company selects the group that will be more receptive to the offer, and finally the definition of the positioning, which reflects the perception that the company transmits to its targeted audience.

Referring to the Marketing Mix, Kotler and Keller [5] defines it as a tool used by the company to reach its marketing objectives. This tool is composed of a set of controllable variables that allow the company to take its idealized strategies and put it concretely into the market [14]. For Wichmann et al. [13] the marketing mix, its instruments, and its function in the value-creation process have evolved considerably over the past decades, and especially in the recent years through technological advances such as Internet of Things (IoT), smart portable devices, and an increased effectiveness in the collection, analysis and application of data, have empowered companies to progressively customize each of the marketing mix instruments to ever smaller segments.

Regarding the marketing plan, it defines the marketing nature of the business and what that organization will do to satisfy its customers' needs in the marketplace [10]. The authors still add that marketing plan is not just a scholar concept of use to academics but is an immensely practical exercise that can mean the difference between success and failure to all types of organizations, and in order of it to be successful, it must be founded in a conceptual framework that provide basis for analysis, execution, and evaluation [10].

3   Research Methodology

3.1   Data Collection

In this study, data was collected by the application of four semi-structured interviews with discussion topics, so the respondent could have the autonomy to dialog and provide a discourse as complete as judged necessary to cover the subject. This method was chosen to enable extent, in-depth understanding and detailed information gathering [7]. Thus, two interview guides were formulated in accordance with the topics necessary for the elaboration of the marketing plan and are divided in the following four dimensions: Organizational Diagnoses, Central Marketing Strategy, Marketing-Mix, Future expectations.

To assist ensuring accuracy and validity, the interviews were recorded, in concordance with Sutton and Austin [11] who say using a recording device will remind the researcher of critical situational factors discussed during the interview and prevent data loss during the analysis stage.

3.2 Data Analysis

The data treatment of the audio recordings started by the transcriptions of the contends, followed by the fragmentation of phrases and sentences, that allowed for further categorization, necessary for the development of this marketing plan.

To perform the categorization, the researcher distributed the fragmented transcriptions in consonance with the four dimensions (organizational diagnoses, central marketing strategy, marketing-mix and future expectations) of the questions. Therefore, to perform the distribution, the first criterion followed was to address each fragment correspondingly to the theme of the question it was answered. Secondly, since the questions were open, in the cases where the fragments of answer deviated from the theme initially discussed, the distribution occurred in a manner that the researcher judged helpful to understand the overall subject.

It is the combination of the information and insights provided by the three experts, along with the secondary data analysed, the theoretical framework here reviewed, and the interview of the owner himself, which aims to obtain information that would allow a coherent internal analysis, is the foundation for the completion of this study's objectives.

The sample is composed by a total of four elements, being the first of them the owner of the company, and the three others are experts from different fields that impact or have a strong connection with the business. Therefore, a Lawyer, a Psychologist and a Real Estate Agent were selected, considering that the knowledge and experience of these professionals will assist, from different perspectives, the comprehension of the business, and provide insights from each of their specialization areas, regarding the surrounding environment of the company. Malhotra and Birks [6] consider that this type of sampling, selected by the convenience of the researcher, proves to be suitable for several situations, including market tests to determine the potential of a new product. According to Brown [1], the lower statistically significant of this sample is sufficient, since this is an exploratory research by nature.

4 Results

4.1 Interview Insights

From the examination of the recorded content, insights regarding the overall environment of the business, strategic choices, services features, and futures expectations were organized. Figure 1 shows these main insights and ideas that were later integrated in the basis of the marketing plan.

4.2 Structure of the Plan

Based on the evaluation of the interview's contents, the following steps were carried out in the marketing plan for the company Probate for Dummies: first, an investigation of the surrounding situation of the company, with an analysis of its internal and external environment. Thereafter, the marketing objectives were defined, and the marketing strategies outlined in the form of segmentation, targeting, brand positioning. Then, a mix of marketing was framed, matching the previously defined strategy, and finally an operational action plan, regarding the marketing objectives, was developed.

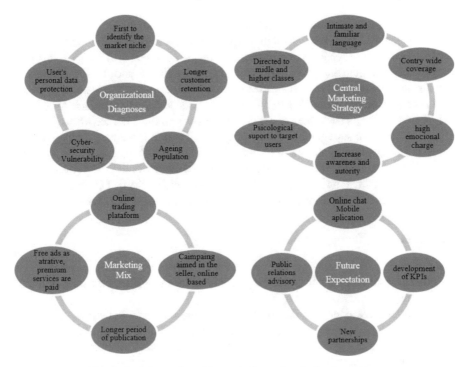

Fig. 1. Insights gathered for each dimension during interviews

5 Marketing Plan

5.1 Situation Diagnosis

Regarding the Macroenvironment, the PESTEL analysis shows that the legal scenario and bureaucracy of the probate subject are a significant opportunity for the company. Moreover, the increasingly number of people using internet are favourable to business development, however followed by a threat related to the cyber-security.

In relation to the Microenvironment, the characterization of the costumers of the company is of a Brazilian citizen, men and women, mainly urban resident, that recently lost a family member who left assets (heritage) that need to be transferred, through a legal/administrative process called Probate. In terms of suppliers, the website needs a hosting company, the online domain, and the platform for editing the website, in this case, Hostgator and Wordpress. As for the competitors, it is necessary to consider that the brand is positioned among the actual online real estate and second-hand assets trading, which have in Brazil three main brands called Vila Real, ZAP Imóveis and OLX.

Considering the Porter's five forces analysis, it can be presumed that the level of attractiveness of the market for online trading of inventory goods is medium high, if the company develops the appropriate marketing strategy. Thus, one possible approach to try to reduce the effects main forces that have potential to affect the company, might be the development of partnership and affiliation with other companies which offer complementary services or products that might interest the customer, such as online insurance, accountants, psychological support, among other to be further identified. Thus, the result of this synergies might help mitigate the effect of these forces.

Concerning its Internal environment, the company's mission is to bring a humanitarian look and support in the moment of intense emotional charge, and provide clarification on bureaucratic demands, as well as offering emotional support to the customer. Regarding the organizational vision, the company aims to become the first and most important website in helping its customers in generating value from the inherited assets.

Lastly, the main strength identified in the SWOT matrix is the fact of being the first website to act over this niche, on the other hand, as an online business, it is important to take actions to sustain the security and integrity of the service online. Moreover, besides in scope of the site being so restricted to a particular niche, the fact of Brazil having a population of over 220 million people implies in a compelling absolute number of users.

5.2 Marketing Strategy Definition

The central marketing strategy for Probate for Dummies was defined in terms of the market segmentation, selection of target market and the definition of the positioning for the company.

Concerning the demographics segmentation, the company aims in a segment of consumers with no specific gender, from classes B and C (middle-high and middle) that are aged between twenty-five to fifty-five years old, since usually people within this age range are more susceptible to need the services and have the highest rates of internet usage in Brazil. Moreover, regarding the psychographics and behavioural segmentation, the consumer that is going to need the services is defined as someone that uses internet and is going through a heritage process or has doubts about the process and is willing to advertise the inherited goods. Ultimately, in relation to the geographic segmentation, since the company is internet based, it is intended to serve costumer from all over Brazil.

The company's market targeting consists of people from B and C class, from 25 to 55 years old, mainly living in the bigger cities of Brazil, who have lost a family member which has left assets that need to be transferred, through an administrative or legal process called the Probate/heritage/Inventory, and are willing to sell those inherited assets, and use the internet to search for answers about how to do this.

In relation to the product identification, it can be defined as an online marketplace for trading heritage assets that focus on features that better suit the particular needs of the online sale of a heritage asset. Therefore, the brand will offer new and specialized options of the service, the main one being the longer time that the heritage process takes in Brazil. Finally, the points of differentiation to be outlined are the intimate and familiar language of the content of the website, along with the focus on the well-being of people that is already going through a loss of a loved one, encompassing whenever possible, the psychological aid for the familiars how need it, and create a mutual beneficial/win-win relationship, linked to the recognition of the emotional charge associated to this targeted customer context.

5.3 Marketing-Mix

Resulting from the strategic guidelines developed and the Marketing objectives determined, a marketing mix composed by the 7P's was defined.

To create value for the targeted consumer, the product to be developed can be defined as an online platform for buying and selling heritage goods. This platform provides the opportunity for sellers and buyers to interact with one another and perform transactions and deals. For the classified ads, the company provides the service of listings of featured ads and free ads. Featured ads will have at least three variations of offerings: 3 month featured ads, 6 month featured ads, 1 year featured ads. In parallel, another source of income for the website is the advertising banners and affiliate sales.

In addition, a freemium model of prices for its listings in going to be employed. The logic of this model the free main offer, but after that, the company offers paid options. It means those users who want their ads to be highlighted and listed at the top of search results, will have to pay fees for their listings.

The promotion strategy adopted aims in creating awareness in the market about the possibility of buying and selling assets of heritage online, since there was not a stablished business of this kind in Brazil before. To shape the message more appropriately for each customer, the communication strategy was divided in two distinct personas, the first one being the Potential Customer, and the second persona is the Current Customer. It was chosen to have a campaign focused on the sellers, because it is considered that if the brand can attract sellers, then it becomes attractive to buyers, as they know they will find more products available. The message of the promotional campaign emphasises how effortless it is to post an online ad of a heritage asset, reinforcing the key features that distinguishes the brand against its competition. This campaign aims not only to sell a product/service, but also to create awareness on consumers and show presence in the market. Regarding its placement, Probate for Dummies is present on a website platform, and additionally there is a second distribution channel prospected, that is going to be a free mobile application.

The brand's communication mix consists of three digital channels techniques, as well as the development of public relations advisory, which makes possible to reach the offline market when applicable (Table 1).

Table 1. Communication Mix

Target	Communication Technique	Effort
Potential Customer	Digital Marketing	Content Writing (Blog)
Potential Customer	Digital Marketing	Social Media Marketing
Potential Customer	Digital Marketing	Search Engine Optimization
Potential Customer	Online Advertising	Display advertise (per click)
Current Customer	Online Direct Marketing	E-mail Marketing
Current Customer	Online Direct Marketing	Online Chat
Mass Media	Public Relations	Press Relation Advisory

In relation to the physical evidence, there are four relevant elements that can serve as differentiators in signaling the intended positioning of the brand, the first being the brand's name, which is self-explanatory, followed by the Blog's writing style, that follows the reader's emotional characteristics, the content structuring of the blog, and the overall appeal of the website.

The website will have two main processes in place, both beginning with the buyer or seller searching on internet about the probate procedure, this step is direct related to the organic traffic generated by the content published in the articles of blog.

Then, the process of the buyer interested in the listings in the trading platform, it is designed to be as simple as possible. Buyers can search the products based on asset categories, geographic region, price, among others. In the event of the buyer being

Table 2. Marketing Actions and Budget

OBJECTIVE	WHAT TO DO	WHEN	COST
O1 - Reach at least 100 visits a month	**A1.** Search trending Key words on Google search console	Quarterly	€ -
	A2. Upload 2 articles monthly with at 500 words	Jan to Jun	€ -
	A3. To apply updates to the security of the website	Jan to Dec	€120,00
O2 - Expand visits by 5% monthly	**A4.** Continuous improvement of SEO strategies	Every 2 months	€120,00
	A5. Apply Google Adwords/Adsense/Facebook Ads	Jul to Dec	€300,00
	A6. Upload 4 articles monthly with at least 500 words	Jul to Dec	€ -
O3 - Reach 5% Conversion Rate for the featured Ads	**A7.** Development of and improvement of the Trading Platform. Testing and experiments	Jul to Dec	€1 000,00
	A8. Offer free ads and 3 types of featured ads	Jul to Dec	€ -
O4 - Public relations advisory	**A9.** Send press releases to online/offline vehicles	Quarterly	€ -
O5 - Diversify the services provided	**A10.** Creating a section in the plataform for Emotional and Psicological support for clients	Long-term: 2nd yr	€700,00
	A11. Implementation of an online chat, for taking doubts of costumers	Long-term: 2nd yr	€700,00
O6 - Build new partnerships	**A12.** Search for companies/professionals that might Fit in the new services	Long-term:3rd yr	€ -
	A13. Negotiating partnership conditions	Long-term: 3rd yr	€ -
	A14. Development of new features with the new partners, on the website	Long-term: 3rd yr	€500,00

Total   €3440,00

interested in a product, the contact with the seller is possible either by clicking on "Make an Offer" button, or by the "Chat with the seller" option.

Lastly, to sell a product, the user must click on "Submit a Free Ad" button, enter the product details, include the title of the add, photographs, and other details of the product. Once the ad is posted, a message which contains a link for "pay and sell faster" option is sent via email.

5.4 Operational Plan

In accordance with the marketing objectives defined and considering the strategy to be adopted in this plan, several actions and targets were defined to be carried out. The Table 2 shows each objective and its corresponding actions, as well as the timing and the costs foreseen for the implementation.

6 Conclusions, Limitations and Future Research Lines

Among the main results unfolded by this study, it was found that the best opportunity regarding the environment of the company lays on the high volume and size of the heritage assets market in Brazil, along with the fact of being the first company to specialize itself in this market niche. Such opportunities underpin the company's decision to pursue a market penetration strategy through a new online platform. This strategy is financially sustainable since the time taken to carry out an inventory lawsuit in Brazil is an average greater than two years, which makes the customer retention period longer. Moreover, the focus on content marketing must be considered a priority for the brand, and along with communication campaigns in social media, are the strategy selected to reach to the targeted clients. On the other hand, the cyber-security of the website was exposed as a vulnerability to the business, followed by the fact of it being a unknown brand. However, mitigation measurements for both cases were included in the operational plan created for the business.

Therefore, by accounting all data acquired, that allowed a comprehensive understanding of the business, its environment, and stakeholders' points of view, which allowed the constitution of the marketing plan for this company, it is understood that the research objective was entirely accomplished. Besides the reliable knowledge developed for this company, this research also contributes to other new and small online enterprises, that are in the early stages of its operation, and need an example of methodological framework about a practical strategic and marketing plan development.

In spite of all the mentioned, there are limitations in this study that could be addressed in future researches, in particular the adoption of a sample composed by specialists who practice their professional activities in the same Brazilian state, namely the state of São Paulo, and even though this state is the most populous and richest in Brazil, it does not capture opportunities from other states, so in the future, having all twenty-seven Brazilian federative units represented in the study can present new opportunities for readjusting the company's strategy, or even creating new products.

Acknowledgment. UNIAG, R&D unit funded by the FCT – Portuguese Foundation for the Development of Science and Technology, Ministry of Science, Technology and Higher Education. Project no. UIDB/04752/2020.

References

1. Brown, R.B.: Doing Your Dissertation in Business and Management: The Reality of Research and Writing. Sage, Thousand Oaks (2006). https://doi.org/10.4135/9781849209069
2. Darmawan, D., Grenier, E.: Competitive advantage and service marketing mix. J. Soc. Sci. Stud. (JOS3) **1**(2), 75–80 (2021)
3. Kartajaya, H., Kotler, P., Setiawan, I.: Marketing 4.0: Moving from Traditional to Digital. John Wiley & Sons, Hoboken (2016)
4. Kartajaya, H., Setiawan, I., Kotler, P.: Marketing 5.0: Technology for Humanity. John Wiley & Sons, Hoboken (2021)
5. Kotler, P., Keller, K.: Marketing Management, 15ª edn. Pearson Education, London (2016)
6. Malhotra, N.K., Birks, D.F.: Marketing Research: An Applied Approach, 2nd edn. Prentice Hall, Hoboken (2007)
7. Merriam, S.B., Tisdell, E.J.: Qualitative Research: A Guide to Design and Implementation, 4th edn. Jossey-Bass, Hoboken (2016). https://digitalcommons.unl.edu/edpsychpapers/214/
8. Mustafa, A.B., Samsudin, J.B., Abdullah, U.K.B.: Service marketing (2021)
9. Naim, A.: Applications of marketing framework in business practices. J. Mark. Emerg. Econ. **1**(6), 55–70 (2021)
10. Stevens, R.E., Loudon, D.L., Wrenn, B., Warren, W.E.: Marketing Planning Guide. CRC Press, Boca Raton (2021). https://doi.org/10.1201/9781003249597
11. Sutton, J., Austin, Z.: Qualitative research: data collection, analysis, and management. Can. J. Hosp. Pharm. **68**(3), 226–231 (2015). https://doi.org/10.4212/cjhp.v68i3.1456
12. Westwood, J.: How to Write a Marketing Plan, 4th edn. Kogan Page, New York (2013)
13. Wichmann, J.R., Uppal, A., Sharma, A., Dekimpe, M.G.: A global perspective on the marketing mix across time and space. Int. J. Res. Mark. **39**(2), 502–521 (2022). https://doi.org/10.1016/j.ijresmar.2021.09.001
14. Wirtz, J., Lovelock, C.: Essentials of Services Marketing, 3rd edn. Pearson, London (2018)

The Role of Digital Marketing in the Process of Musical Events' Participation

Elvira Vieira[1]([✉]) [iD], Manuel Fonseca[2] [iD], Ana Pinto Borges[3] [iD],
and Bruno Miguel Vieira[4] [iD]

[1] IPVC- Polytechnic Institute of Viana do Castelo and UNIAG - Applied Management Research Unit and ISAG – European Business School and Research Center in Business Sciences and Tourism (CICET-FCVC), Porto, Portugal
elvira.vieira@isag.pt

[2] IPVC- Polytechnic Institute of Viana do Castelo and UNIAG - Applied Management Research Unit, Porto, Portugal
manuelfonseca@esce.ipvc.pt

[3] ISAG – European Business School and Research Center in Business Sciences and Tourism, (CICET-FCVC) and Center for Research in Organizations, Markets and Industrial Management (COMEGI), Porto, Portugal
anaborges@isag.pt

[4] ISAG – European Business School and Research Center in Business Sciences and Tourism (CICET-FCVC), Porto, Portugal
bruno.miguel@isag.pt

Abstract. This paper intends to study the importance of digital marketing to attract participants to musical events. We use the case of NOS Primavera Sound, and we analysed the 2022 edition, which contemplated 1126 valid responses. To reach our objective we ran a logit model and verified that the sociodemographic characteristics present a different impact on digital marketing tools (when compared to other means of communication). The results of this study established that participants can be influenced to participate in the music festival through various digital marketing tools while being sensitive to the two characteristics themselves.

Keywords: Social Networks · Music Festival · Digital Marketing Communication · NOS Primavera Sounds

1 Introduction

The growth of the economic representativity of live music concerts emerged during the decline of the media economy. The changes in the artists' business model itself dictated a loss of revenue through music sales and the consequent need to define other sources of income, namely through live concerts. In parallel and consequently, the diffusion of networked digital media has contributed to the redefinition of promotional strategies associated with this type of events.

By resorting to new media of digital nature, the conventional forms of content had to be reinvented and, consequently, the distribution channels and revenue streams

A. Rocha et al. (Eds.): WorldCIST 2023, LNNS 802, pp. 111–119, 2024.
https://doi.org/10.1007/978-3-031-45651-0_12

di- versified [1]. Therefore, digital tools as means of communication for an event have been gaining ground. It is important to understand the significance of digital marketing and specifically the role of social media in the context of digital marketing to promote a musical event.

We have studied the specific case of the 2022 edition (from the 9th to the 11th of June) of the NOS Primavera Sound festival held in the Porto (Parque da Cidade). Two years after the interruption, caused by the COVID-19 pandemic, the Nos Primavera Sound Festival welcomed more than 100 thousand people in the three days of the event, witnessing the urgency of a long-awaited return. Nick Cave and the Bad Seeds, Gorillaz, Pavement, Beck, Interpol, and Tame Impala were some of the many bands and artists present in the event. The music was distributed across five stages (Nos Primavera Sound, Super Bock, Cupra, Binance and Bits); Many different languages could be heard in this particular location, considering that many foreigners were attracted by the event. In a few words, the NOS Primavera Sound is a festival of reference in the context of European festivals.

Considering the main goal of this work, the following research question was formulated: what is the role of digital marketing in the decision-making process of musical events' participation?

2   Literature Review

Events are a growing phenomenon on a global scale and can function as a tourism product, whether they are large-scale or small-community-level cultural events. In other words, they are opportunities for artistic, heritage, and cultural entertainment in the communities where they take place [2].

Festivals and events play an important role in the context of tourism marketing. The events are a relevant promoter of tourism and are a very important element in the development and marketing plans of most destinations [3]. Also, according to the author, a music festival is an event that represents a social phenomenon found in almost all cultures, representing an opportunity for leisure, social and cultural development, as well as new experiences. The music festivals are defined as a type of event arising from non-routine occasions that pursues leisure, cultural, personal, or organizational goals and whose purpose is to celebrate, entertain or challenge the experience of a group of people [4]. Music festivals are now considered an important subset in the universe of cultural events and have earned interest from different fields of study, given their universality and popularity of the experiences provided [5].

Festivals and events influence the community where they are located, since they provide diversified activities to their participants (locals and tourists), promoting the image of the locality due to their social and cultural significance and generating a boost in economic activity [6]. In addition to the social and economic effect, music festivals contribute to building and developing links between communities [7] and also drive cultural and ethnic development and social integration between groups [8].

Social media have revolutionized the communication strategies between brands and consumers. Financed mainly through advertising, social media have extraordinary potential in spreading information and values, increasing competitive advantages for companies, and sharing knowledge, among others. One of the main goals of event marketing is to help attendees learn more about the event and experience something new. Thus, social media offers a great opportunity to interact with event organizers, marketing managers, and participants. It is in this context that there is an expectation that event managers ought to pay attention to social media and include them in their event marketing plans [9]. In addition, potential event participants can be influenced by sharing other people's experiences [10].

Many studies have conducted empirical analyses on the use of social media in music events promotion, yet there is a consensus on the importance and usefulness of social media in most marketing plans, regardless of the industry they belong to.

The use of positive images of events on social media can increase motivation to attend them [11]. At the same time, when using social media as a communication channel for events, event marketers and organizers should understand how other people's opinions and perceptions affect intention and behavior formation.

In general, music festivals are quite active in using social media, engaging their audiences throughout the consumer decision journey [12]. The authors emphasize that social media are making the "evaluation" and "recommendation" stages of the decision journey more relevant to music festival-related marketers.

The target audiences of major live music festivals tend to be more proactive in seeking information and disseminating their experiences through social media, and to pay more attention and give greater credibility to content created by other users, rather than what is produced by event organizers [13]. Regardless of this fact, organizers should consider the use of social media as a key factor in integrated marketing communication strategies, aiming to increase the brand equity of their festivals by optimizing the effectiveness and efficiency of such communications.

The consumers who engage with their favorite music festival brands using social media have stronger relationships with those brands compared to consumers who do not interact through social media [14]. As a consequence, investments in social media programs can provide marketing benefits when they succeed in facilitating customer brand interactions. Additionally, the relationships created from social media lead to positive results, for example, in word-of-mouth contexts.

Events are present in the main social networks available, therefore festivals' marketing managers are not only able to assess customers' needs, but also to have easy access to customers' feedback, namely the levels of their satisfaction or dissatisfaction, as well as the reasons for such opinions [15]. Thus, the author argues that the use of social networks in the context of music festivals should always be an integral part of the set of promotional actions to be carried out.

Regarding the audience that most attends this type of music event, it should be noted that young people's travels and their decisions regarding the options for a particular music festival are considerably influenced by social media, so the role of digital marketing can be significant in reaching these generations [16]. Thus, the content shared on festivals' social media can make some locations better known, shaping their image and, consequently, attracting more tourists.

All the studies consulted corroborate the positive perspective of the use of social networks as a way to promote and communicate events in general and music festivals in particular.

3 Methodology and Data Analysis

For data collection, questionnaires were applied during the three days of the event, leading to 1126 valid answers (June 9, Thursday we collected 359 answers, June 10, Friday) 350 answers and June 11, Saturday) 417 answers). This edition of the event was attended by around 100,000 participants. In this sense, given the population of 100 thousand participants, the sample should contemplate 660 complete answers, with a confidence interval of 99% and a sampling error of 5%, so it can be concluded that the sample is representative of the population under study.

For this study, we only used the i) socio-demographic characteristics of the respondents (gender, age, marital status, education level, and nationality) and the ii) main means of communication through which the respondent was aware of the event (social networks of NOS Primavera Sound, social networks of family and friends, social networks or website of one of the poster's bands, social networks of a sponsoring brand, advertisement/internet advertising/social network, online media (information websites), e-mail / newsletter, radio or television, word-of-mouth, media (newspapers, magazines) and posters/ mupis).

For data analysis, we started with descriptive statistics to describe the profile of the respondent and the main means of communication of the event. After that, for the econometric model, in order to explain the role of digital media to promote the event, we applied logistic regression. The logistic regression (also called logit regression) estimates the parameters in which the coefficients follow the linear combination [17]. In this regression the dependent variable must assume a binary variable, while the independent variables can each be a binary variable (two classes, coded by an indicator variable) or a continuous variable [18]. In our model, the dependent variable "digital marketing communication" was transformed in the binary choice, "yes" and "no". Concerning the option "yes" we considered the following options: social networks of NOS Primavera Sound, social networks of family and friends, social networks or website of one of the poster's bands, social networks of a sponsoring brand, advertisement/internet advertising/social network, online media (information websites), e-mail/newsletter. For the option "no" we contemplated the variables: radio or television, word-of-mouth, media (newspapers, magazines), and posters/ mupis. The econometric model included gender, age, marital status, education level, and nationality as independent variables. For the analyses, we used the SPSS (version 21) and STATA (version 14).

4 Results

In scope of the sample description (see Table 1), we observe that 50.7% were female. In terms of age, the average age was 29.9 years old (focusing on age ranges between 15 and 35 years). Most of the respondents were single (75.6%), and we also observe that they had higher levels of education, as 73.5% had at least completed a degree. In terms of working conditions, we highlight that 55.0% were paid employees, 14.8% were self-employed, 23.3% were students and 5.5% were unemployed. In the scope of nationalities, we concluded that 33 different nationalities were identified at the venue. Spain, the United Kingdom, Germany, France, Italy, Brazil, and Switzerland stood out as the main countries of origin.

Table 1. Sample of sociodemographic characteristics of the public (n = 1126)

Variable	Description	%
Gender	1- Female	50.7
	0- Male	47.1
	Prefer not to say	2.2
Age	1 – Less than 18 years old	2.8
	2 – Between 18–25 years old	34.5
	3 – Between 26–35 years old	40.8
	4 – Between 36–50 years old	18.0
	5 – Over 50 years old	3.9
Marital status	1 – Single	75.6
	2 – Married	20.5
	3 – Divorced	3.6
	4 – Widow	0.3
Academic qualifications	1 – Elementary studies	3.1
	2 – Secondary studies	23.5
	3 – Bachelor's Degree	46.2
	4 – Master's/PhD degree	27.3
Work status	1 – Paid employment	55.0
	2 – Self-employed	14.8
	3 – Unemployed	5.5
	4 – Retired	1.0
	5 – Domestic activities	0.4
	6 – Student	23.3
Nationality	1 – Portugal	63.4
	0 – Foreigner	36.6
Nationalities	German	10.9
	Brazilian	8.5
	Spain	18.9
	French	11.7
	English	18.0
	Italy	6.1
	Swiss	2.2
	Other	23.8

Source: Own elaboration

Considering our sample, it could be observed that it has a high level of literary qualifications, the most representative age group is between 26 and 35 years old, and that 36.6% of the audience was composed of foreigners.

Within the context of the framework that the respondents indicated stating their knowledge of the event, it is observed that the digital media have a representation of 80.0% in the main means through which the respondent was aware. More specifically, the social networks of NOS Primavera Sound (30.2%), social networks of family and friends (27.4%) and the social networks or website of one of the poster's bands (10.9%) are the main means of dissemination. On the opposite side, we found that non-digital media have a representation of 20.0% in the main means through which the respondent was informed about it, with word-of-mouth having the highest percentage of this group (12.4%) (Table 2).

Table 2. The main means of communication through which the respondent learns about the event

Group	Variable	%
Digital Marketing Communication	Social networks of NOS Primavera Sound	30.2
	Social networks of family and friends	27.4
	Social networks or website of one of the poster's bands	10.9
	Advertisement/internet advertising/social network	5.2
	Social networks of a sponsoring brand	3.7
	Online media (information websites)	2.2
	E-mail/Newsletter	0.4
Non-digital media	Word-of-mouth	12.4
	Radio or television	2.4
	Posters/mupis	3.8
	Media (newspapers, magazines)	1.4

Source: Own elaboration

The top 3 communication framework through which the respondent learns about the event are: social networks of NOS Primavera Sound, social networks of family and friends and word-of-mouth. The following table presents the result of the econometric model estimated by logistic regression. It shows the model's coefficient and also the marginal effects, because the first ones are not directly interpretable [18]. Through the results presented in Table 3, we observe the variables that influence event communication in terms of the framework of digital marketing (social networks, advertisement/internet advertising, online media, and e-mail/newsletter) when compared to other ways of communication (radio or television, word-of-mouth, media (newspapers, magazines) and posters/mupis). We pointed out that gender is statistically significant in the model, which means that digital marketing is more effective to communicate an event for the female public (when compared to male) by 1.7%. We also noted that as age increases, the effective role of digital media to communicate the event decreases by 1.2%. In other words, digital marketing works better for a young public to communicate and recommend the participation in an event. Respondents with higher levels of education (bachelor's, master's, or doctorate), compared to those with no academic qualifications (elementary studies), are more likely to learn about the event through digital media. Regarding nationalities, the Portuguese respondent is less likely to learn about the event through a

Table 3. Econometric model to explain the role of digital marketing in event communication, logistic regression

Variable	Coefficient	Average marginal effects
Gender	0.569 (0.248)***	0.170
Age	−0.018 (0,001)**	−0.012
Marital status		
Single	−	−
Married	0.431 (0.622)	0.012
Divorced	0.723 (0.856)	0.201
Widow	0.652 (0.785)	0.275
Academic qualifications		
Elementary studies	−	−
Secondary studies	0.765 (0.987)	0.129
Bachelor's Degree	0.354 (0.013)***	0.112
Master's/PhD degree	0.162 (0.012)**	0.024
Work status		
Paid employment	−	−
Self-employed	0.335 (0.532)	0.061
Unemployed	0.367 (0.843)	0.090
Retired	0.266 (0.522)	0.521
Domestic activities	0.422 (0.552)	0.654
Student	0.221 (0.328)	0.070
Nationality		
Other	−	−
Portuguese	−0.766 (0.351)***	0.052
Constant	1.743 (0.826)***	

Notes: Standard errors in parentheses. Significant at: * $p < 0.10$ level;** $p < 0.05$ level; *** $p < 0.01$
Source: Own elaboration

digital medium by 5.2%. The foreign public is captivated to participate in events through digital tools. The marital status and work status variables have no statistical significance.

5   Conclusions

Digital marketing plays a key role in the promotion and dissemination of information today. With the right techniques, we can achieve much better results than just using offline promotion methods.

Our results show that most participants are young (ages range from 15 to 35 with an average of about 29) and highly educated, and those facts combined with the percentage of respondents who heard about the event through social media (80%), suggest that participants are regular users of these platforms, so the use of digital marketing strategies can help us reach them for the dissemination of future events.

Something that should also be considered, is the fact that the results confirm a greater influence of digital marketing on women when compared to men. Although the results show a lower influence of digital marketing on participants from other countries when compared to Portuguese participants, the high number of countries represented (33), highlight digital marketing strategies as the best way to spread the word and reach a wider range of likely participants.

In conclusion, the results emphasize the urgency to privilege the use of digital marketing to communicate a music event to increase the demand for participants.

Acknowledgment.   This work is supported by the Fundação para a Ciência e Tecnologia (FCT) under the project number UIDB/04752/2020.

References

1. Holt, F.: The economy of live music in the digital age. Eur. J. Cult. Stud. **13**(2), 243–261 (2010). https://doi.org/10.1177/1367549409352277
2. Small, K.: Social dimensions of community festivals: an application of factor analysis in the development of the social impact perception (SIP) scale. Event Manag. **11**(1–2), 45–55 (2007). https://doi.org/10.3727/152599508783943219
3. Getz, D.: Event tourism: definition, evolution, and research. Tour. Manag. **29**(3), 403–428 (2008). https://doi.org/10.1016/j.tourman.2007.07.017
4. Shone, A., Parry, B.: Successful Event Management: A Practical Handbook, 2 edn. Thomson Learning, London (2004)
5. Getz, D.: The nature and scope of festival studies. Int. J. Event Manag. Res. **5**(1), 1–47 (2010)
6. Pine, B.J., Gilmore, J.H.: The Experience Economy. Harvard Business Review Press, Boston (2011)
7. Pegg, S., Patterson, I.: Rethinking music festivals as a staged event: gaining insights from understanding visitor motivations and the experiences they seek. J. Conv. Event Tour. **11**(2), 85–99 (2010). https://doi.org/10.1080/15470141003758035
8. Yeoman, I., Robertson, M., Ali-Knight, J., Drummond, S., McMahon-Beattie, U.: Festival and Events Management. Routledge, London (2012)

9. Lee, W., Xiong, L., Hu, C.: The effect of Facebook users arousal and valence on intention to go to the festival: applying an extension of the technology acceptance model. Int. J. Hosp. Manag. **31**(3), 819–827 (2012). https://doi.org/10.1016/j.ijhm.2011.09.018
10. Litvin, S.W., Goldsmith, R.E., Pan, A.: Electronic word-of-mouth in hospitality and tourism management. Tour. Manag. **29**(3), 458–468 (2008). https://doi.org/10.1016/j.tourman.2007.05.011
11. Harb, A.A., Fowler, D., Chang, G.J., Blum, S.C., Alakaleek, W.: Social media as a marketing tool for events. J. Hosp. Tour. Technol. **10**(1), 28–44 (2019). https://doi.org/10.1108/JHTT-03-2017-0027
12. Hudson, S., Hudson, R.: Engaging with consumers using social media: a case study of music festivals. Int J. Event Festiv. Manag. **4**(3), 206–223 (2013). https://doi.org/10.1108/IJEFM-06-2013-0012
13. Llopis-Amorós, M.P., Gil-Saura, I., Ruiz-Molina, M.E., Fuentes-Blasco, M.: Social media communications and festival brand equity: millennials vs Centennials. J. Hosp. Tour. Manag. **40**, 134–144 (2019). https://doi.org/10.1016/j.jhtm.2019.08.002
14. Hudson, S., Roth, M.S., Madden, T.J., Hudson, R.: The effects of social media on emotions, brand relationship quality, and word of mouth: an empirical study of music festival attendees. Tour. Manag. **47**, 68–76 (2015). https://doi.org/10.1016/j.tourman.2014.09.001
15. Oklobdžija, S.: The role and importance of social media in promoting music festivals. In: Synthesis - International Scientific Conference of IT and Business-Related Research, pp. 583–587. Belgrade, Singidunum University, Serbia (2015). https://doi.org/10.15308/synthesis-2015-583-587
16. Süli, D., Martyin-Csamangó, Z.: The impact of social media in travel decision- making process among the Y and Z generations of music festivals in Serbia and Hungary. Turizam **24**(2), 79–90 (2020). https://doi.org/10.5937/turizam24-24678
17. Wooldridge, J.: Introductory Econometrics: A Modern Approach, 5th edn. Cengage Learning, Mason (2013)
18. Tolles, J., Meurer, W.: Logistic regression relating patient characteristics to outcomes. JAMA **316**(5), 533–534 (2016). https://doi.org/10.1001/jama.2016.7653

"Conversas Made in CO": An Audio Podcast Featuring Alumni, Building a Collective Memory and Contributing to the Brand Identity

Alexandra Leandro[1,2,3](), João Morais[1,2,4], and Estela Silva[1]

[1] Polytechnic Institute of Coimbra, Higher School of Education, Coimbra, Portugal
aleandro@esec.pt
[2] Research Group in Social and Human Sciences (NICSH), Coimbra, Portugal
[3] University of Minho, Communication and Society Research Centre (CECS), Braga, Portugal
[4] Nova University (IC-NOVA), Lisbon, Portugal

Abstract. One of the newest trends in marketing is an obvious interest in technological applications, as in virtually all fields, albeit focusing on human needs and aspirations. This framework can be viewed as a junction where Relationship Marketing and Marketing 5.0 have led to. When it comes to Higher Education Institutions (HEIs), a fierce global competition has pushed them to think strategically about marketing themselves. Although there is a visible discomfort to "market" educational institutions, there is also a pressing need to do so. Academics have suggested clarifying the use of "marketing" instead of "marketization", and to think of HEIs as brands that must draw from Relationship Marketing philosophy and create value through maintaining closeness to their stakeholders. This paper is the result of a recent initiative taken by two faculty members that created a podcast of interviews with *alumni* from a graduate degree. This experience has been so rewarding and, unlike many other academic endeavors, it compelled the authors to learn more about Educational Marketing (being the authors Marketing professors themselves) and analyze the impact of the podcast to date. The preliminary insights, after having six episodes online, is that it constitutes a repository of voiced memories by former students, contributing to the collective narrative; there is a common experience shared by students from different years, that helps summarize the strengths of the bachelor degree; *alumni* constitute a precious, accredited, discourse that may inspire current students; the initiative of the podcast has brought interesting exchanges of knowledge and even deepened ties within and outside the HEI. As it was a pioneer experience in this particular context, it also brings visibility to the Institution. Taking upon one of the golden rules of Relationship Marketing: it has been a win-win situation and it has been an innovative tool to strengthen the bonds with former and current students.

Keywords: Educational Marketing · Podcast · Higher Education Branding · Relationship Marketing · Marketing 5.0

A. Rocha et al. (Eds.): WorldCIST 2023, LNNS 802, pp. 120–129, 2024.
https://doi.org/10.1007/978-3-031-45651-0_13

1   Introduction

"Conversas Made in CO"[1] ("Talks Made in CO" - CO being Comunicação Organiza-cional[2] / Organizational Communication, the name of the graduate degree) is an audio podcast created by two faculty members from Coimbra School of Education with the primary purpose of taping short conversations with Organizational Communication's *alumni*. In about 15 min, the three participants talk about the career of the interviewee, with a semi-flexible script. The creation of this podcast is rooted in the hope of cre-ating a source of knowledge about potential careers for the students, and, of course, to enhance the pride of belonging to the institution. Other effects could be creating a more detailed database of *alumni*, networking, renewing contacts and emotional ties. Although this endeavor started off as a 'mere' idea of reconnecting with former students and learn more of what they developed after getting their degree - both academically and professionally – the authors believe that if it continues, it may constitute a kind of a "treasure chest" filled with stories from people that were "made in CO". This audio conversation, even though individually, can contribute to the narration of the collective history of this bachelor's degree. As a communication product, it can be analyzed from different points of view, depending on which place to drop the anchor. *Alumni* can be viewed as a crucial stakeholder group. The authors argue that their importance may dim as they graduate, but they are the most important ambassadors that the institution can have. Their voice carries the brand of the institution outside. Also, they can be potential candidates for postgraduate studies, which makes them a relevant target to reach as the institution grows and diversifies its formative offer. On the other hand, being viewed as an important stakeholder, this podcast can also be looked upon as a communication tool to build engagement, sense of belonging, pride of being "CO". Furthermore, the podcast can also be a *marketing* tool to attract new students, new generations of "Made in CO". So, organizational culture, communication, marketing, are all valid areas to move in, as the authors frame this podcast theoretically. The authors opted to look at it from an educational marketing point of view, as well as taking the marketing 5.0 lesson of putting humanity in the core of marketing and technology applications [15]. The use of podcasts has been increasing in diverse fields of knowledge, information, and entertainment. In education, podcasts have been used for either conveying syllabus in an audible form, or to complement the curricular activities with relevant content for the students [17]. The aim of this project is more related to the latter, as its content could be of added value for the current students, prospective students and, of course, *alumni* and faculty members.

---

[1] Please check the podcast on Spotify: https://open.spotify.com/show/4kTKo4jWf25ZjU5i4K wjOL?si=0af0ffeb568e4b1f.

[2] 'Comunicação Organizacional' is a graduate degree within the School of Education's formative offer (Polytechnic Institute of Coimbra, Portugal). For further information, please visit www. esec.pt.

2   Theoretical Context

2.1   Educational Marketing

While designing a condensed history of Marketing, Kotler et al. [15] reach the newest wave of Marketing, calling it 5.0, and describing it as a merger of 3.0 and 4.0 stages. These recent stages are all about transitioning from analog to digital, from transactional to relational. These authors, whilst reflecting upon Generation Z's approach to marketing, state that its main concerns are bringing positive changes to humanity and improving the quality of life, using as much technology as one can [15, pp.52–53]. That is, this generation is focused on how technology can better the human experience, people being at the core of marketing strategies. This new wave is probably the right time to give 'educational marketing' a true opportunity. Up until recent years, the concept of applying marketing strategies, lexicon and techniques to the HE world was appalling and hindered a much-needed systematic research of the specific field [19]. "The marketing field is still to be developed and adapted for the HE sector, without probably never being applied in the same way as in the business sector." [19, p.42]. It is more or less consensual that the business way is not the way for education: but the weight of words should not prevent evolution towards a closeness between the institution and its main stakeholders. Marketing, in its core, is about being of value for all parties involved in a given relationship (either commercial or other). Furthermore, relationship marketing, marketing 5.0 and all forms of the contemporary framework of this discipline, are actually very innate for education, contrary to transactional marketing, as Farrell [11, p.177] uncovers: "So far the focus has been on a 'transactional' view of marketing. This is because there may be the danger that heads [*responsible managers of schools*], feeling under pressure to formally start marketing their institution, may imagine that this is what 'proper' marketing is all about. Ironically, in doing so they may overlook relational marketing. This is ironic because to do so would be to ignore a type of marketing which schools have always understood.". Helgesen [12] argues that the perspective of relationship marketing is, indeed, the adequate one for educational marketing. The fundamentals of relationship marketing are taking care of retention and development of actual customers as a priority, instead of focusing only on acquiring new customers. The relationship marketing approach will indicate the way for the HEI to work on satisfaction, reputation, and loyalty of students and managers should focus on getting data to add value for their primary "client", students. Also, "marketers need to know what creates student value so that they can craft appropriate marketing campaigns." [12, p.70]. Stachowski [22] and Nicolescu [19] agree that competition has increased, both in strength and volume, for HEIs as well, and that justifies marketing. Nicolescu [19], however, points out two features about educational context that makes it hard to apply marketing theories as business does. The fact that, in most countries, HE is a public sector, non-profit, and not suitable to think of marketing as a profit-seeking set of techniques; the other relates to the fact that HE is a service, so service marketing should be the framework, instead of 'product marketing', and one should look at educational marketing with all the specificities that services imply. Oplatka [20] also contributes, based upon a study of the perceptions of teachers' involvement in marketing their institutions, that a business model of marketing is not appropriate, and it would also help to 'translate' some terms to be more suitable for the educational setting:

educational marketing should be based on service sector or non-profit organizations, student should be seen as a citizen instead of a customer, and marketing be essentially viewed as "sharing of information to match educational resources to student needs" [20, pp.18–19]. Nevertheless, Nicolescu [19] deconstructs the main marketing themes for educational context, moving over the sensitive issue of applying certain terms to people, students, teachers. In fact, students, *alumni* and employers can be seen as the primary and secondary 'customers' of the HEI. If one takes the stakeholder approach, society as a whole is also a beneficiary of HE; students are certainly the primary stakeholder, and so on. From a marketing perspective, segmentation and targeting in HE functions for their primary "customers" (students), but all other stakeholders' needs must be considered, and it is not easy to assess them through the usual segmentation methods. As far as marketing-mix is concerned, rather than the original Product, Placement, Promotion and Price, it makes more sense to adapt the service marketing-mix, or, as Kotler and Fox [14] suggested: Programme, Price, Place, Promotion, Processes, Physical Facilities and People, a more intricate mix that encompasses the complexity of all service marketing decisions. In conclusion, referencing Tahir et al. [24]: educational marketing is important because it will help institutions reach their maximum level of coverage of all aspects of the social network that matters; also, it propels the institution to get to know students' characteristics better, and by doing so, it develops its sense of the benchmark; and, more important for this case, a marketing approach will open the institution to new ways to reach their audiences and connect with them.

2.2 Higher Education Institutions as Brands

HEIs are becoming increasingly aware of the development of marketing programs and are beginning to realize the importance of acceptance, reputation, and visibility itself [9]. These are entities that face challenges at the level of institutional branding due to a high degree of uniformity and consequently the difficulty to differentiate themselves and create unique images [19]. It is in this perspective of brand management that HEIs try to bet on their strengths, position themselves in order to differentiate from the others [7, 26] and try to gain acceptance in the market. The relationship between brand, institutional image, and reputation of the HEIs then become concepts to be considered [1].

The sector is discussing a new logic that combines brand management, corporate communication, identity, image, and reputation management, making HEIs more aware of the link between what they represent in terms of values and characteristics, and how they are perceived in the market [8]. Despite this sensitivity, it is a view still not consensual among all agents, confronting the perspectives of reputation and institutional image and how they can be related with those inherent to the management of a HEI brand in a marketing approach [19]. Although coincident, the concepts are necessarily different when applied to HEIs *vis-à-vis* the business sector. Traditionally, reputation and institutional image have always been seen by those in charge as their own being, the way that best and that really sells [25]. Therefore, they have an interpretation that reflects the reality of the institution that differs from marketing because when they are high, they are often linked to minimum sales and lower student acceptance of the educational offer [13]. The reasoning leads Bulotaite [6] to state that in HE, the purpose of branding should not be to sell products and services, but rather to communicate the corporate identity as a way of attracting potential students and enhancing loyalty. For that author, an educational brand must build, manage, and develop associations, emotions, faces and images because they are brands with the potential to create stronger feelings than most brands.

2.3   Podcasting for *Millennials* and Gen Z: Audio Makes a Comeback

Whoever thought audio was dead, better think again. *Millennials* and Generation Z, or people born 1980–1994, and 1995–2010, have their ears glued to podcasts. Berry [5] even claims we are living a new golden era of audio, with very accessible platforms and means to anyone to produce audio content and divulge it on several gateways on the internet. A profoundly democratic technology, as Balzen [3] stated, that enables individuals and organizations to create content that may not fit the working agenda of the mainstream media. Reis and Ribeiro [21] describe podcasts as suited for both big audiences and small niches, as well as a medium for a multitude of areas, "from politics to public relations, from education to organizational communication, from culture to sports" [21, p.1] and other topics like music, crime, inclusion. The fact that podcasting doesn't need professional equipment to actually function means that it can serve as a megaphone for all voices to be heard. Moreover, Markman and Sawyer [16] highlight that this phenomenon happens under a context of a collaborative and participatory culture. This phenomenon has impacted Portugal too. An OBERCOM[3] survey referring to 2021 showed that 41,4% of internet users had listened to at least a podcast during the month prior to the survey.

A steady growth from 34,3% in 2019 and 38,4% in 2020. In the same report, OBER-COM compares this data to other countries, and Portugal stands well above the mean (31%). Focusing on the younger generations, United States' data reveal that *millennials* (also known as generation Y) are the biggest consumers of podcasts, followed by generation X (born 1960–1979) and generation Z[4]. The HE students and *alumni* are mainly within the younger generations. Beyond any statistics, there's an obvious interest in audio podcasts that could be observed in their conversations, inputs in classes, references. One of the most important facets of being a teacher is being able to engage students, creating touchpoints that mitigate the generation gap and motivate them for learning. Nevertheless, this doesn't come easy for HEIs. As Drake stated [10], referring to the much similar use of social media, HEIs face a dilemma: "On the one hand, the university represents the entire institution, the receiver of their tuition, the administrators that create the rules and regulations of their study program, a kind of disciplinarian and enforcer. On the other hand, the university represents a collection of educators, the professors who nurture the learning and motivate the students to achieve and reach their maximum potential. […] Any university or educational institution must establish its voice or brand via social media, and simultaneously, faculty members may be leveraging social media

---

[3] OBERCOM - Reuters Institute for the Study of Journalism. Data retrieved from https://pop casts.pt/blog/podcasts-em-portugal-consumo-numeros-e-estatisticas-de-2021/ (27th of october of 2022).

[4] A study by eMarketer (February 2022). Data retrieved from https://www.insiderintelligence.com/content/spotify-tries-solve-its-gen-z-podcasting-problem (27th of october of 2022).

in the engagement of students in a particular area of study." [10, p.6]. Besides, one can argue, as do Aguiar, Carvalho and Maciel [2], that the use of podcasts as a learning tool is highly connected with the established value of oral/audio communication as important vectors for increasing motivation and cognition. Among other reasons, the fact that students can pause, listen anytime, anywhere, a high-flexibility feature, gives podcasts a sure place as a viable pedagogical platform [18].

2.4   The Importance of *alumni* in Creating HEI Brands

Considered one of the stakeholders to every HEI [27], *alumni* are also viewed as products of the education institutions, former and prospective customers, if one considers postgraduate studies available for enrollment. And although they might experience a different sense of the HEI from the current students – changes that are introduced in the institution's mission, leadership, dynamics, *etc.* [27], they are still a voice to be reckoned with when it comes to the establishment of a HEI's reputation and image. The same authors [*ibidem*] refer to the need for contemporary HEIs to have a more intentional approach to brand strategies that include a shared brand meaning across all the main stakeholders, where *alumni*, students and employers are contemplated. Stephenson and Yerger [23] concluded, in their research, that one of the most effective ways to make alumni positive ambassadors for the HEI is to feed their sense of belonging to that institution, as well as how it differentiates from other HEIs and how attractive it is. These drivers will propel brand advocacy among *alumni*, so it is paramount to work on brand building and managing considering this stakeholder, keep communicating with them after their degree completion.

3   The Case Study of "Conversas Made in CO": Preliminary Insights

For all that was discussed previously, educational marketing is full-on hybrid today. Every HEI aims to build and feed its brand value both offline and online and create a *persona* that "speaks the same language" as their students. Although the podcast "Conversas Made in CO" was not an institutional initiative, the HEI's President, as well as the Communication responsible, and most of the faculty members connected to the degree, were highly positive and enthusiastic about it. This paper's main objective is to reflect upon this podcast as a motivational tool for current students, *alumni*, and faculty members. With the objectives of the podcast in sight, a semi-flexible script was designed, considering four main "fixed" questions, posed to the interviewee after a short introduction done by one of the hosts: 1. Where did you do your internship? / What did you do after graduation (academically, professionally)?; 2. What do you believe to have taken away from this degree?; 3. Leave a message for current Organizational Communication students; 4. Who do you think would be a good guest for this podcas.t? (reference of other *alumnus*). Recently, another question was added, due to some feedback from colleagues: 5. How do you foresee the future in our area (communication, marketing, public relations, advertising, human resources)? What are the hot issues now?. Having 6 episodes online (the pilot plus 5 regular episodes, from May 2022 to October 2022), preliminary insights can be provided about the outcomes of the podcast. Specifically for the hosts, and although this is a *pro-bono* part of the job, it has been highly rewarding to be able to reconnect with some *alumni*, reminisce, and it a boost in the love for the job, as the authors get to see firsthand people who have been - even if just a bit - touched by their work. It has also sparked some interesting conversations with colleagues, in and out of the institution. For current students, it has taken a while for them to 'warm up' to it, but now the authors start to get feedback in person, and it certainly looks like an element of pride of belonging to this degree. Less subjective, some conclusions can also be drawn from the podcast itself, from data analytics to the content of the answers provided by the 6 guests so far. The data retrieved from the RSS page of the podcast contains the basic analytics for the podcast: there have been more than 500 downloads of the podcast since it started and, until October 2022, there were 37 followers. Considering that the population is roughly 250 current students and approximately 1400 former students (along almost 30 years of existence), this number is a good indicator of acceptance. For the interviews' analysis, a content analysis was conducted, supported by Bardin's works [4]. This author stated that content analysis is a set of techniques that analyze communications using systematic and objective procedures to describe the content of the messages [4]. The intention of this method is to infer knowledge from communicative products,

Table 1. Summary of answers of interviewees

What do you believe to have taken away from this degree?	Leave a message for current Organizational Communication students
Bibliography and knowledge (still current) for day-to-day application; Practical assignments, oral presentations; ability to expose ideas; confidence; Transversal vision across different topics	The ability to carry knowledge into practice, and knowing how to do it in a professional way is a differentiating factor
Complete course; Tools/skills to work in various areas; Workload prepares for 'real' life	The will to learn and the humility to know that we don't have all the answers; Go on doing complementary training; Be careful when following trends - specifically digital; Taking risks, doing new things, not being afraid
360° view of what you can do in an institution; Comprehensive course; Ability and flexibility in different areas	The need to have good marketing and communication skills, because without them digital will not work
Entrepreneurship; Ability to work to achieve success; Quality of the professors and their professional experiences; Foundations of 'classic' marketing	Beware of the danger of confusing digital marketing with the personal management that we do of our networks - 'blindness' to digital; Take advantage of the most classic things your training gives you in terms of marketing because it will be an opportunity for differentiation
Curricular Internship - first contact with the market; International curricular internship Practical assignments and projects every school year; Learning background in various areas	Be aware of the triangle: market, competition, consumer; Available and open to different things; explore and understand what really attracts you the most; open mind. Be aware of the people who influence us and who can be good examples; Experimenting; Networking
Hands-on; Many practical assignments and many oral presentations	Reading the markets; Don't be afraid to explore other areas; All experiences are valuable; Don't be afraid to take risks

that can be retrieved both quantitively and qualitatively [*ibidem*]. On what concerns the answers the authors got from the interviewees; they are summarized in a main outlook in Table 1. The focus of the analysis was on the answers whose content is highly connected with the degree's identity (questions 2 and 3). Although the first question is left out of this analysis, it is worth highlighting that having an internship in the last semester of the degree is always indicated as one of the strengths of this course. The other questions were not analyzed due to the highly individual content of their answers. To analyze the content of the answers to those questions (2 and 3), firstly, a full transcription of the episodes was conducted. Afterwards, the core answers were highlighted, to separate unnecessary discourse or secondary text. The remaining content is relevant and related to the question that was posed in the first place.

From these first episodes can be drawn that the most likable features of the bachelor's degree are a transversal view of all Organizational Communication fields and being a highly practical, hands-on, course. On what concerns 'leaving a message for current students', the main messages are to take chances, experiment, and to be aware of hollow applications of trends, such as digital marketing and social media uses.

4 Conclusions

From this first set of episodes, a main take away is that the podcast "Conversas Made in CO" constitutes a repository of voiced memories by former students, contributing to the collective narration of the history of the course and the HEI at large. Although these interviewees have enrolled in the course in different years, there is an obvious common experience that they share, which helps summarize the strengths of the degree. *Alumni* have an accredited discourse and embody the finest ambassadors for current students (as inspiration) and in the market (as references). The initiative of the podcast has brought interesting exchanges of knowledge and even deepened ties within and outside the HEI. As it is a pioneer experience in this specific universe, it also brings visibility to the Institution. Taking upon one of the golden rules of Relationship Marketing: it has been a 'win-win' situation and an innovative tool to strengthen the emotional bonds with former and current students. The more the students, whether still in the HEI or having graduated, associate themselves with the organizational identity of their *alma mater*, the more loyal they will be, which ends up reinforcing the HEI brand. As the theoretical review suggests, Educational Marketing is all about relations, identification, involvement, and delivery. This reasoning brings us to the junction of this particular area of Marketing. When it comes to HEIs, a fierce global competition has pushed them to think strategically about marketing themselves. Although there is a visible discomfort to "market" educational institutions, there is also a pressing need to do so. Academics have suggested clarifying the use of "marketing" instead of "marketization", and to think of HEIs as brands that must draw from Relationship Marketing philosophy and create value through maintaining closeness to their stakeholders. The "talks" with *alumni* - 'Conversas Made in CO' - may be a good contribution for just that. The study has some limitations, namely not measuring the impact on listeners' behavior, their opinion, and suggestions for new content, as well as how the podcast can arouse a sense of belonging and the impact on the institutional brand. It is also not possible to know the totality and characteristics of the people who listen to it. The suggestions for the future are precisely to fill the limitations pointed out above. Understand who are the listeners, what is their opinion about the podcast, how it can condition their future behavior and have suggestions for the development of new content. It will also be interesting to measure the impact that this initiative has on the perception of the Organizational Communication (CO) course and the institutional brand.

Funding. This work is financed by national funds through FCT – Fundação para a Ciência e a Tecnologia, I.P., under the project UIDB/00736/2020 (base funding) and UIDP/00736/2020 (programmatic funding).

References

1. Aaker, D., Keller, K.: Consumer evaluations of brand extensions. J. Mark. **54**, 27–41 (1990)
2. Aguiar, C., Carvalho, A.A.A., Maciel, R.: Podcasts na Licenciatura em Biologia Aplicada: Diversidade na Tipologia e Duração. In: Carvalho A. (org.) Actas do Encontro sobre Podcasts, pp. 22–38. Centro de Investigação em Educação, Braga (2009)
3. Balzen, R.: Podcasting is THE Democratic Medium. Discover Pods (2017). https://discoverpods.com/podcasting-democratic-medium/. Accessed 29 Dec 2022
4. Bardin, L.: Análise de conteúdo, p. 70, 4th edn. Edições, Lisboa (2013)
5. Berry, R.: Just because you play a guitar and are from Nashville doesn't mean you are a country singer': the emergence of medium identities in podcast. In: Llinares, D., Fox, N., Berry, R. (eds.) Podcasting New Aural Cultures and Digital Media, pp. 15–33. Palgrave Macmillan, London (2018)
6. Bulotaite, N.: An institutional tool for branding and marketing, higher education in Europe. University Heritage, vol. XXVIII, No.4, pp. 449–454 (2003)
7. Chapleo, C.: Exploring rationales for branding a university: should we be seeking to measure branding in UK universities? J. Brand Manag. **18**, 411–422 (2011). https://doi.org/10.1057/bm.2010.53
8. Chapleo, C.: Interpretation and implementation of reputation/brand management by UK university leaders. Int. J. Educ. Adv. **5**(1), 7–23 (2004)
9. Del Rio-Cortina, J., Cardona-Arbelaez, D., Simancas-Trujillo, R.: Proposal of a theoretical model of branding for the positioning of the university brand. Revista Espacios **38**(53) (2017)
10. Drake, P. D.: A Study of digital communications between universities and students. Dissertations, p. 673 (2017). https://irl.umsl.edu/dissertation/673
11. Farrell, F.: Postmodernism and Educational Marketing. Educational Management & Administration. SAGE Publications 0263-211X (200104), vol. 29, no. 2, pp. 169–179 (2001)
12. Helgesen, Ø.: Marketing for higher education: a relationship marketing approach. J. Mark. High. Educ. **18**(1), 50–78 (2008). https://doi.org/10.1080/08841240802100188
13. Hemsley-Brown, J.V., Oplatka, I.: Universities in a competitive global marketplace: a systematic review of the literature on higher education marketing. Int. J. Public Sect. Manag. **19**, 316–338 (2006)
14. Kotler, P., Fox, K.: Strategic Marketing for Educational Institutions, 2nd edn. Prentice-Hall, Englewood Cliffs (1995)
15. Kotler, P., Kartajaya, H., Setiawan, I.: Marketing 5.0 - Tecnologia para a Humanidade. Actual Editora, Lisboa (2021)
16. Markman, K.M., Sawyer, C.E.: Why pod? Further explorations of the motivations for independent podcasting. J. Radio Audio Media **21**(1), 20–35 (2014)

17. McCarthy, S., Pelletier, M., McCoy, A.: Talking together: using intercollegiate podcasts for increased engagement in marketing education. Mark. Educ. Rev. **31**(2), 125–130 (2021). https://doi.org/10.1080/10528008.2021.1875849
18. Menezes, C., Gamboa, M.J., Brites, L.: Dez minutos de conversa: podcasting como recurso de formação multidimensional. Comunicação Pública **16**(31), (2021). https://doi.org/10.34629/cpublica.65
19. Nicolescu, L.: Applying marketing to higher education. Manag. Mark. **2**(4), 35–44 (2009)
20. Oplatka, I.: Teachers' perceptions of their role in educational marketing: Insights from the case of Edmonton, Alberta. Canadian Journal of Educational Administration and Policy, Issue #51 (2006)
21. Reis, A.I., Ribeiro, F.: Os novos territórios do Podcast. Comunicação Pública **16**(31), (2021). https://doi.org/10.34629/cpublica.251
22. Stachowski, C.A.: Educational marketing: a review and implications for supporting practice in tertiary education. Educ. Manag. Adm. Leadersh. **39**(2), 186–204 (2011)
23. Stephenson, A., Yerger, D.: Does brand identification transform alumni into university advocates? Int. Rev. Public Nonprofit Mark. **11**, 243–262 (2014). https://doi.org/10.1007/s12208-014-0119-y
24. Tahir, A.G., Rizvi, S.A.A., Khan, M.B., Ahmad, F.J.: Keys of educational marketing. J. Appl. Environ. Biol. Sci. **7**(1), 180–187 (2017)
25. Temple, P., Shattock, M.: What does Branding mean in higher education? In: Stensaker, B., D'Andrea, V. (eds.) Branding in Higher Education. Exploring an Emerging Phenomenon, EAIR Series Research, Policy and Practice in Higher Education, pp. 73–82 (2007)
26. Whelan, S., Wohlfeil, M.: Communicating brands through engagement with 'lived' experiences. Brand. Manag. **13**(4/5), 313–329 (2006)
27. Wilson, E., Elliot, E.: Brand meaning in higher education: leaving the shallows via deep metaphors. J. Bus. Res. **69**, 3058–3068 (2016)

Data Mining and Machine Learning in Smart Cities

Qualitative Data Analysis in the Health Sector

Maria Veloso[1], Marta Campos Ferreira[1,2(✉)] (iD), and João Manuel R. S. Tavares[1] (iD)

[1] Faculdade de Engenharia, Universidade do Porto, R. Dr. Roberto Frias, 4200-465 Porto,
Portugal
up202100658@edu.fe.up.pt, {mferreira,tavares}@fe.up.pt
[2] INESC-TEC, R. Dr. Roberto Frias, 4200-465 Porto, Portugal

Abstract. In the health sector, the implementation of qualitative data research is very important to improve overall services. However, the use of these methods remains relatively unexplored when compared to quantitative analyses. This article describes the qualitative data analysis process that is based on the description, analysis and interpretation of data. It also describes a practical case study and the use of NVivo software to assist in the development of a theory-based qualitative analysis process. This article intends to be a step forward in the use of qualitatively based methodologies in future research in the health sector.

Keywords: Grounded Theory · Qualitative Research · Case Study · Healthcare · NVivo

1 Introduction

Most studies developed in the health science sector use the quantitative method, which is logical, experimental and mathematical, with a preference for the extensive phenomenon, which cultivates an alleged objectivity and neutrality, is hypothetical-deductive, replicable and generalized [1]. Denzim and Lincoln state that qualitative research is an investigative field of study that can be considered a large umbrella that covers different approaches used to describe, understand, and interpret experiences, behaviors, interactions and social contexts. Qualitative approaches to health also encompass different theories and study models, such as ethnography, case studies, oral history, document analysis, among others [2, 3].

The scarce use of the qualitative method in health research, especially by professionals with more specific training such as doctors and engineers, is caused by the lack of knowledge in human sciences [4]. This makes it difficult to learn more subjective topics because these courses require students to acquire skills to perform complex procedures required to act in the face of challenges presented in the day-to-day life. Living and working conditions influence the way in which people think, feel and act about health. Thus, it is essential to understand the social determinants that drive the lives of these people. Qualitative approaches seek to understand this reality that the numbers indicate but do not reveal.

A. Rocha et al. (Eds.): WorldCIST 2023, LNNS 802, pp. 133–142, 2024.
https://doi.org/10.1007/978-3-031-45651-0_14

In the qualitative research cycle, the stage that is more complex for those with an essentially technical background is data analysis [5]. Criticism of qualitative research developed by those who do not have a background in the humanities or social sciences refer mainly to the superficiality with which they approach reality, the inability to debate empirical data ad failures in the consistent and in-depth application of the theory [6, 7]. This article presents a synthesis of theoretical and practical knowledge on the analysis of textual data with the objective of helping the researcher in their qualitative studies related to the health sector.

2   General Considerations About Qualitative Data Analysis

The analysis of qualitative research data aims to understand, confirm or not the research assumptions, answer formulated question and thus expand the knowledge on the investigated topic. There are several qualitative data analysis techniques that can be used. However, nothing prevents each researcher from creating a new one, adapting existing techniques, or improving them. In any analysis technique, interpretation is the main action of the research. It is present throughout its process and constitutes the essential part of the analysis. During data collection, the analysis is already taking place, unlike in quantitative studies where analysis only starts after the completion of the field research. It is the pre-analysis that allows the use of the saturation criterion to determine the sample size in qualitative studies. Often, after collecting the data, the information is not enough to come to a conclusion. In these cases, researchers must return to the field to gather the missing information. These comings and goings to the field are characteristics of qualitative investigations. They allow, through the interpretation of data, the reach of greater proximity of knowledge or reality, surpassing the instance of common sense. This stage is subjective to the researcher's interpretative capacity and not only on the data collected.

In the interpretative work, one should get the most ideas out of the text, comparatively analyze the new ideas that appear, what confirms and what reject the initial assumptions, which leads to thinking more broadly. A frequent questioning about the results of qualitative studies is their representativeness and validity. These are related to the ability to understand the meaning of the phenomenon studied. In qualitative analysis, quantity is replaced by intensity, by deep immersion. The number of people is not the most important aspect but analyzing the issue from several perspectives and understanding the social fact that is being investigated. There is also no concern for generalization.

The purpose of the qualitative approach is not to count opinions and people. Instead, the purpose is to explore the spectrum of opinions and different representations about an issue. While experiences may seem unique to the individual, representations of such experiences do not arise from individual's minds alone. To some extent, they are the result of social processes. When analyzing the data, the researcher must be careful not to get carried away by conclusions that apparently seem clear and transparent. The more familiar the researcher is with what he/she is researching, the greater the illusion that the results are obvious.

The ability of a qualitative researcher to be flexible in the fieldwork is very important. When the researcher is too attached to methods and techniques, he/she may end up forgetting the meanings present in the data, disregarding important aspects of the field, due to the restrictions of questioning of the methodological procedure. In quantitative research, a neutral posture is required, under penalty of incurring bias in the investigation, which would invalidate the knowledge produced. In qualitative studies, the attitude is different as it is not possible to conceive that there are studies with absolute neutrality because they deal with human beings. The researcher's task is to recognize the bias to prevent its interference in the conclusion.

2.1 Processing of Qualitative Data

Some structuring terms that support qualitative research must be known and contained in qualitative analysis: the nouns, experience, common sense and social action and the verbs understand and interpret [6]. The experience is what the being experiences in the world, the actions it performs. It expresses itself in language and is mediated by culture. The experience is the product of personal reflection on the experience, that is, what it represents for that person. The same experience can be lived differently by two individuals. Common sense is the set of knowledge arising from the experiences of individuals and consists of opinions, beliefs ways of thinking, acting, feeling, and relating. Human action is what individuals do to build their lives in the conditions they find. The verb to understand means exercising the ability to put oneself in the other's shoes, considering the subject's uniqueness and subjectivity in the historical and social context in which it is inserted, and interpreting is based on understanding, on the elaboration of possibilities about what is understood.

The treatment of qualitative data can be didactically divided into 3 interconnected steps: description, analysis, and interpretation. In the description, researchers work so that the opinions of the different informants are preserved as faithfully as possible. The analysis seeks to go beyond what is described. A systematic path that seeks in the testimonies the relationships between the factors is traced. It produces the decomposition of a data set, looking for the relationships between the parts that compose it. One of its purposes is to expand the description. The interpretation can be a follow-up to the analysis and can be also developed after description. Its goal is the search for meanings of speeches and actions to reach understanding or explanation beyond the limits of what is described and analyzed.

These qualitative data processing are not mutually exclusive, nor do they have clear boundaries between them. They are just perspectives on the treatment of qualitative data that may not formally co-exist. In qualitative research, according to Minayo [6], the interpretation is the starting point, because it starts with the actors' own interpretations, and is the arrival point, because it is the interpretation of the interpretations.

1) Data Description

Data can be grouped and classified into different structures such as by type of interlocutor, such as professionals, patients, managers and educators, by location of data collection, such as schools, health services and neighborhoods, or by type of instrument collections, such as interviews, groups, field diaries and reports. The textual data that come from recording must be checked against the original recording, when it is not the interviewee who transcribes them. As the transcripts are read repeatedly, markups are introduced to highlight the ideas that come to mind. The data collected must be carefully guarded as they are unique and irreplaceable. After five years, they can be destroyed so that they are not misused.

2) Analysis

The analysis begins with a careful reading of the textual data, with a comprehensive rereading to be better familiarized with its content, have an overview, and grasp the particularities present. From this exhaustive reading, it is possible to identify the main body of data and separate what is not directly related to the interest of the study. In the dialogue that is established between the researcher and the interlocutor, other subjects often are discussed can be discarded from the analysis because they are outside the objective of the investigation. From the identification of the main body of the text, it can be organized by relevant themes and start categorization process.

When a relevant topic is identified, which can be, for example, of the items in the interview script, the text is cut and pasted, organizing in the same "drawer" or file the statements of all the interviewees about it. At this moment, another stage of reading begins, this time from the file with the clippings and collages of the speeches, in an attempt to understand what message, the interlocutors are giving on the topic, what meaning the listed subject has for them. The researcher can now create classification categories. For this, one must: observe the structure of relevance of the text, what is common in the narratives and what is divergent; make comparisons between groups; seek ideas that behind the text, that is, go beyond the speeches and facts described; identify and problematize the explicit and implicit ideas; seek broader (sociocultural) meanings attributed to ideas; dialogue with information from other studies on the subject and theoretical framework of the research. This classification must be anchored to the theoretical approach adopted by the researcher and be contextualized. To understand the context of the speeches, it is not enough to analyze the narrative. It is necessary to reconstruct the social and historical conditions of production that gave rise to it.

3) Interpretation

The last phase of analysis is interpretation. It is about the elaboration of a synthesis between the theoretical dimension and the empirical data: a dialogue is made between the theoretical foundation adopted, information from other studies and the narratives of those surveyed to seek broader meanings. Triangulation can be used as a methodological resource for better interpretations, which is processed through the dialogue of different methods, techniques, sources and research [8]. The organization of the data is observed with the use of imagination for a better understanding of the subject, proposing concepts and theories that provide new and useful meanings and used to the community. The reinterpretation of the social actors about the social facts is made, revealing models underlying the ideas, illuminating obscure points, building new theories and establishing new concepts and knowledge. Interferences can be made from the theoretical-conceptual approach adopted in the study. The interpretation must have a guide to respond to research objectives, seeking a broader understanding of the topic under study in which meaning found is no longer of the individual but of the group. This is not common sense and must be based on the theory exposed in the introduction of the work. When you only reach a conclusion that you already had before, you didn't need to do research.

2.2 Use of the Computer for Data Analysis

The existing software on the market do not perform the analysis for the researcher, they are only instruments to help organize the data. New concepts and theories are built in the field, that is, they emerge as the data is analyzed. Some software tries to be smart, such as, for example, asking the researcher question when a certain line appears in the text, similar to the already established categories.

Some examples of data analysis software:

1) ALCESTE: Uses factorial method, analyzes the frequency of a certain word appear in the speech of individuals;
2) NVIVO: Allows users to organize and analyses content from interviews, group discussions or articles.

3 Methodology

In this section, it will be discussed the methodology used to analyze a case study. The study aims to understand customers experience while booking medical appointments and how this experience can be improved.

To understand the phenomenon, the method used is Grounded theory. Grounded theory is a set of methodological procedures for conducting research that tend to explore understudied phenomena or experiences and situations [9]. The reason why this method was chosen with relation to any other is that grounded theory allows the researcher to understand and explain a phenomenon without first having a preconceived notion about the topic or an initial theory, preventing the researcher bias from affecting the outcome of the research.

When conducting research using the grounded theory approach, the researchers should have no preconceived ideas, all analysis must be driven by data, and the theorist should use constant comparison method (compare data with data). The main objective of this theory is to allow the data to drive your analysis.

3.1 Processing of Qualitative Data

The method chosen to collect the data for the research is interview. There are many advantages of collecting the data by interviewing. Interview is generally a very pleasurable and memorable experience for both the interviewee and the interviewer. The fact that it's such a pleasurable and general experience contributes to gathering more in-depth and more valid data. Interviews will help to observe and pay attention to verbal and non-verbal cues as well as generally the participants' emotions. This allows to recognize whether the participant wants to talk to or maybe they want to change the subject. So, this links to emotions and behaviors in general. One can see whether the participants feel uncomfortable. For example, the researcher decides to lead the discussion in a slightly different direction. It is very important to be responsive during the interview.

Flexibility is another major advantage of interview. One can make sure that the interview takes place in a place and time that is very convenient to the participants which is important. Influencing directly in the will of the participants to remain in the study.

The type of interview chosen was the semi-structured. In a semi-structured interview, the interviewer writes an interview guide, which is a document where is outlined the topics of the interview and the questions that the researchers want answered by the interviewees. However, this is just an interview guide. So, the fact of having one does not mean that the researcher has to stick to every single point outlined in that guide, but the guide is meant to help the investigator stay relatively focused. Interview guides often include some prompts and some questions that generally aim to encourage the participants to develop their thoughts further. The whole point of having this interview guide is to have structure but at the same time to remain extremely flexible and responsive to the interview situation.

1) Design and development of an interview guide

The first stage of an interview is the introduction. So, the introduction is the phase where you introduce the whole the whole context and the whole study. Usually what is advised at this stage is to briefly outline the main aims of the study to the participants, explain again what the researcher is doing and what is going to be talked about. Sometimes people at this stage have the participants sign the informed consent form. However, generally the main point of the introduction is to introduce the study to your participant but also to introduce some general rules. The investigator will talk about situation and what is going to happen for the next 45 min or the next hour or whatever time it will take to interview the participants. It is important to highlight that there are no right or wrong answers. Participants need to understand that sometimes the researcher does not comment because he/she is not expected to comment. It may in fact damage, or it may affect the validity of the findings because the investigator does not want to suggest to the participants what his/her opinion is. Instead, the researcher wants the participants not to feel like they said something wrong, or they said something that you agree with.

The second part of the interview is the main part, where the interviewer starts asking the questions and developing them further. Interviewers should ideally try to ask the question in a way that reflects the chronology of the real events. If there are any controversial questions or some very complex questions it requires the participants not only to remember something very well but also to open up and discuss controversial things or personal things. It is crucial to start with some questions that are simpler. In that manner, the researcher takes the opportunity to wait until the participants start to feel more and more comfortable and more importantly around the camera or audio recording equipment.

Finally, there is the closing stage to closure. To finish the interview what is usually advised is that the investigator should thank the participant and then ask them if they have anything else to add or if they have any questions [10].

3.2   Data Analysis

The software used to analyze the data collected from the interviews was NVivo because it allows researchers to analyze qualitative data more effectively [11].

Before starting using NVivo, it is important to read the interviews transcripts and prepare a coding table based on the criteria that is considered important from the transcripts. Figure 3 shows a table with the codes created based on the interviews. After reviewing the transcript, it is important to add every related information under the codes. The purpose of coding is to create a table of contents that will help understand the data to the extent that the researcher can reflect on the findings and understand what message the data is transmitting.

The transcript of the interviews was divided in 8 codes. First, the researcher was trying to understand how frequently the participants go to the doctor. After that, it was important to know how the participants book their doctor appointments and finally understand if there is interesting in booking appointments via the internet.

3.3 Data Interpretation

After creating the codes and by using the visualization options available in NVivo it was easier to draw conclusions from the data. The results will be discussed in the next section.

4   Results and Discussion

The purpose of doing qualitative data analysis is to understand phenomena based on the data available. In this study, the researcher was trying to understand how the process of scheduling medical appointments is and how could this process be improved. For the intended purposed, a group of 10 students from the university of Porto, faculty of engineering, where interviewed and the results drawn from the interviews are going to be discussed next.

Based on the data analysis, it was clear that the participants go to the hospital at least twice a year to do a routine checkup. The average was once a year as can be seen in Fig. 1.

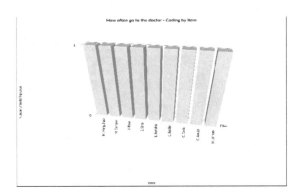

Fig. 1. Chart representing how often participants go to the doctor.

On the x axis, one can see the name of the participants and on the y axis the average time they go to the doctor per year.

Knowing that they go to the hospital for routine checkup at least once a year, the next question was made to understand how these appointments were booked. It was concluded that 53% book their appoints over the phone. While 38% prefers to schedule in person. Some believe that it is more reliable if they go in person, also sometimes there are long wait lines on the phone. When questioned about the idea of booking appointments online, 100% of the participants considered it as a good option. Some argued that "… it's more practical, because a person doesn't have to go to the place (in case it's in person) or if it's by phone, what happens is that we have to wait to be contacted."; while others said "I would like that because then I wouldn't have to wake up super early to get an appointment and wait for my time. It's more convenient". Words such as practical, convenient, faster were commonly used as can be seen from the word cloud (Fig. 2).

Fig. 2. Word cloud of participants' opinions about online booking appointments.

Codes	Search Project				
⊙	Name		⌃ Files	Refer	
⊟ ○	How do you book your appointments		10	10	
	○ online		1	1	
	○ in person		5	5	
	○ over the phone		7	7	
⊟ ○	How do you prefer to book online appo		10	14	
	⊟ ○ mobile app		5	5	
	○ Reasons		2	3	
	⊟ ○ website		5	6	
	○ reasons		1	1	

Fig. 3. Codes created based on the interviews.

When asked about whether they would prefer to schedule the online appointment via a website or a mobile app, there was a tie in the opinions (50/50). Some argued that a mobile phone would be better for its practicability. Other argued that "The number of apps that multiple services suggest us to install makes it almost impossible to manage the phone's memory". However, all participants were very comfortable in using both technologies (cellphone and computer). It was asked if they would like to receive an email, a phone call, or a text message with a reminder of the scheduled appointment and the opinions were mixed. The majority (60%) preferred to receive an email with a reminder because they don't usually pay attention to their phones during the day. It is easier for them to read an email than to read a text message. Finally, it was questioned if the participants would agree with sharing their personal information online to book the appointment and the answer was unanimous. Every participant was in agreement to share their personal information online disregarding the concern for data privacy. One participant even argued "Yes, because when I usually make an appointment, whether in person or over the phone, I also have to give this data."

From the analysis of the data, one could infer that it is very common for students to go to the doctor at least once a year and usually the appointments are scheduled over the phone. However, it is not very practical having to call to book an appointment making the idea of an online booking system very attractive to the students. Because some of them would not have to go in person or call and wait for long time until booking the appointment. It would be ideal to have an online service that could be accessed via a mobile app or via a website. The total booking process should not last more than 10 min otherwise, it would not be practical.

In Portugal, the Serviço Nacional de Saúde (SNS) allows all portuguese citizens to schedule appointments via internet [12]; however, during the interview 80% of the participants did not know about this option. Those that knew, found out via word of mouth or by calling to schedule an appointment and being redirected to the website. Both participants that have used online booking systems before mentioned that "…the actual system has thing to improve. It is effective if it is a routine appointment, scheduled well in advance." They believe that the already existent system is efficient and user-friendly. They would recommend it to other users as well.

It is clear therefore that although online booking services are available most people still don't know about them. However, they are very interested in using an online booking system for medical appointments because it would be very practical. The user has the flexibility to book the appointment whenever he or she wants to, no waiting list, and reduction on mistakes.

5   Conclusion

There are several paths to follow in the analysis of qualitative data, but all require rigor in the application of procedures. Qualitative analysis does not dispense with the use of formal techniques for data processing, as well as quantitative methods. Unlike the latter, in qualitative research the analysis takes place from the beginning of data collection, during and after the completion of the fieldwork. When reporting the results, it is very important for the researcher to describe all the steps followed with transparency, informing possible biases and limitations of the study. How the data was collected and recorded, the sampling criteria and the ethical aspects involved should be described. The path taken into de decomposition of the analyzed material must be clear, as well as the interpretations and their theoretical, supported by evidence and in dialogue with the updated literature.

References

1. Demo, P.: Ciências sociais equantidade. In: Demo, P. (Org.). Ciências sociais e qualidade (1985)
2. Amis, J.: Book Review: Amis. In: Denzin, N.K., Lincoln, Y.S. (eds.). The landscape of qualitative research. Sage. Organizational Research Methods, Thousand Oaks, CA, vol. 14, pp. 239–242 (2010). (2008). https://doi.org/10.1177/1094428109332198
3. Victora, C.G., Knauth, D.G., Hassen, M.D., Knauthk, D.R., Knauth, H.F.E.: Pesquisa qualitativa em saúde: uma introdução ao tema (2000)
4. Taquette, S.R., Minayo, M.C., Rodrigues, A.O.: The perceptions of medical researchers on qualitative methodologies. Cad Saude Publica 31(4), 722–32 (2015). https://doi.org/10.1590/0102-311x00094414
5. Gomes, R., Souza, E.R., Minayo, M.C.S., Silva, C.F.R.: Organização, processamento, análise e interpretação de dados: o desafio da triangulação. In: Minayo, M.C.S., Assis, S.G., Souza, E.R., (eds.) Org, Avaliação por triangulação de métodos: abordagem de programas sociais, Fiocruz, pp. 185–221 (2005)
6. Minayo, M., Assis, S., Souza, E.: Avaliação por Triangulação de Métodos. Abordagem de Programas Sociais (2006)
7. Gomes, M.H., Silveira, C.: Sobre o uso de métodos qualitativos em Saúde Coletiva, ou a falta que faz uma teoria. Revista de Saúde Pública 46(1), 160–165 (2012)
8. Gomes, R.: Analysis and interpretation of qualitative research data. In: Minayo, M.C.S. (ed.), Organizer, Social Research: Theory, Method and Creativity, 25th Edition, Vozes, Petrópolis, pp. 79–107 (2007)
9. Walker, D., Myrick, F.: Grounded theory: an exploration of process and procedure. Qual. Health Res. 16(4), 547–559 (2006). https://doi.org/10.1177/1049732305285972
10. de la Croix, A., Aileen, B., Terese, S.: How to… do research interviews in different ways. Clín. Teach. 15(6), 451–456 (2018)
11. Edwards-Jones, A.: Qualitative data analysis with NVIVO. J. Educ. Teach. 40(2), 193–195 (2014). https://doi.org/10.1080/02607476.2013.866724
12. Marcar consultas. https://www.sns24.gov.pt/servico/marcar-consultas/. Accessed 11 June 2022

Gamification in Mobile Ticketing Systems: A Review

Marta Campos Ferreira[1,2](✉) ⓘ, Diogo Gouveia[1], and Teresa Galvão Dias[1,2] ⓘ

[1] Faculdade de Engenharia da Universidade do Porto, R. Dr. Roberto Frias, 4200-465 Porto, Portugal
{mferreira,tgalvao}@fe.up.pt, up202003081@edu.fe.up.pt
[2] INESC-TEC, R. Dr. Roberto Frias, 4200-465 Porto, Portugal

Abstract. This review is an analysis of the literature on public transport and mobile ticketing systems and their gamification. The review is divided into three main topics: (i) Behavioral Change in relation to Public Transport, (ii) Gamification, and (iii) Gamification in Public Transport and Mobile Ticketing. This study shows the diversity of the theme of gamification applied to the transport sector and demonstrates its potential to attract and retain more customers for more sustainable means of transport.

Keywords: Game-based · Public Transport · Smart Cities · Smartphones · Behavioral Change

1 Introduction

In today's society, complex transport networks and the lack of seamless options are barriers to the use of public transport services [1] while modern mobile ticketing services allow to travel in a convenient and smooth way, enhancing the customer experience [2]. Even though some cities implemented mobile ticketing solutions on their public transport network, the adoption of this type of services by the general population has seemed to have achieved limited success [3].

As for the use of gamification in marketing, it is seen has an effort to engage users and improve user experience. Gamification is defined as the implementation of game elements in a non-game context [4] for users to enjoy using a product or service and despite the enormous potential it holds, its application in mobile payment solutions is very limited and not widely used [5].

The aim of this study is to understand how gamification can be introduced into public transport mobile ticketing systems, in order to promote the use of sustainable means of transport and the use of mobile ticketing solutions. For this, a literature review is carried out on behavioral change regarding public transport, gamification in general and gamification applied to mobile ticketing solutions.

A. Rocha et al. (Eds.): WorldCIST 2023, LNNS 802, pp. 143–152, 2024.
https://doi.org/10.1007/978-3-031-45651-0_15

2   Survey Methods

In order to gather the necessary literature to produce an accurate analysis of the multiple topics at hand, a survey was conducted using the following queries on Scopus.

- (Gamification AND "Public Transportation" OR "Public Transport"), which returned 17 documents.
- ("Behavioural Change" AND "Gamification") AND (LIMIT-TO (SUBJAREA, "ENGI") OR LIMIT-TO (SUBJAREA, "COMP")), which returned 113 documents.

The first query is focused on the main topic of this work, combining the concept of gamification with public transport. The second query is more directed to sources that can help explain how the use of gamification can affect people behavior.

3   Survey Results

The literature review is divided into three: (i) Behavioural change regarding public transport which will introduce the topic and explain it, (ii) Gamification which will explain the concept and how to implement it successfully and (iii) Gamification in public transport and mobile ticketing which present successful cases of applied gamification inside the public transport sector.

3.1   Behavioural Change Regarding Public Transport

This section of the review will proceed to analyze and review the current state of the art surrounding the topic of behavioural change regarding public transport and how gamification has affected this subject.

It is known that the task of behavioral change is a very difficult subject to tackle without proper preparation, especially in the transport sector, as people's transport patterns [6] are very difficult to modify and change because these patterns are usually embedded in each person's lifestyle, identity and habits, and the change produced from the change campaign can influence the person's well-being [7, 8].

There are several barriers to this change. The first barrier found are the obstacles that appear during the change process, and these can vary in types and forms, such as political or bureaucratic [8]. The second barrier is the population itself, as stated earlier in [7, 8], it is very difficult to change the values and lifestyle of a single person, let alone the population, but it is a possible agenda that it can be achieved even in the face of the greatest difficulties.

Among the many theories that explain how it is possible to change the decisions people make, some see these changes as a result of internal factors (values, attitudes, personal norms), while others see them as a result of external factors (financial and social) or both [6, 9].

The authors [6] classify behavioral science interventions to increase public transport use into three broad categories: communication-based approaches, anti-bias approaches, and technology-based approaches. This review will focus exclusively on the aspect of technology-based approaches, as this is where gamification falls within that spectrum.

Within technology-based approaches, two alternatives are given: feedback and gamification. The first explains that feedback that is given through applications can change behavior [10], which works by providing useful information that can be used to evaluate progress towards a goal [6]. One of the experiments done with this approach was carried out in Japan, where, through the feedback provided by the participants on the "daily diaries of your transport" [11, 12], the researchers were able to analyze and repackage the information in different groups in order to change the behavior of the participants "leveraging insights" [6]. This reorganization of information would then lead to changes in participant behavior. This study led to a decrease of about 15% in the use of the family car and a 4% increase in the use of public transport [11]. The second alternative will be explained and analyzed in detail in the next section.

3.2 Gamification

This section of the review will proceed to analyze and review the current state of the art surrounding the topic of Gamification and all of its variances.

Gamification is an emerging trend that has been growing very rapidly in recent years within multiple sectors, such as business, marketing, education, communication, literature and transport. There is a range of different definitions that appear in the literature found for this topic. The authors [13] define games as a system in which players engage in an artificial conflict, defined by rules, which results in a quantifiable result. Another definition is given by [13] who proposes that all games have six main characteristics: rules; variable; quantifiable results; results loaded with value; player effort; player investment; and negotiable consequences, regarding the real life effects.

In all definitions, there is a common thread: all games are defined by rules, structure, uncertain outcomes, conflict, representation and resolutions, and from these aspects of games, each in its own proportion, different games arise.

Gamification can also be used to describe two additional concepts: (1) the creation or use of a game for any non-entertainment context and/or purpose and (2) the transformation of an existing system into a game [14]. In both cases, games are incorporated into an existing system, while replacing and adding to the structure of the system itself, or the system is converted strictly into a game. Various action words and phrases have been widely used to refer to them instead of gamification, such as "Gamified", "gamify" and "the gamification of" [14].

In gamified experiences, there are four different parties involved: players, designers, spectators and observers [15], and each has a significant and important role to play within the sphere of gamification. The main aim behind gamification is to introduce the use of game elements and aspects into non-game applications and concepts, it can also be referred to as the enrichment of well-known gaming software to invoke immersive gaming-like experiences [16].

Gamification also tries to work by providing users with different activities through "gamified" aspects that are added to the platforms being used, in order to lead to the achievement of goals and objectives defined by the developer.

As stated by [17] gamification involves applying game elements with a very specific intention in mind. Although the introduction of "gameplay, playful interaction and playful design", it always comes to the point that gamification does not need to be in a serious context, but requires the system not to be a game, otherwise it would not be gamification, but just a game.

The gamification of a system can be implemented in many ways, formats and forms. From introducing various game elements into the system, to just a simple leader board. The authors [18] gathered a very complete collection of information about the type of game elements present in the whole subject of Gamification. The authors divide game elements into three main types: social, competitive, and a mixture of the other two. Social game elements provide the system with a new social element, such as emotion, feedback, gift, social discovery, social pressure, or social status. The competitive game elements provide the system with a competitive side, such as badges, challenges, competition, points, unlockable or rewards. Elements of social and competitive games give the system both a competitive side and a social side. Elements of social and competitive games give the system a competitive side and a social side, such as exploration, collection, honor system, leaderboard or virtual economy.

While there are many complex and small ways to gamify software, or any kind of experience, it's still a very complex process that requires time and work to get right [16]. One of the challenges that comes with running and implementing a gamified experience for game designers is the individuality that players, whether users or employees of a company, bring to the table, and understanding this variability is difficult and necessary to create engaging experiences [19]. This comes with a set of "rules" that most designers need to follow in order to create an exceptional gamified experience that affects people's behavior and not just entertain as the main objective, thus creating another layer of complexity that challenges designers to create challenges. Engaging as context provides operational requirements that limit the unlimited design space that games typically have [16].

A great example of a successful application is Duolingo, an application that provides users with a new way to learn a new language while playing a "game". Although the main focus of Duolingo is for the user to learn a new language, it does so by providing the user with different means to learn while having fun doing it. Some of the gamified elements that Duolingo provides are the daily goals, the day streak, a fully implemented challenge system, a progression system, achievements/rewards to complete and collect, customization of your avatar, power-ups, a heart system that functions as a life bar, a leaderboard where you can compare yourself and compete with other "Duos" (name given to platform users) and even a friends list where you can see the progress made by your peers [20]. As shown in the study by [19], another example of a gamified experience is the launch of the famous rapper Jay-Z's book Decoded, which incorporated gamification elements into the release of the book instead of the old fashioned and traditional book release which are not particularly interactive. Another case that demonstrates a successful implementation of gamification is Freshdesk, which is a helpdesk software program which aimed at improving employee productivity and customer satisfaction [19].

But not all cases are successes, many cases of gamification fail because not a lot of thought was put into the process and how it would affect the "players". A famous case was the Disneyland and Paradise Pier hotels that in order to increase productivity installed monitors that functioned like scoreboards in order to increase productivity margins, but this process had a different outcome from what was expected, employees felt even more pressured seeing their performance measured, and felt that Disney had create an "electronic whip" in order to control their workers into working faster and harder so their performance did not lower.

The next section details how gamification can be implemented in public transport and mobile ticketing systems.

3.3 Gamification in Public Transport and Mobile Ticketing

This section of the review details the current state of the art around the topic of gamification implemented in public transport and mobile ticketing. In our society, gamification is an aspect that has been growing a lot in recent years and one of the areas that has affected it is the public transport sector, which has been a factor that has allowed the sector to develop even more in the world. Future. Next, some cases will be presented to demonstrate how this change affected this sector and its future.

Wikibus is a system that allows passengers to share information about public transport and real-time occurrences of vehicles, stops and bus routes that depend on the contribution of the crowd to maintain their services [21]. Wikibus collects information about bus stops, lines, vehicles, companies and events in real time (for example, if the bus was late, had an accident, etc.). Wikibus can be used by everyone, where anonymous users can just find and read information, while verified users can create, read and update information.

To gamify Wikibus, the authors used the G.A.M.E approach, which is a four-phase plan that gathers, analyzes, models, and executes the gamified experience. To obtain the best results from the approach, two preliminary tests were carried out to find out about the main problems that end users had with the application (gathering). After the collection process, the authors concluded that the two main problems present were the difficulty of understanding how to find and how to contribute new content and that they were analyzed using the 3C collaboration model: communication, cooperation and coordination [21, 22]. So, to fix these issues and gamify the app, the authors decide to design how the gameplay, play and fix the issues with the Wikibus app would be. Finally, the developers prototyped the changes in Wikibus and carried out usability tests to verify that the implementation was effective, which according to the final test results, more people were able to complete the test (40% to 56%) and more than 80% thought the gamified version was more reliable than the original [21].

Trafpoint Trafpoint is a system that was created with the aim of getting more people to travel by public transport and is divided into four elements, where the first is a mobile application that registers people boarding and leaving the bus, where people can accumulate miles for each trip and share their position on the leaderboard created for the app [23]. The second element, is an application that counts all passengers getting on and off the bus using video and motion detection, and then all the collected information is sent to the Trafpoint backend [23]. The third and fourth element is the backend analytics, where the data is processed. Trafpoint is a promising application, but has not yet been implemented on a large scale, but reports show that Trafpoint can be adapted to other types of transport, such as airplanes and trains [23].

Viaggia Rovereto is a mobile routing application that allows users to register and plan their daily itinerary. So, to create a proof of plan, an experiment was carried out that lasted 5 weeks and was divided into 3 phases, where the first phase (1 week) was the baseline and users had to familiarize themselves with the App [24].

The second phase (2 weeks) added new functionality (Sustainable Mobility Recommendations) without any gamification, where the app gave the user alternative routes to use and then ranked them based on sustainability policies.

The third and final phase (2 weeks), gamification was added to the features of phase 2, which added three types of points: Green Points, Health Points and Park&Ride points. Added badges, collections, leaderboards, and certificates for achievements (see Fig. 1) [24].

According to the data obtained in the study, after implementing the application's gamified elements, Viaggia Rovereto encouraged its players to use the application more frequently and to use the sustainable alternative routes recommended by the application where there was an increase of 17.9% in the selection of new sustainable routes [24].

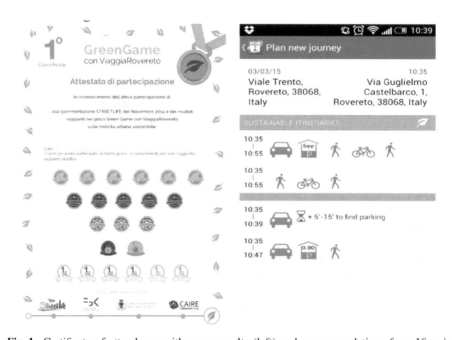

Fig. 1. Certificate of attendance with game results (left) and recommendations from Viaggia Rovereto App (right) adapted from [24].

Stand Up Heroes!, or SUH for short, came to try to solve a problem present in the peak hours where the public transport vehicles are very crowded and there is nothing to do but stand up, so that users are inside of any kind of public transport, their avatar inside the SUH app starts to evolve and get "stronger". Users can also compare avatars with other players and help each other find treasure, which leads to cooperation between users and motivation to keep standing [25].

SUH is a game system that was implemented in an application that is divided into two subsystems, the first a transport-and-standing estimation subsystem and the second an avatar and game representation subsystem [25].

The first subsystem, which estimates transport using the accelerometer and GPS, tries to get a user's current location and compare their position either at a bus station or a train station, and then checks the speed to see if they are moving in transport. Public. Then, to estimate whether a user is standing or not, the subsystem uses the orientation of the mobile device's gravity acceleration [25].

The second subsystem runs an RPG (Role Playing Game) that allows users to control and develop their own avatar. Within the game, the avatar explores a "dungeon" and the game shows the progress of the exploration which is explored further the longer the user is on his feet. The gamified experience also offers monsters, loot, treasures, health points, skills and equipment that allow the user to change their avatar at will [25].

To check if the app was a success, a study was carried out which showed that the gamified experience can motivate users to keep using the app even more and to stay on their feet. The only negative point raised was that there were few items for the "players" to collect, because at the end of the experiment most of the participants had gathered most of the available items [25].

The **Collaborative Public Transport Optimization** or CPTO for short was an application that was developed to try and change the habits of Trikala city dwellers to use public transport more and change their transport habits [26].

Within the CPTO, the concept of the serious game framework was implemented, which aims to connect different public transport users and create an ecosystem, and reward appropriate usage behavior through virtual and real incentives.

Given the variety of users involved [26], four games were developed: Virtual Bank (VB), Competition Game (CG), Snake and Ladders (S\&L), and Passenger Game (PG), in order to adapt to the different needs and expectations of users.

In the VB game, the user can get coins that can later be used to buy bus tickets, parking spaces and be used in the application world itself (all other applications) [26].

In the Competition Game, users can enroll in open competitions where they can see the evolution of their score through the analysis of the user's performance, and at the end of the competition a detailed analysis of the scores, each user and awards are displayed in the application [26].

In the S&L game, the user plays the classic Snakes and Ladders game with other users of the application. To advance, a virtual die is rolled each time a real-world event happens and the user's position advances in the slots [26].

In the PG game, it is just a simple interactive space shooter that depends on the gaming skills of the user playing. The game itself consists of two levels with challenges that the player needs to overcome using energy collected from the real world [26].

4 Discussion

The literature review carried out allowed to verify that there are several ways to implement a behavioral change campaign [6], ranging from simple approaches based on communication to more complex approaches to bias-busting and based on technology.

But that alone is not enough to change behavior, and the team responsible for the campaign must develop different ways of approaching the situation in order to have multiple strategies to have a better chance of changing the population in question.

As shown in [6], gamification is one of the technology-based approaches that entails implementing game elements in a non-game context and this can be done with a multitude of variations between each possible iteration of this process. From the implementation of a simple narrative in an application about the ocean to a more complex approach involving the implementation of a social network with players, badges, trophies, collections, challenges. There are many other ways to gamify the experience for users to enjoy.

But the success of a gamified experience depends on the users using it and building the community around the experience. Although they are not the ones who develop the experience itself, intrinsically they contribute the most since the experience develops around them, their habits, their choices and above all their needs.

One of the best gamified experiences to date is Duolingo, where the user can learn a new language while "playing" the language game with lanes, badges, leaderboards, and even the latest competitions. But not all experiences go as planned, for example, during the implementation of a leaderboard system at Disney hotels, employees saw this as another way for the corporate side of the business to control them and ensure that they maintain their performance in relation to their peers.

Some examples of a good gamified experience within the public transport sector are the Stand Up, Heroes! project from Japan that gamified the process of standing up during peak hours on public transport with an RPG-like game (Role Playing Game) with character development, dungeons and multiplayer capability [25]. Also, Viaggia Rovereto project in Italy that made users think about their transport plans and gamified their experience by implementing badges, certificates of public transport use that showed an increase of about 18% in the selection of new sustainable routes [24].

5   Conclusion

More than ever, it is important to encourage the use of sustainable modes of transport, such as public transport. This article presents a review of the state of the art on changing people's behavior and how this behavior can be influenced through gamification. Success cases implemented in public transport are also explored.

Gamification is a very diverse topic that can be implemented in different ways, for example using leaderboards, narratives, social aspects that connect users, prizes, challenges, customization. Some of the most successful cases of gamification are: Duolingo, Freshhdesk, Viaggia Rovereto.

It is believed that adopting gamification can help attract and retain more customers on public transport, however there is still a lot of research work to be done.

References

1. Ferreira, M.C., Dias, T.G., Cunha, J.F.: An in-depth study of mobile ticketing services in urban passenger transport: state of the art and future perspectives. In: Smart Systems Design, Applications, and Challenges, pp. 145–165. IGI Global (2020). https://doi.org/10.4018/978-1-7998-2112-0.ch008

2. Ferreira, M.C., Dias, T.G., Cunha, J.F.: Anda: an innovative micro-location mobile ticketing solution based on NFC and BLE technologies. IEEE Trans. Intell. Transp. Syst. **23**, 6316–6325 (2022). https://doi.org/10.1109/TITS.2021.3072083
3. Dahlberg, T., Guo, J., Ondrus, J.: A critical review of mobile payment research. Electron. Commer. Res. Appl. **14**, 265–284 (2015). https://doi.org/10.1016/j.elerap.2015.07.006
4. Deterding, S., O'Hara, K., Sicart, M., Dixon, D., Nacke, L.: Gamification: using game design elements in non-gaming contexts. Conf. Hum. Factors Comput. Syst. Proc. 2425–2428 (2011). https://doi.org/10.1145/1979742.1979575
5. Putri, M.F., Hidayanto, A.N., Negara, E.S., Budi, N.F.A., Utari, P., Abidin, Z.: Gratification sought in gamification on mobile payment. In: ICICOS 2019 - 3rd International Conference on Informatics and Computational Sciences Accel. Informatics Computer Research Smarter Social Era Industry 4.0, Proceedings (2019). https://doi.org/10.1109/ICICoS48119.2019.8982424
6. Kormos, C., Sussman, R., Rosenberg, B.: How cities can apply behavioral science to promote public transportation use. Behav. Sci. Policy **7**, 95–115 (2021). https://doi.org/10.1353/bsp.2021.0004
7. Chatterjee, K., et al.: Commuting and wellbeing: a critical overview of the literature with implications for policy and future research. Transp. Rev. **40**, 5–34 (2020). https://doi.org/10.1080/01441647.2019.1649317
8. Davies, N.: What are the ingredients of successful travel behavioural change campaigns? Transp. Policy **24**, 19–29 (2012). https://doi.org/10.1016/j.tranpol.2012.06.017
9. Hoffmann, C., Abraham, C., White, M.P., Ball, S., Skippon, S.M.: What cognitive mechanisms predict travel mode choice? A systematic review with meta-analysis. Transp. Rev. **37**, 631–652 (2017). https://doi.org/10.1080/01441647.2017.1285819
10. Sunio, V., Schmöcker, J.D.: Can we promote sustainable travel behavior through mobile apps? Evaluation and review of evidence. Int. J. Sustain. Transp. **11**, 553–566 (2017). https://doi.org/10.1080/15568318.2017.1300716
11. Fujii, S., Taniguchi, A.: Reducing family car-use by providing travel advice or requesting behavioral plans: an experimental analysis of travel feedback programs. Transp. Res. Part D Transp. Environ. **10**, 385–393 (2005). https://doi.org/10.1016/j.trd.2005.04.010
12. Taniguchi, A., Hara, F., Takano, S., Kagaya, S., Fujii, S.: Psychological and behavioral effects of travel feedback program for travel behavior modification. Transp. Res. Rec. 182–190 (2003). https://doi.org/10.3141/1839-21
13. Salen, K., Zimmerman, E.: Rules of Play: Game Design Fundamentals. MIT Press, Cambridge (2004)
14. Seaborn, K., Fels, D.I.: Gamification in theory and action: a survey. Int. J. Hum. Comput. Stud. **74**, 14–31 (2015). https://doi.org/10.1016/j.ijhcs.2014.09.006
15. Robson, K., Plangger, K., Kietzmann, J.H., McCarthy, I., Pitt, L.: Is it all a game? Understanding the principles of gamification. Bus. Horiz. **58**, 411–420 (2015). https://doi.org/10.1016/j.bushor.2015.03.006
16. Morschheuser, B., Hassan, L., Werder, K., Hamari, J.: How to design gamification? A method for engineering gamified software. Inf. Softw. Technol. **95**, 219–237 (2018). https://doi.org/10.1016/j.infsof.2017.10.015
17. Deterding, S., Dixon, D., Khaled, R., Nacke, L.: From game design elements to gamefulness: defining "gamification. In: Proceedings of the 15th International Academic MindTrek Conference: Envisioning Future Media Environments MindTrek 2011, pp. 9–15 (2011). https://doi.org/10.1145/2181037.2181040
18. Klock, A.C.T., Gasparini, I., Pimenta, M.S., Hamari, J.: Tailored gamification: a review of literature. Int. J. Hum. Comput. Stud. **144** (2020). https://doi.org/10.1016/j.ijhcs.2020.102495

19. Robson, K., Plangger, K., Kietzmann, J.H., McCarthy, I., Pitt, L.: Game on: engaging cus-
    tomers and employees through gamification. Bus. Horiz. **59**, 29–36 (2016). https://doi.org/
    10.1016/j.bushor.2015.08.002
20. Shortt, M., Tilak, S., Kuznetcova, I., Martens, B., Akinkuolie, B.: Gamification in mobile-
    assisted language learning: a systematic review of Duolingo literature from public release of
    2012 to early 2020. Comput. Assist. Lang. Learn. 1–38 (2021). https://doi.org/10.1080/095
    88221.2021.1933540
21. Brito, J., Vieira, V., Duran, A.: Towards a framework for gamification design on crowdsourcing
    systems: the G.A.M.E. approach. In: Proceedings of the - 12th International Conference on
    Information Technology-New Generations (ITNG 2015), pp. 445–450 (2015). https://doi.org/
    10.1109/ITNG.2015.78
22. Vieira, V., Fialho, A., Martinez, V., Brito, J., Brito, L., Duran, A.: An exploratory study on
    the use of collaborative riding based on gamification as a support to public transportation.
    Proc. - 9th Brazilian Symp. Collab. Syst. SBSC 2012. 84–93 (2012). https://doi.org/10.1109/
    SBSC.2012.32
23. Johannessen, M.R., Berntzen, L.: Smart cities through implicit participation: using gamifi-
    cation to generate citizen input for public transport planning. Innov. Public Sect. **23**, 23–30
    (2016). https://doi.org/10.3233/978-1-61499-670-5-23
24. Kazhamiakin, R., et al.: Using gamification to incentivize sustainable urban mobility. In:
    2015 IEEE 1st International Smart Cities Conference on ISC2 2015 (2015). https://doi.org/
    10.1109/ISC2.2015.7366196
25. Kuramoto, I., Ishibashi, T., Yamamoto, K., Tsujino, Y.: Stand up, heroes!: gamification for
    standing people on crowded public transportation. In: Marcus, A. (eds.) Design, User Experi-
    ence, and Usability. Health, Learning, Playing, Cultural, and Cross-Cultural User Experience.
    DUXU 2013. LNCS, vol. 8013, pp. 538–547. Springer, Berlin, Heidelberg (2013). https://
    doi.org/10.1007/978-3-642-39241-2_59
26. Drakoulis, R., et al.: A gamified flexible transportation service for on-demand public transport.
    IEEE Trans. Intell. Transp. Syst. **19**, 921–933 (2018). https://doi.org/10.1109/TITS.2018.279
    1643

Analyzing Quality of Service and Defining Marketing Strategies for Public Transport: The Case of Metropolitan Area of Porto

Marta Campos Ferreira[1,2(✉)] , Guillermo Peralo[1], Teresa Galvão Dias[1,2] ,
and João Manuel R. S. Tavares[1]

[1] Faculdade de Engenharia, Universidade do Porto, R. Dr. Roberto Frias, 4200-465 Porto,
Portugal
{mferreira,tgalvao,tavares}@fe.up.pt, up202102419@edu.fe.up.pt
[2] INESC-TEC, R. Dr. Roberto Frias, 4200-465 Porto, Portugal

Abstract. The aim of this work is to determine, based on a market research, the level of passenger satisfaction with public transport services, in order to support better marketing decisions. This survey involves dimensions such as the level of satisfaction with timetables and frequency, vehicle conditions, driver attitudes and behavior, fares and information made available to passengers. The study was applied to the case of public transport in the Porto Metropolitan Area, Portugal, and aims to help define recommendations to improve the quality of service and define more effective marketing strategies.

Keywords: Grounded Theory · Quantitative Data

1 Introduction

One of the biggest problems faced by many countries around the world is the rapid growth in the use of private vehicles, which is associated with the reduction in the use of public transport. This situation has negative impacts such as traffic congestion, increased use of fossil fuels and air pollution.

Therefore, it is important to study the perception and behavior of public transport users, in order to obtain valuable information to develop policies and strategies to encourage the use of public transport.

Empirical studies on public transport in different cities around the world show that it is important to focus on people's subjective opinions about public transport, rather than an objective analysis of the quality of service [1, 2, 8].

In this context, it is intended that this work serves as a guide to understand which aspects are subjectively valued by people with regard to public transport. For this purpose, a study was carried out in the city of Porto, Portugal, in order to draw the intended conclusions. With the information presented, some improvement points and marketing strategies are discussed may encourage greater use of this service.

This article is organized as follows: the next section presents the methodology that was followed. Section 3 presents the main results derived from the data collected and analyzed. Section 4 discusses the results and Sect. 5 presents the main conclusions.

© The Author(s), under exclusive license to Springer Nature Switzerland AG 2024
A. Rocha et al. (Eds.): WorldCIST 2023, LNNS 802, pp. 153–161, 2024.
https://doi.org/10.1007/978-3-031-45651-0_16

2 Methodology

As a basis for starting to study the population's perception of public transport in Porto, a survey was carried out with 231 people of different ages, genders and social conditions, so that the information obtained was as objective as possible.

The questionnaire consists of 3 separate sections (see Appendix A):

1. The first section seeks to ask people general questions to identify them within the groups and from there begin to analyze the data.
2. In the second section, questions are asked about how people normally use public transport and what kind of services they usually use.
3. In the third, a total of 33 questions on different topics are asked to assess people's level of satisfaction with different aspects. Topics addressed are: ease of understanding in access and use, price, quality of stations/stops, waiting and displacement times, quality of vehicles, relationships with other people and satisfaction with the routes covered by vehicles.

With all the data obtained from the respondents, a statistical analysis was carried out in order to have a clearer view of the dimensions where it is possible to improve and the factors that may be influencing the population with regard to the use of public transport.

The analysis is divided into three distinct parts:

- Study of the correlations between the sociodemographic variables that define the respondents

  A series of questions are asked to respondents in order to categorize them into different groups and thus analyze whether there is any relationship between these groups and the use of public transport. For this purpose, Pearson's correlation coefficient was calculated and conclusions drawn from the results. Specifically, respondents were asked about their age, gender, level of education and employment status.

- Study of the type of use and mobility services used by the population.

  It explores the kind of use people make of public transport and what kind of services they usually use. Questions are asked about frequency of use, usual travel time, waiting time at stops, type of tickets used and where they are obtained, payment method for tickets, what type of public transport services are used and which channel is used. Used to access these services. In this way, it is possible to obtain a more accurate picture of how public transport is normally used, as well as the services associated with it.

- User satisfaction

  User satisfaction levels are analyzed in relation to different points. The most unfavorable data are chosen to propose solutions to increase satisfaction levels, and the most satisfactory data are chosen to highlight them and attract people through them. To assess this point, 33 questions were asked (see Appendix A) on different aspects of the service (ease of understanding and use, prices, experience at stations and in the vehicle…). Participants are asked to rate on a scale from 1 to 5 how much they agree with the statement presented to them. With this information it is possible to have an idea of which aspects need to be improved and which can be used as strengths.

3   Results and Data Analysis

The questionnaire was released online and a total of 231 responses were collected. The sociodemographic characteristics of the survey respondents are shown in Table 1.

Table 1. The sociodemographic characteristics of the survey respondents.

Variable	Category	Nr of responses (%)
Gender	Male	40.44%
	Female	59.56%
Age	Under 18 years old	1.3%
	Between 18 and 25 years old	48.3%
	Between 26 and 45 years old	21.7%
	Between 46 and 65 years old	19.7%
	Over 65 years old	1.7%
Level of education	Secondary education (12th year)	23.9%
	Bachelor's Degree	28.3%
	Master's	21.7%
	PhD	26.1%
Employment status	Student	60.9%
	work for others	35.2%
	Self-employed	1.7%
	Unemployed	1.3%
	Retired	0.9%

Table 2. Pearson's correlation coefficient related to sociodemographic variables of respondents.

	Age	Frequency of use	Education level	Employment status	Gender
Age	1				
Frequency of use	−0,401984511	1			
Education level	0,764408383	−0,355383563	1		
Employment status	0,63773463	−0,30702164	0,648555893	1	
Gender	0,11963918	−0,10332888	0,099049719	0,04827921	1

3.1   Relationship Between the Variables that Define the Respondents

The results obtained by calculating Pearson's correlation coefficient in the different sections analyzed are presented in Table 2.

Regarding the frequency of use of public transport, some conclusions can be drawn:

- Use of public transport and age:
  It is a weak negative correlation, that is, the older the person, the less likely they are to use public transport. Even so, it's not a high enough number to be considered a determining factor.
- Use of public transport and level of education:
  Once again, this is a weak negative correlation, that is, the higher the level of education, the lower the tendency to use public transport. However, the value is not high enough to be a significant ratio.
- Use of public transport and professional status:
  Again, there is a weak negative correlation, i.e. employed or self-employed people are less likely to use public transport. The value is too low to be relevant.
- Use of public transport and gender:
  This is extremely small and therefore excludes the possibility of any relationship between gender and public transport use.

The results obtained here show a tenuous relationship between education, professional situation and age and use of public transport, but so tenuous that it cannot be considered significant enough to do something about it. Therefore, it is considered that the variables selected to classify people are not related to the use of public transport.

3.2   Type of Use of Public Transport and Services Used

The percentages obtained are calculated according to the answers given to the different questions. From the results obtained, it appears that most people tend to make the same use of public transport and its services.

The following conclusions can be summarized:

- 95% of people use public transport for trips longer than 15 min. In other words, it is generally not used for short trips. Most people use it for medium trips (between 15 and 30 min).
- In 96% of cases, the waiting time at the station/stop is less than 20 min, with the usual waiting time being between 5 and 10 min.
- 49% of people buy tickets at vending machines and 16% at Andante stores. The rest of the people buy them spread across different services.
- Most people use the monthly pass. Another high percentage of people choose to use occasional tickets. Around 10% use the Anda App or Andante 24.
- 98% of people pay with a credit card or cash. Only 2% use a cell phone.
- The most used additional services are map consultation, timetable consultation and mobile applications. The other services are used in low proportion.
- The most common way of accessing these services is through the website (41%), followed by stops and stations (27%) and mobile applications (21%).

3.3 Level of Satisfaction with Different Aspects

Of all the aspects evaluated in the survey, we are going to focus on those that have a high level of satisfaction, to highlight them and use them as a marketing strategy; and those with unfavorable levels of satisfaction, to try to improve them and make the service more pleasant and attractive.

Therefore, we established as a satisfaction criterion that the average score assigned to the question is greater than 4; and as a criterion of low satisfaction that the average score assigned is less than 3.

Next, the average score obtained in each question for the different aspects to be evaluated is presented in the form of a bar graph.

Most questions have an average score between 3 and 4, that is, they present a moderate level of satisfaction.

The aspects considered with a high degree of satisfaction are:

- Easily understandable schedules (average = 4.01351351)
- Ticket/mobile validation on the easy reader (average = 4.27477477)

    There are other points that come close to the established parameter of high satisfaction, which are:

- Easy purchase of tickets (average = 3.9954955)
- Feeling of safety in the vehicle (average = 3.86936937)

    Analyzing these aspects, it is possible to observe that, in general, users tend to classify the ease of access and use of public transport very positively.

The aspects considered of low satisfaction are:

- The stations/stops are comfortable (average = 2.86486486)
- Vehicles always arrive on time (average = 2.77927923)
- There is always a free place to sit (average = 2.17567568)
- Waiting time at suitable stops/stations (average = 2.93243243)
- Satisfaction with the frequency of public transport (average = 2.82432432)

    It can be seen that the most negative aspects that people perceive about public transport have to do with the discomfort they feel both in waiting and on the trip, as well as with the time spent on the trip.

4 Discussion

It is worth noting, first of all, that, as can be seen in this case, it is more important to look at the subjective opinion of consumers than the objective opinion of the service. For example, in the case of waiting time at stops, something that was classified negatively, we found that in 96% of cases the waiting time is less than 20 min, and in 20% less than 5 min. These are results that a priori could be considered favorable, but that the population considers not to be so.

In order to encourage the use of public transport and make it more attractive, two paths will be followed: firstly, solutions will be offered for the negative aspects of public transport, so that they can be improved and so that people have a more positive view of it. The service; secondly, ways will be sought to highlight the good points of public transport, so that they are what people think of when they think of public transport.

4.1   Improve the Negative Aspects

What people value most about public transport is comfort at stops and inside the vehicle, and waiting time at stops and inside the vehicle.

The solution, which is probably the most economical and serves to improve all these points, is to slightly increase the frequency of vehicle arrivals. By doing so, 4 points will improve directly: waiting time at stops/stations will decrease, improving the perception of this aspect, satisfaction with the frequency of public transport will increase, vehicles will arrive more on time and there will be more free spaces on trips, as vehicles will be emptier. In addition, indirectly, the perception of the comfort of the stops will also improve, since the waiting time will be more bearable and will lead to a less negative view. If comfort still does not improve with this solution, you can also consider investing in improving the quality of the stations and stopping to make people more comfortable.

Another possible solution would be to change the location and number of stops to try to make them more efficient and thus improve waiting times. However, this option can be much more expensive. In addition, the location and number of stops are currently evaluated positively, so changing the location and number of stops can be harmful in other respects.

4.2   Highlighting the Positive Aspects

As seen above, the positive points of public transport are the ease of understanding the schedules, the ease of validating and purchasing tickets, and the safety felt inside the vehicle.

In the marketing strategy to be used, these points should be highlighted so that people associate public transport with something easy and very accessible to everyone.

Anything that also undermines the perception of the private car in favor of the public car would also be sensible. For example, highlighting the time that can be spent looking for parking and in traffic jams would be beneficial, as it would make the perception that using public transport more efficient in many cases, also improving the perception of negative qualities.

5   Conclusion

The study of the population's subjective opinion about public transport is essential to know which areas of this service are considered negative and which are considered positive. With this information, strategies and policies can be designed to make public transport a better service and encourage its use.

To find out which aspects people value most, a survey was carried out in the city of Porto. With the answers obtained, conclusions are drawn that allow to have a clear point of view on which points should be improved and which should be highlighted.

Appendix

A. Structure of the Questionnaire
The questionnaire is divided into three sections:

1. Variables that define individuals:

First, a series of questions are asked about different characteristics and qualities that allow people to be grouped into different groups.

The questions asked in this area are:
a) gender
b) Age
c) Level of education
d) Employment status
e) Nationality
f) Frequency of use of public transport

2. Type of use and services used by the population

In this case, the objective is to know how the population uses public transport and what type of services associated with them are most used. So, the questions asked are:
a) What means of transport do you use most often?
b) What is your average number of transfers when using public transport?
c) What is the average duration of your trips?
d) What is the average waiting time at stops/stations?
e) What kind of banknotes do you buy most often?
f) Where can you buy tickets for public transport?
g) What means of payment do you use to pay for the tickets?
h) What kind of additional services related to public transport do you usually use?
i) What type of channels do you usually use to access the services of your public transport operator?

3. User satisfaction

To measure people's level of satisfaction with public transport, 33 questions were asked on different topics related to public transport, which give an idea of the population's assessment of these groups.

Ease of Access and Use
1. I find the timetables easy to understand
2. I find the transport network maps easy to understand.
3. I think the rates are easy to understand
4. I find it easy to buy tickets
5. It is easy for me to validate the ticket/mobile phone on the ticket reader.
Price
6. I think the occasional ticket price is adequate.
7. I think the monthly passes are priced appropriately.
Quality of stations/stops
8. I think stations/stops are always clean.
9. I find the stations/stops comfortable.
10. I feel safe at stations/stops

11. Overall, I am satisfied with the quality of stations/stops.

Quality of vehicles

12. I believe that vehicles are always clean.

13. I consider modern vehicles

14. I find vehicles comfortable

15. I find the smell inside vehicles pleasant.

16. I believe that the noise level inside the vehicle is adequate.

17. I believe the vehicle temperature is adequate

18. I always have a seat available

19. Overall, I am satisfied with the quality of the vehicles.

Relationships with other people

20. I feel safe inside the vehicle

21. I consider the behavior of the other passengers to be adequate

22. I find bus drivers friendly.

23. I find bus drivers polite.

24. I believe that bus drivers should be careful when driving.

25. I think reviewers are understanding.

26. I think reviewers are polite.

Waiting and travel times

27. I believe that vehicles always arrive on time.

28. I consider the travel time adequate.

29. I consider that the waiting time at stations/stops is adequate.

30. I am satisfied with the frequency of public transport.

Satisfaction with the routes covered by vehicles.

31. I am satisfied with the number and location of stations/stops.32. I am satisfied with the connections between public transport lines and operators.

Overall

33. I am overall satisfied with the public transport service in the Porto Metropolitan Area.

References

1. dell'Olio, L., Ibeas, A., de Oña, J., de Oña, R.: Qualidade do serviço de transporte público: factores, modelos e aplicações. Elsevier
2. Dong, H., Ma, S., Jia, N., Tian, J.: Compreender a satisfação dos transportespúblicos no pós-pandemia da COVID-19. Elsevier
3. Maha, A., Bobâlcã, C., Tugulea, O.: Estratégias para a melhoria da qualidade e eficiência dos transportes públicos. Elsevier
4. Gijsenberd, M.J., Verhoef, P.C.: Avançar: O papel do Marketing na Promoção da Utilização dos Transportes Públicos. Associação Americana de Marketing
5. de Oña, J., de Oña, R.: Qualidade de Serviçonos Transportes Públicos Baseados em Inquéritos de Satisfação do Cliente: Uma Revisão e Avaliação das Abordagens Metodológicas
6. Ibraeva, A., de Sousa, J.F.: Comercialização de transportes públicos e fornecimento de informação ao público. Elsevier
7. Morton, C., Cauldield, B., Anable, J.: Customer Perceptions of Quality of Service in Public Transport: Evidence for bus transit in Scotland

8. Yarmen, M., Sumaedi, S.: Qualidade de Serviço Percebida dos Passageiros dos Transportes Públicos Juvenis (2016)
9. van Lierop, D., Badami, M.G., El-Geneidy, A.M.: O que influencia a satisfação e a lealdadenos transportes públicos? Uma revisão da literatura
10. Irtema, H.I.M., Ismail, A., et al.: Percepções Passageiros sobre a Qualidade do Serviço: Transporte Público em Kuala Lumpur

Estimating Alighting Stops and Transfers from AFC Data: The Case Study of Porto

Joana Hora[1,2] , Marta Campos Ferreira[1,2(✉)] , Ana Camanho[1,2] , and Teresa Galvão[1,2]

[1] Faculdade de Engenharia, Universidade do Porto, R. Dr. Roberto Frias, 4200-465 Porto, Portugal
`{jhora,mferreira,tgalvao}@fe.up.pt, acamanho@reit.up.pt`
[2] INESC-TEC, R. Dr. Roberto Frias, 4200-465 Porto, Portugal

Abstract. This study estimates alighting stops and transfers from entry-only Automatic Fare Collection (AFC) data. The methodology adopted includes two main steps: an implementation of the Trip Chaining Method (TCM) to estimate the alighting stops from AFC records and the subsequent application of criteria for the identification of transfers. For each pair of consecutive AFC records on the same smart card, a transfer is identified considering a threshold for the walking distance, a threshold for the time required to perform an activity, and the validation of different boarding routes. This methodology was applied to the case study of Porto, Portugal, considering all trips performed by a set of 19999 smart cards over one year. The results of this methodology allied with visualization techniques allowed to study Origin-Destination (OD) patterns by type of day, seasonally, and by user frequency, each analyzed at the stop level and at the geographic area level.

Keywords: Origin-Destination · Trip Chaining Method · Public Transportation · Automatic Fare Collection

1 Introduction

Until the end of the 20th century, data for the management of Public Transport was mainly gathered from surveys. Aligned with the technological advent of the last decades, Public Transport operators started to invest in ITS technologies and applications, such as Automated Vehicle Location, AFC, Mobile Data Terminals, Automatic Passenger Counter, wireless communications, among others. These new sources of data bring forward a huge potential for the development of innovative optimization techniques, the continuous improvement of Public Transport systems, in addition to uphold the enhanced understanding of impacts arriving from decision-making processes. However, for most cases, the full potential from such large data sets remains unexploited.

This work aims to extract reliable OD matrices from entry-only AFC data. To that end, a methodology combining the TCM and the identification of transfers is proposed. The adequacy of the methodology is demonstrated with its application to the case-study of Porto.

A. Rocha et al. (Eds.): WorldCIST 2023, LNNS 802, pp. 162–171, 2024.
https://doi.org/10.1007/978-3-031-45651-0_17

The TCM has been applied to Public Transport systems worldwide. The TCM is an algorithm that allows to infer the alighting-stops from entry-only AFC data. Although TCM studies incorporate differences in some assumptions and approaches adopted, they share the two grounding assumptions proposed in the seminal work of Barry et al. (2002): (1) most passengers will start the next trip of the day at or near the alighting location of their previous trip, and (2) most passengers end the last trip of the day at or near the boarding location of their first trip of the day. A review of main assumptions used in TCM studies is provided in Hora et al. (2017).

The TCM has been applied to the city of Porto in previous works by Nunes et al. (2015) and by Hora et al. (2017). However, these two works were developed at the disaggregated level of trips, which means that they estimated OD trips without identifying transfers amid them. This work considers trips at an aggregate level, meaning that transfers amid trips are identified and removed, this way allowing the analysis of OD flows.

Some studies have proposed the implementation of the TCM aligned with the identification of transfers. To that end, additional assumptions have been proposed including transfer walking distance thresholds (Trépanier et al. 2007, Munizaga et al. 2012, Alsger et al. 2015), transfer time thresholds (Nassir et al. 2011, Munizaga and Palma, 2012, Alsger et al. 2015), and transfer network feasibility conditions (Munizaga and Palma, 2012). This work contributes to this body of literature, by providing a clear methodological description of the TCM and transfer identification criteria adopted, and by reporting the results obtained for the implementation in the case-study of Porto. This work also implements visualization techniques applied to the analysis of OD matrices, which help in understanding mobility patterns at the stop level and at the geographic area level.

2 Methods

The methodology to estimate destinations and transfers from entry-only AFC data is depicted in Fig. 1. It starts by sorting AFC records by smart card ID and then chronologically. Therefore, the AFC records are analyzed sequentially in this order in the following steps.

2.1 Trip Chaining Method

The TCM is not applied for smart cards that only have one daily AFC record. Consequently, for those cases there is no alighting stop estimation and the algorithm proceeds to the next smart card ID in the dataset.

When a smart card has two or more AFC records in the same day, the algorithm proceeds to estimate the alighting locations of each trip-leg by applying the TCM.

The TCM implemented in this study follows the work described in Hora et al. (2017), and considers the following assumptions:

(1) The majority of passengers will start the next trip-leg close to the alighting-stop of their previous trip-leg;
(2) The majority of passengers end the last trip-leg of the day close to the boarding-stop of the first trip-leg of the day;

For each day

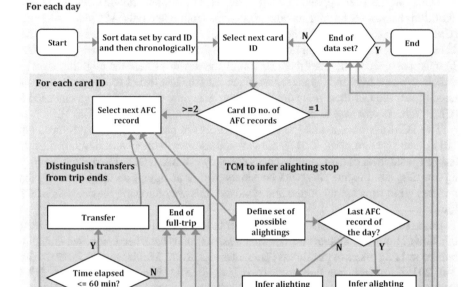

Fig. 1. Diagram of the methodology followed to estimate destinations and transfers, which includes the TCM followed by the criteria for the identification of transfers.

(3) By default, passengers travel within the Public Transport system;

(4) Passengers travel outside the Public Transport system when there is a single trip-leg on a day;

(5) Passengers travel outside the Public Transport system when the walking-distance between consecutive trip-legs exceeds 3 km;

(6) When an alighting-stop cannot be inferred, the corresponding trip-leg is discarded;

(7) Passengers can only alight in the sequence of downstream stops on the directed-route boarded;

(8) The alighting-time is not inferred.

Accordingly, knowing the boarding-stop, route and direction, the algorithm specifies all stops not yet traveled in that directed route as possible alighting-stops. From this set of possibilities, the TCM estimates the alighting stop as the one with the shortest walking distance regarding (1) the boarding-stop of the subsequent AFC record if it exists, or (2) the boarding-stop of the first AFC record of the day for the last daily record (assuming the passenger would travel back home at the end of the day).

When the estimated walking distance is higher than the threshold of 3 km, we assume that the passenger traveled out of the Public Transport system (e.g., picking up a ride or other alternative transportation mode). In these cases, the alighting stop estimation is discarded. This assumption is adopted both for consecutive AFC records and for the case of the last AFC record of the day.

2.2 Identifying Transfers

The next step of the methodology comprises the identification of transfers. This identification concerns distinguishing if the alighting stop of an AFC record corresponds to a transfer within a sequence of trip-legs, or if it corresponds to the destination of a trip.

The main assumptions adopted to identify transfers between two consecutive AFC records are detailed next:

(9) Passengers do not walk more than 400 m to transfer;
(10) Passengers do not wait more than 60 min to transfer;
(11) A transfer occurs between consecutive trip-legs;
(12) A transfer occurs between trip-legs on the same smart card;
(13) Passengers do not transfer to the same route;
(14) When passengers travel out of the Public Transport system, the next trip-leg starts a new full-trip;

The thresholds adopted in (9–10) were considered adequate to the case study of Porto. Assumptions (11–12) establish basic rules to identify transfers and are self-explanatory.

Assumption (13) considers that passengers will only perform two consecutive trips legs in the same route when executing two distinct trips. That is, a passenger does not perform a transfer to board the same route he was already traveling, even if in the opposite direction, unless by mistake. This way, if a passenger boards the same route in the consecutive AFC record, the algorithm considers that the passenger is performing a new trip and not a transfer.

Finally, assumption (14) addresses situations in which passengers travel by means of transportation alternatives to the Public Transport system under consideration. When the estimation of the alighting-stop of trip-leg is discarded (i.e., the passenger traveled out of the Public Transport system), the following trip-leg is always considered as the beginning of a new trip.

3 Results

This section presents the results obtained by implementing the methodology described in the case study of Porto. The Public Transport system considered encompasses the joint activity of the bus and tram operator Sociedade de Transportes Colectivos do Porto, S.A.

(STCP), and the metro operator Metro do Porto (MP). The network considered in this study prefaces 2,672 stops. The metro system has six routes and 81 stops, the bus system has 71 routes, and the tram system has three routes (tram and bus share stops).

This study considered a sample of 19,999 randomly selected smart cards. All AFC records from this set of smart cards were analyzed over the entire year of 2013. The open-source database software PostgreSQL was used to select and sort data. The TCM algorithm was implemented in C++, using a 3.4 GHz Intel Core i7 processor and 16 GB of Random Access Memory (RAM).

Table 1 provides a summary of the results obtained with the application of the methodology to the case study of Porto. The set of 19,999 smart cards included a total of 902,819 AFC records. From this total, the methodology excluded 82,080 AFC records corresponding to cases of one AFC record in the day, 23,011 AFC records corresponding to trips performed out of the Public Transport system, and 97,548 AFC records corresponding to transfers made within full-trips.

Table 1. Results from applying the methodology to the case study of Porto, by month.

Month	All AFC Records	Assumption (4)	Assumption (5)	Transfers
1	78,455	7,023	2,085	15,813
2	68,930	6,176	1,843	13,404
3	74,444	6,925	1,903	14,676
4	80,984	7,222	2,136	16,010
5	86,985	8,180	2,146	16,871
6	70,800	7,165	1,898	13,568
7	76,169	6,882	1,946	15,452
8	61,745	5,874	1,547	12,806
9	73,551	6,569	1,843	14,694
10	86,323	7,136	2,087	16,762
11	76,169	6,521	1,816	13,820
12	68,264	6,407	1,761	12,209
Total	902,819	82,080	23,011	176,085

Note that a transfer is identified amid two AFC records, but its accounting is not duplicated. For example, one trip-leg from A to B, followed by a trip-leg from B to C, and where a transfer was identified at B, results in a OD flow from A to C. This way, the methodology allowed the identification of 700,180 OD flows.

Table 1 also provides detail on the proportion of AFC records for each month. All proportions were calculated from the total number of AFC records (i.e., 902,819). It is possible to see that August presents the lowest volume of AFC records, followed by February and by December, which relates to summer and other holidays such as

Carnival and Christmas, noting, however, that February is a shorter month. Months with the highest volume of records are May followed by October.

The set of OD flows was further analyzed considering three distinct segmentations: type of day, seasonality, and passenger frequency. Each segmentation classifies OD flows into categories. Table 2 provides the summary of each segmentation analysis, with the detail on all categories considered.

Table 2. Summary of the segmentation analyzes to the set of OD flows. Proportions calculated on the total number of OD flows (i.e., 700,180).

Segmentation	Category	No. of days in the year	Average no. of daily trips	No. Smart cards	Annual OD flows	% of total OD flows
Type of day	Workdays	251	2,361	–	592,670	84.65
	Sundays/Holidays	63	758	–	47,771	6.82
	Saturdays	51	1,171	–	59,739	8.53
Seasonality	Regular service	257	2,030	–	521,660	74.50
	Reduced service	108	1,653	–	178,520	25.50
Passenger frequency	Frequent riders	–	–	1,354	551,416	78.75
	Occasional riders	–	–	13,227	148,764	21.25
Total		365	1,918	14,581	700,180	100.00

Figures 2, 3 and 4 include the OD matrices obtained for each segmentation analysis. Each category is analyzed with two OD matrices, one corresponding to a perspective at the level of the geographic area (left), and the second corresponding to a perspective at the level of the stop (right).

In order to allow for a direct comparison between all OD matrices shown in this section, they were all built over the same scale ratio, which is a proportion to the total of 700,180 OD flows estimated. Accordingly, all OD matrices share the same scale of proportional dots, where the smallest dot indicates that at least 0.01% of total trips (around 70 trips in the year) were made in that OD pair, and the largest dot indicates at least 25% of total trips (around 175,045 trips in the year) were made in that OD pair.

All OD matrices in Figs. 2, 3 and 4 follow a symmetric pattern towards the diagonal, which is consistent with regular daily commuting patterns from an overall perspective. Matrices at the stop level show a higher density of trips in the first 800 stops, resulting in the visualization of a denser square positioned in the lower-left corner of these matrixes. These stops are almost all positioned within the geographic area C1, which is the city center. From these set of stops, it is worth mentioning that stops from the metro are at first positions, which translates in a small but even denser zone within the referred square. In parallel, we can see that the corresponding matrices built at the geographic area level always have the most significant dot in the area C1. This larger dot encompasses all trips with origin and destination within C1.

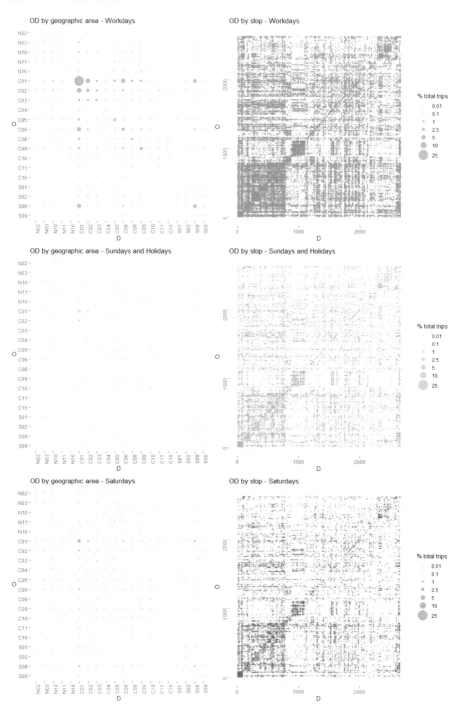

Fig. 2. OD matrixes by type of day, at the stop and geographic area levels. Proportions calculated on the total number of OD flows.

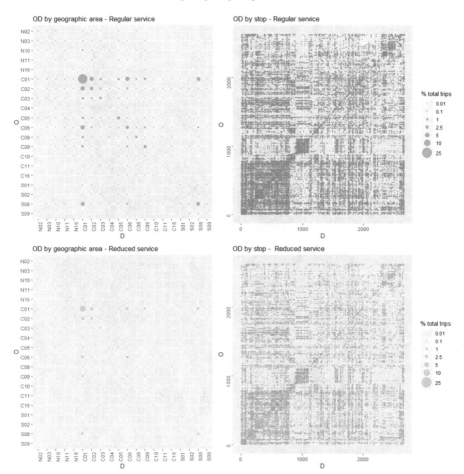

Fig. 3. OD matrixes by seasonality, considering regular and reduced services (reduced service includes summer and school holiday periods), at the stop and geographic area levels. Proportions calculated on the total number of OD flows.

All trips starting within another zone and finishing within C1 are visualized in the dots aligned vertically to C1 in these matrices. Analogously, all trips starting within C1 and finishing within other areas are visualized in the dots aligned horizontally to C1. The size of these dots aligned vertically and horizontally with C1 is significant in all categories, showing that area C1 not only is the densest regarding internal travel; it is also the area with the most significant flow of movements to other zones.

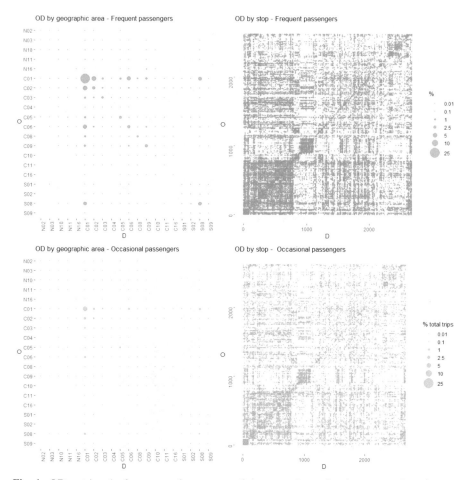

Fig. 4. OD matrixes by frequency of passengers (frequent and occasional passengers), at the stop and geographic area levels. Proportions calculated on the total number of OD flows.

4   Conclusion

This work proposes a methodology to estimate alighting-stops and transfers from entry-only AFC records. The methodology has two main steps, the first considers the implementation of the TCM, and the second step considers a set of criteria to identify transfers performed amid full-trips. This methodology was applied in the case study of Porto, consisting of a Public Transport system integrating metro, bus, and tram. The data covers all AFC records from 19,999 randomly selected smart cards, considering the entire year of 2013.

The implementation of the methodology was reported, and its results were analyzed. Around 9% of AFC records were discarded from relating to single records in the day, around 2.5% AFC records were discarded by not complying with the 3 km assumption, and around 11% of AFC records were identified as transfers.

The methodology allowed to retrieve a data set of OD flows. The set of OD flows was also analyzed in this work considering three segmentation analyses: (1) by type of day considering the categories of workdays, of Sundays and holidays, and of Saturdays; (2) by seasonality considering the categories of regular service days and of reduced service days; (3) passenger frequency considering the categories of frequent and of occasional passengers. The visualization of OD matrices was provided for each category at the stop level and at the geographic area level (considering the geographic zones currently used by the operators). The segmentation analyses showed significant differences in the volume of trips of the different categories, while a similar OD design was observed across all cases.

Future work can include a comparison of the results obtained with this methodology with other literature on this research topic. Future work can also encompass the incorporation of sophisticated statistical methods into this methodology to perform sensitivity analysis on the threshold values adopted, to identify mobility patterns at the level of uniform demand periods.

References

Alsger, A.A., Mesbah, M., Ferreira, L., Safi, H.: Use of smart card fare data to estimate public transport origin–destination matrix. Transp. Res. Rec. **2535**(1), 88–96 (2015)

Barry, J.J., Newhouser, R., Rahbee, A., Sayeda, S.: Origin and destination estimation in New York City with automated fare system data. Transp. Res. Rec. **1817**(1), 183–187 (2002)

Hora, J., Dias, T.G., Camanho, A., Sobral, T.: Estimation of origin-destination matrices under automatic fare collection: the case study of Porto transportation system. Transp. Res. Procedia **27**, 664–671 (2017)

Munizaga, M.A., Palma, C.: Estimation of a disaggregate multimodal public transport Origin-Destination matrix from passive smartcard data from Santiago, Chile. Transp. Res. Part C: Emerg. Technol. **24**, 9–18 (2012)

Nassir, N., Khani, A., Lee, S.G., Noh, H., Hickman, M.: Transit stop-level origin–destination estimation through use of transit schedule and automated data collection system. Transp. Res. Rec. **2263**(1), 140–150 (2011)

Nunes, A.A., Dias, T.G., Cunha, J.F.: Passenger journey destination estimation from automated fare collection system data using spatial validation. IEEE Trans. Intell. Transp. Syst. **17**(1), 133–142 (2015)

Trépanier, M., Tranchant, N., Chapleau, R.: Individual trip destination estimation in a transit smart card automated fare collection system. J. Intell. Transp. Syst. **11**(1), 1–14 (2007)

Camera Movement Cancellation in Video Using Phase Congruency and an FFT-Based Technique

Abdorreza Alavi Gharahbagh[1], Vahid Hajihashemi[1], J. J. M. Machado[2],
and João Manuel R. S. Tavares[2(✉)]

[1] Faculdade de Engenharia, Universidade do Porto,
Rua Dr. Roberto Frias, s/n, 4200-465 Porto, Portugal
[2] Departamento de Engenharia Mecânica, Faculdade de Engenharia,
Universidade do Porto, Rua Dr. Roberto Frias, s/n, 4200-465 Porto, Portugal
tavares@fe.up.pt

Abstract. One of the interesting fields in video processing is motion detection and human action detection (HAR) in video. In some applications where both objects in the scene and the camera may be moving, camera movement cancellation is very important to increase accuracy in extracting motion features. HAR systems usually use image matching/registration algorithms to remove the camera movement. In these methods, the source (fixed) image frame is compared with moved image frame, and the best match is determined geometrically.

In video processing, due to the existence of a set of frames, one can correct errors using previous data, but at the same time, it is needed a fast frame registration algorithm. According to the above explanations, this article proposes a method to detect and minimize camera movement in video using phase information. In addition to having the acceptable speed and the ability to be implemented online, the proposed method, by combining texture and phase congruency (PC), can significantly increase the accuracy of detecting the objects in the scene. The proposed method was implemented on a HAR dataset, which includes camera movement, and its ability to compensate for camera motion and pre-serve object motion was verified. Finally, the speed and accuracy of the proposed method were compared with a number of the latest image registration methods, and its efficiency in terms of camera movement cancellation and execution time is discussed.

Keywords: Camera Movement Cancellation · Phase Congruency (PC) · Video · Image Registration · Fast Fourier transform (FFT)

1 Introduction

Image registration, image matching, and video stabilization are some of the most common applications of image matching. However, there are similarities and differences between these applications. For example, Image registration is used in cases where panoramic images are used to form a complete image [1]. A match between two images, and possibly their rotations or scales, occurs when components from one image are searched for in the other image, and exactly similar components are found [2]. The video stabilization systems are designed to stabilize camera motion in videos acquired with a shaky camera. This work aims to remove the camera movement from the input video in order to extract a more accurate motion vector from the moving objects in a HAR system [3].

It is important to note the speed and accuracy simultaneously and also to consider the difference between an object's movement and the camera's movement. The matching of two images requires matching fixed parts to obtain a better model of moving objects and their location by removing the camera movement. In image registration, features such as colour and edge are used in key points of images to match each other and find the involved shift, rotation, or scale. The PC method is commonly used to find edges in images. The nature of PC differs from that of derivatives and gradients, and its error usually is lower since it does not depend on the image's brightness. Using edges and important parts in matching, which is found with PC, will improve the accuracy. Considering suitable efficiency of PC compared to gradient and adequate run time, this feature, combined with colour features, is used in this work to eliminate camera movement in HAR systems. The following cases demonstrate the innovation of this work:

- Using PC to extract edges more accurately and use the result in image registration;
- A modified registration algorithm to make sure PC during calculations between frames is optimally used, and the key points of the frames can be found with a low run time;
- A motion vector comparison before and after removing the camera movement is performed to validate the proposed method.

The following parts of this article are as follows, after introduction, a review of the literature and history of related works is given. In the next section, the proposed system is described in detail. The fourth section is dedicated to the used dataset, obtained results, and comparison of the proposed method against previous works, and the final section gives the conclusions.

2   Literature Review

According to the wide existent applications, various types of research have been done in fields of image registration. Among these fields, one can mention medical imaging, remote sensing, and computer vision [4]. The current work is related to the fields of remote sensing and computer vision, which can be divided into different categories such as Area based, Feature point based, Learning based, Visible to infrared, Cross spectra, and Cross temporal [4]. The proposed approach belongs to the Feature point based group. In the type of data processing, image registration methods are divided into two categories: Area based pipeline and Feature based pipeline. The area based pipeline is divided into two general groups, traditional framework and methods based on deep learning.

There are three traditional components in usual image registration systems: the measure metric, the transformation model, and the optimization method. In Feature based systems, being the proposed method of this subgroup, features are used to find important image regions. A feature based image registration method generally includes feature extraction, feature description, and feature matching steps. In the Feature description, descriptors such as float descriptor, binary descriptor, and learned descriptor are commonly used, and in feature matching, methods such as graph matching, point set registration and indirect matching are used. In the feature extraction, which is the main focus of the current work, features such as corner [5], blob [6], line/edge [7], and morphological feature [8] are used. Usually, the groups of line or region features are also converted into a set of points so that it is easier to compare and match [9,10]. All the above features are extracted based on information like gradient, intensity, second order derivative, contour curvature [11,12], or learning based [13,14]. Based on the authors' knowledge, PC has previously been used in matching radar images and deep learning based image registration systems [15], but this feature has not been yet used for camera movement cancellation in HAR systems.

3   Proposed Method

The block diagram of the proposed method for matching images and removing camera movement is shown in Fig. 1. The main advantage of the proposed method is the usability of the extracted features for the next steps of a HAR system. In addition, the proposed system can be used as a preprocessing step in any HAR system and increase its accuracy. According to Fig. 1, PC and Image Registration are the most important steps in the proposed system. In this study, the PC technique is used to extract important regions and edges of the input image, while the image registration technique is used to determine camera movements. The dominant result is presented in terms of a geometric transformation corresponding to the camera movement. The important parameters in the proposed system are the speed and, simultaneously, the accuracy of the camera movement estimation. In the following, the PC and then the image registration block is described.

3.1   Phase Congruency

In conventional algorithms for extracting image edges and borders, gradients are usually used. The image gradient can specify edges and corners using the derivative of intensity. On teh other hand, the Local Energy Model uses the concept of phase in the Fourier series to identify edges and transitions.

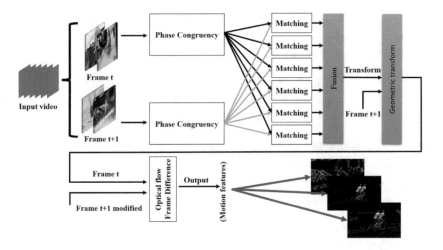

Fig. 1. Block diagram of the proposed method.

The initial formula of PC is the following relationship [16]:

$$PC_1(x) = \frac{|E(x)|}{\sum_n A_n(x)} \tag{1}$$

where $A_n(x)$ is the set of all Fourier transform domains that are added together, and $|E(x)|$ measure the resultant vector at any point. PC can also be expressed in the form of the following relationship:

$$PC_1(x) = \frac{\sum_n A_n \left(\cos \left(\phi(x) - \bar{\phi}(x) \right) \right)}{\sum_n A_n(x)} \tag{2}$$

In this regard, An is the coefficients of the Fourier series, ϕ is the phase of the coefficients and $\bar{\phi}$ is the average phase of the total Fourier series coefficients at each point. Since this relationship is highly sensitive to noise, Kovesi [17] proposed a modified robust to noise PC relationship, which is:

$$PC_1(x) = \frac{\sum_n W_n(x) \lfloor A_n(x) \left(\cos \left(\phi_n(x) - \bar{\phi}(x) \right) - \left| \sin \left(\phi_n(x) - \bar{\phi}(x) \right) \right| \right) - T \rfloor}{\sum_n A_n(x) + \varepsilon} \tag{3}$$

where W_n is the importance weight of each frequency in PC, T is the image noise, and $\lfloor \; \rfloor$ operator indicates that if the enclosed quantity has a positive value, it is equal to itself, and if negative, it is zero. The 2D PC as a Local Energy Model has been proposed [17] for identifying the edges and corners of an image. Xiao and Zhengxin [18] showed that this feature is very close to the human visual system. In 2D mode, different directions (6 basic directions) are usually considered, and Fourier values are extracted from oriented 2D Gabor wavelets. In this way, the output will be a relative expression of the edges and corners at all angles. In order to decrease error, the maximum and minimum values of the PC can be calculated at each point, and then a judgment about the point type can be made using the maximum and minimum values. In 2D mode, the modified formulas of PC are as follows:

$$
\begin{aligned}
a &= \sum \left(PC\left(\theta\right) \cos\left(\theta\right) \right)^2 \\
b &= 2 \sum \left(PC\left(\theta\right) \cos\left(\theta\right) \right) . \left(PC\left(\theta\right) \sin\left(\theta\right) \right) \\
c &= \sum \left(PC\left(\theta\right) \sin\left(\theta\right) \right)^2
\end{aligned}
\tag{4}
$$

where $PC(\theta)$ refers to the PC determined at orientation θ, and the sum is calculated over all orientations.

3.2   Image Registration

Considering the pros and cons of different methods in image registration, phase based methods which use the Fourier transform of the input image and phase correlation or PC are used in this work [4]. Among these methods, [19,20] can be mentioned. In the proposed method, the method [19] in combination with PC is used due to the more appropriate speed and greater adaptation to the intended application. Finally, the output of this method is examined in different orientations, and the output is determined. The PC for the six main orientations will first be calculated for both consecutive frames separately. All six orientations are calculated in PC, so the processing time for this step is the same as the calculation of PC.

After this step, a geometric transformation is obtained for each orientation. Ideally, it is expected the transformations to be the same, but in practice, these transformations are not the same due to various errors. In the image registration step, the proposed method first selects transformations whose correlation scores are higher than 0.015, and then mathematical averaging is performed among the transformations. In the field of rotation angles, if the absolute value of an angle is greater than 150°, it is rounded to the nearest number by adding or subtracting 180° (period of rotation angle calculation), and then averaging is done. The final geometric transformation is determined after the shift and rotation been determined. This work ultimately leads to the best possible result while reducing possible errors.

4   Dataset and Results

4.1   Dataset

The 2017 DAVIS Challenge on Video Object Segmentation dataset was used to evaluate the effectiveness of the proposed method [21]. This dataset is a collection of different video frames, and a mask is defined for them with the purpose of video segmentation. In this work, it was important to evaluate the camera's movement and remove it. The number of sequences in this dataset is 150 and number of frames is 10459.

4.2   Comparison Criteria

Unlike other image registration algorithms, this work uses a different evaluation metric since it aims to extract the motion features more accurately by removing the camera movement. According to this objective, the exact detection of the object is not important, and the goal is to reduce the motion value of fixed components in the output of motion extraction methods. According to this explanation, two simultaneous evaluation criteria were defined:

$$Cancellation\ Ratio = \frac{\displaystyle\sum_{In\ all\ static\ pixels} Frame\ difference}{\displaystyle\sum_{In\ ideal\ matched\ frames} Frame\ difference} \tag{5}$$

$$Matched\ Ratio = \frac{\displaystyle\sum_{In\ the\ motion\ area} Frame\ difference}{\displaystyle\sum_{In\ ideal\ matched\ frames} Frame\ difference} \tag{6}$$

The cancellation ratio indicates the amount of noise removal in non-moving areas, and Matched ratio indicates the degree of compliance of the moving area with the ideal state after correcting the camera movement. As a result of this process, a better algorithm has a lower cancellation ratio and a higher matched ratio. The ideal value of the cancellation ratio is equal to zero, and the ideal value of the matched ratio is equal to one. Figures 2 and 3 show the results of corrections made by the proposed method using two motion extraction methods: Optical flow and Frame difference, compared to the without correction, the correction made by [19] on grey images and usual PC of two frames. Figure 2 shows the output image of optical flow between two frames without correction, where some parts of the wall and path are visible.

In the case where two images were matched using the grey image of two frames or the PC of two frames using the method [19], there is less image noise than in the case without error correction, but the result is still not ideal. In the proposed method, first, all six directions of PC are checked in two frames, and finally, the appropriate geometric transformation is selected. After cancelling the camera movement using desired geometric transformation, the optical flow result is much more accurate and less noisy.

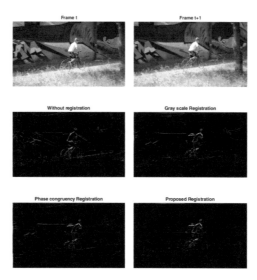

Fig. 2. Optical flow results of the proposed method compared to the case without camera motion cancellation, with image registration using [19] and image registration using PC.

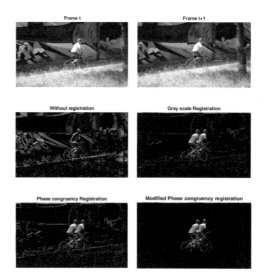

Fig. 3. Frame difference in the proposed method compared to the case without camera motion correction and the method [19] results in gray images and PC of two frames.

In all three cases where the images have been corrected, after applying the geometric shift, the redundant edges of the image frames were removed in the processing so that there are no errors in the edges of the images. Figure 3 shows the result of the Frame difference for different modes. A comparison of optical flow and frame difference illustrates the effectiveness of the proposed method in this case, as optical flow corrects some noise and errors, whereas frame difference clearly shows any mismatch. In this situation, the effectiveness of the proposed method is much more specific than other methods.

Table 1. The average result of two evaluation criteria on the whole image dataset [21].

Cancellation ratio				Matched ratio			
W.C	G.C	P.C	Proposed Method	W.C	G.C	P.C	Proposed Method
3.12	2.42	2.61	1.004	0.43	0.48	0.47	0.53
W.C: Without camera movement cancellation.							
G.C: Camera movement cancellation using grey images and [19] method.							
P.C: Camera movement cancellation using PC images and [19] method.							

Table 1 shows the average results obtained on all image dataset based on two defined metrics. In the cancellation ratio, the worst result is caused by the uncorrected state, which indicates a high noise output. Among the camera motion removal methods, the best result was obtained by the proposed method. It is possible to reduce the cancellation ratio again by adding different steps, but it should be noted that the processing time is very important. In the field of matched ratio, all the methods shown acceptable results, and this can be due to the fact that in the uncorrected state, there is the same noise in moving parts which helps to increase the result of matched ratio. At a final glance, it seems that the proposed method can effectively increase the accuracy of HAR methods by reducing the motion vector error when the scene or the camera is moving.

5   Conclusion

The objective of this research is to increase the effectiveness of motion feature extraction in HAR methods by eliminating camera movement based on PC. The proposed method removes camera movement and adapts successive frames to reduce motion vector error, improves the efficiency of HAR systems, and improves the accuracy of moving object detection. In the proposed method to minimize motion estimation error, the edges of the image are extracted by PC in different orientations. PC is independent of scene intensity, so noise and error of edge extraction are lower gradient-based methods. After that, geometric transformation (shift and rotation) between two frames is calculated in basic orientations between every two consecutive image frames. Ideally, all these geometric transformations should be similar, but there is always an error in practice. In order to identify these errors, transformations with a higher degree of similarity than others are selected based on correlation. The averaging is done between the remained results. Correcting the angle of rotation in values close to zero and removing very small values that are usually errors have been used to decrease errors. Finally, the proposed method was compared with the existing methods, and its efficiency in detecting and removing the camera movement was shown. In addition, if the scene does not move and the camera is fixed, the proposed method works without any error.

Acknowledgements. This article is partially a result of the project Safe Cities - "Inovação para Construir Cidades Seguras", with reference POCI-01-0247-FEDER-041435, co-funded by the European Regional Development Fund (ERDF), through the Operational Programme for Competitiveness and Internationalization (COMPETE 2020), under the PORTUGAL 2020 Partnership Agreement. The second author would like to thank "Fundação para a Ciência e Tecnologia" (FCT) for his Ph.D. grant with reference 2021.08660.BD.

References

1. Li, Y., Wang, J., Yao, K.: Modified phase correlation algorithm for image registration based on pyramid. Alex. Eng. J. **61**(1), 709–718 (2022). https://doi.org/10.1016/j.aej.2021.05.004
2. Ma, J., Jiang, X., Fan, A., Jiang, J., Yan, J.: Image matching from handcrafted to deep features: A survey. Int. J. Comput. Vision **129**(1), 23–79 (2021). https://doi.org/10.1007/s11263-020-01359-2
3. Gharahbagh, A.A., Hajihashemi, V., Ferreira, M.C., Machado, J.J.M., Tavares, J.M.R.S.: Best frame selection to enhance training step efficiency in video-based human action recognition. Appl. Sci. **12**(4), 1830 (2022). https://doi.org/10.3390/app12041830
4. Jiang, X., Ma, J., Xiao, G., Shao, Z., Guo, X.: A review of multimodal image matching: methods and applications. Inf. Fusion **73**, 22–71 (2021). https://doi.org/10.1016/j.inffus.2021.02.012
5. Ye, Y., Wang, M., Hao, S., Zhu, Q.: A novel keypoint detector combining corners and blobs for remote sensing image registration. IEEE Geosci. Remote Sens. Lett. **18**(3), 451–455 (2020). https://doi.org/10.1109/LGRS.2020.2980620
6. Bellavia, F.: Sift matching by context exposed. IEEE Trans. Pattern Anal. Mach. Intell. (2022). https://doi.org/10.1109/TPAMI.2022.3161853
7. Fan, Q., Zhuo, W., Tang, C.-K., Tai, Y.-W.: Few-shot object detection with attention-RPN and multi-relation detector. In: Proceedings of the IEEE/CVF Conference on Computer Vision and Pattern Recognition, pp. 4013-4022 (2020)
8. Shao, Z., Zhou, W., Deng, X., Zhang, M., Cheng, Q.: Multilabel remote sensing image retrieval based on fully-convolutional network. IEEE J. Sel. Top. Appl. Earth Observ. Remote Sens. **13**, 318–328 (2020). https://doi.org/10.1109/JSTARS.2019.2961634
9. Kong, S.-Y., Fan, J.-S., Liu, Y.-F., Wei, X.-C., Ma, X.-W.: Automated crack assessment and quantitative growth monitoring. Comput.-Aided Civil Infrastruct. Eng. **36**(5), 656–674 (2021). https://doi.org/10.1111/mice.12626
10. Berg, A.C., Malik, J.: Shape matching and object recognition. In: Ponce, J., Hebert, M., Schmid, C., Zisserman, A. (eds.) Toward Category-Level Object Recognition. LNCS, vol. 4170, pp. 483–507. Springer, Heidelberg (2006). https://doi.org/10.1007/11957959_25

11. Rani, P.E., Velmurugan, S.: Behavioral analysis of students by integrated radial curvature and facial action coding system using DCNN. In: 8th International Conference on Advanced Computing and Communication Systems (ICACCS), vol. 1, pp. 68–73. IEEE (2022). https://doi.org/10.1109/ICACCS54159.2022.9785056

12. Cheng, T., Juping, G., Zhang, X., Hua, L., Zhao, F.: Multimodal image registration for power equipment using Clifford algebraic geometric invariance. Energy Rep. **8**, 1078–1086 (2022). https://doi.org/10.1016/j.egyr.2022.02.192

13. Wodzinski, M., Ciepiela, I., Kuszewski, T., Kedzierawski, P., Skalski, A.: Semi-supervised deep learning-based image registration method with volume penalty for real-time breast tumor bed localization. Sensors **21**(12), 4085 (2021). https://doi.org/10.3390/s21124085

14. De Silva, T., Chew, E.Y., Hotaling, N., Cukras, C.A.: Deep-learning based multimodal retinal image registration for the longitudinal analysis of patients with age-related macular degeneration. Biomed. Opt. Express **12**(1), 619–636 (2021). https://doi.org/10.1364/BOE.408573

15. Fan, J., Ye, Y., Li, J., Liu, G., Li, Y.: A novel multiscale adaptive binning phase congruency feature for SAR and optical image registration. IEEE Trans. Geosci. Remote Sens. **60**, 1–16 (2022). https://doi.org/10.1109/TGRS.2022.3206804

16. Morrone, M.C., Ross, J., Burr, D.C., Owens, R.: Mach bands are phase dependent. Nature **324**(6094), 250–253 (1986). https://doi.org/10.1038/324250a0

17. Kovesi, P.: Image features from phase congruency. Videre: J. Comput. Vision Res. **1**(3), 1–26 (1999)

18. Xiao, Z., Hou, Z.: Phase based feature detector consistent with human visual system characteristics. Pattern Recogn. Lett. **25**(10), 1115–1121 (2004). https://doi.org/10.1016/j.patrec.2004.03.018

19. Reddy, B.S., Chatterji, B.N.: An FFT-based technique for translation, rotation, and scale-invariant image registration. IEEE Trans. Image Process. **5**(8), 1266–1271 (1996). https://doi.org/10.1109/83.506761

20. Xie, X., Zhang, Y., Ling, X., Wang, X.: A novel extended phase correlation algorithm based on Log-Gabor filtering for multimodal remote sensing image registration. Int. J. Remote Sens. **40**(14), 5429–5453 (2019). https://doi.org/10.1080/01431161.2019.1579941

21. Pont-Tuset, J., Perazzi, F., Caelles, S., Arbeláez, P., Sorkine-Hornung, A., Van Gool, L.: The 2017 Davis challenge on video object segmentation. arXiv preprint arXiv:1704.00675 (2017). https://doi.org/10.48550/arXiv.1704.00675

Audio Event Detection Based on Cross Correlation in Selected Frequency Bands of Spectrogram

Vahid Hajihashemi[1], Abdorreza Alavi Gharahbagh[1], J. J. M. Machado[2], and João Manuel R. S. Tavares[2](✉)

[1] Faculdade de Engenharia, Universidade do Porto, Rua Dr. Roberto Frias, s/n, 4200-465 Porto, Portugal
[2] Departamento de Engenharia Mecânica, Faculdade de Engenharia, Universidade do Porto, Rua Dr. Roberto Frias, s/n, 4200-465 Porto, Portugal
tavares@fe.up.pt

Abstract. Audio event detection (AED) systems have various applications in modern world. Examples of applications include security systems, urban management and automatic monitoring in smart cities, and online multimedia processing. The noise and background sound vary in an urban environment, so frequency domain and normalized features usually show better efficiency in AED systems.

This work proposes a Mel spectrogram-based approach that uses the spectral characteristics of audio signals and cross-correlations to build a dictionary of effective spectrogram frequency bands and their patterns in different audio events. Initially, the proposed approach extracts the Mel spectrogram of audio input. In the next step, a mathematical-statistical analysis is used to specify the effective frequency bands of the spectrogram in each audio event. The pattern of selected frequency bands varies due to the type of event, which can effectively help decrease spectrogram size as input feature, reduce errors and increase the accuracy of different AED methods. The proposed approach was implemented on the URBAN-SED database, and its efficiency was compared against deep learning base state-of-the-art researches in the field. According to the results, about 50% of the frequency bands in the spectrum are useless and can be discarded in the training process of an AED system without any loss in terms of accuracy.

Keywords: Audio event detection · Spectrogram · Cross-correlation · Feature selection · Deep learning

1 Introduction

In machine-human communication systems, it is very important to use sound of the environment in addition to its visual information in order to create a better understanding of the environment. Actually, many events only are recognized by sound, as observers are incapable of observing them. Detecting audio events by processing the sounds of the environment has many applications in the management of urban scenarios and security systems [1]. Audio event detection (AED) in urban scenarios is more complicated than in indoor spaces, because there is more noise in outdoor areas, and the diversity of sound is also greater.

In urban scenarios, things like the possibility of overlapping audio events with each other, sound and multi-path reflection, different weather conditions, and highly variable range of sound when receiving are among the factors that add to the complexity of the work [1,2]. cross-correlation, as a mathematical-statistical operator that evaluates the similarity of time series even in the state of shifting and with different lengths, is very effective in the analysis of time series. Considering the challenges in AED systems and cross-correlation specifications, cross-correlation is used in the current to select effective frequency bands of the spectrogram in an AED system. In summary, the novelties of this research are:

- Using cross-correlation to analyse the event effect on different frequency bands of the Mel spectrogram;
- Using a time domain operator as a frequency feature to use the advantages of both domains simultaneously.

The structure of this article is as follows. In the second section, a summary review of the research in this field is presented. The proposed method is explained in Sect. 3. Section 4 is dedicated to the used database, implementation details, and results, and, finally, the final section is devoted to conclusions.

2   State-of-the-Art

This review focuses on the research in audio signal processing that used cross-correlation. Plenkers et al. [3] used Waveform stacking and cross-correlation to detect seismic events. Plinge et al. [4] used Mel and gammatone frequency cepstral coefficients to extract features in an AED system, using the Bag-of-Features approach for feature analysis. Xugang Lu et al. [5] used spectral patch extraction from Mel band spectra in AED. They used K-means clustering based vector quantization (VQ) for feature analysis and a support vector machine (SVM) classifier. Espi et al. [6] used A high-resolution spectrogram patch and applied it to a convolutional neural network (CNN) for AED.

Pikrakis and Kopsinis [7] used a moving window technique and extracted spectroscopic features. The extracted features are compared with features of other events using the Smith-Waterman algorithm. Farina et al. [8] used spectrograms for Ecoacoustic Event Detection and Identification. They used The Acoustic Complexity Index to measure the information that occurs between two successive frequency bins. Grzeszick et al. [9] proposed an AED system using a triple combination of MFCC, gammatone frequency cepstral coefficients (GFCCs) and perceptual loudness, which is derived from the A-weighted magnitude spectrum.

Adavanne et al. [10] used a deep recurrent neural network for AED, which has as input Log Mel energies. Cross-correlation is one of the layers used in their deep recurrent neural network. Kim and Jin [11] used spectrogram feature to detect audio event in Wireless Acoustic Sensor Networks. The classifier used in the proposed system was a deep recurrent neural network. Lu et al. [12] used the spectrum and cross-correlation information extracted from different inputs of a multichannel microphone array to train a CNN, for audio event detection and localization. Cordourier et al. [13] used Fourier-based spectrograms and applied results to a Convolutional Recurrent Neural Network (CRNN) for audio event localization and detection. Cao et al. [14] used a combination of Generalized Cross Correlation with Log-Mel feature and Intensity vector for audio event localization and detection. Noh et al. [15] determined the location of the sound source using the log Mel-spectrogram and the generalized cross-correlation phase transform.

Nguyen et al. [16] used Log-Mel spectrograms and spatial features such as generalized cross-correlation with phase transform for multi-channel polyphonic AED. Sam-pathkumar and Kowerko [17] used Log-Mel features for AED and intensity vector and generalized cross-correlation (GCC) for sound source localization. The classifier used in the proposed work was CRNN. Ick and McFee [18] used per-channel energy normalization (PCEN), which uses signal spectrogram for AED in urban areas. Nguyen et al. [19] proposed a method to detect the audio events and determine the location of the sound source using Spatial Cue-Augmented Log-Spectrogram Features. According to our search, cross-correlation has been widely used in various studies in order to determine sound source location, but its use directly for AED has been very limited. MFCC, Log-Mel, and Mel spectrograms have been used in many audio processing research, but none of the studies examined the method proposed in this article to analyze frequency bands of spectrograms through cross-correlation.

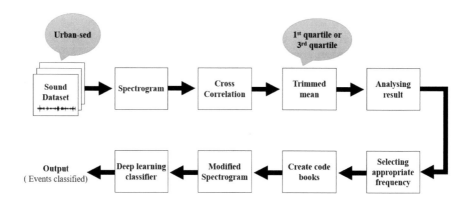

Fig. 1. Block diagram of the proposed method.

3   Proposed Method

The block diagram of the proposed method is shown in Fig. 1. As one can see, the most important steps of the proposed method are the cross-correlation, spectrogram, and deep learning based classifier steps. In the following, the theory of cross-correlation and spectrogram is briefly explained, and then the different steps of the proposed method are explained.

3.1   Cross-Correlation

The cross-correlation function (Φ) for two real signals x and y in continuous and discrete form are defined as:

$$\phi_{xy}(t) = \int_{-\infty}^{\infty} x(\tau - t)y(\tau)\,d\tau \tag{1}$$

with $x \in [0 \quad M-1]$, $y \in [0 \quad N-1]$, and $k \in [-(M+1) \quad (N-1)]$, and: (2)

$$\phi_{xy}[k] = \sum_{j=\max(0,k)}^{\min(M-1+k,\ N-1)} x[j-k]\,y[j] \tag{3}$$

where M and N are the lengths of x and y signals in discrete form, respectively. In the present study, normalized cross-correlation is used to examine the differences and similarities between frequency bands of the spectrogram. In the Normalized cross-correlation, the effect of two signals' energy is removed by adding the denominator:

$$\bar{\phi}_{xy}(t) = \frac{\phi_{xy}(t)}{\sqrt{\phi_{xx}(0)\,\phi_{yy}(0)}} \tag{4}$$

The normalized cross-correlation always has a value between 1 (one) and -1 ($-$ one). The 1 and -1 values indicate complete similarity, and zero means that the two signals are completely uncorrelated. The normalized cross-correlation in the proposed method is used to test whether spectrogram samples with similar frequency bands in an event are related to the same frequency band in other events.

3.2   Mel Spectrogram

In AED systems, frequency features have been widely used. Since the normal spectrum extracted by the Fourier transform does not match the auditory characteristics of the human system, it does not perform well in AED system, and Mel scale characteristics are usually used instead [20,21]. In the design of Mel features, the human auditory system was used. Since Mel is designed to match the human auditory system, it performs much better than normal frequency scale in AED. A Mel spectrogram is calculated by obtaining equal frequency intervals based on the Mel criterion. Usually, a triangular filter is designed for each interval after knowing the frequency of the start and end of intervals. Since each filter in each frequency interval will have an output at different times, one will have a 2D output in which the columns are time and rows are Mel frequencies. The 2D output can be shown as an image equivalent to the audio spectrogram in the Mel domain for an audio file. Figure 2 shows an output of the Mel spectrogram in the forms of a 3D plot and an image. In this work, the importance of each frequency band of Mel spectrogram in an AED system is analyzed separately by cross-correlation, and a unique frequency pattern is found for each event.

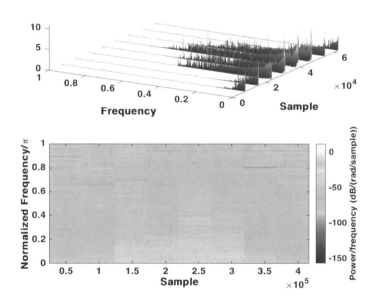

Fig. 2. Mel spectrogram for a sample sound.

3.3   Proposed Method in a Glance

As shown in the block diagram of Fig. 1, in the training phase, the sounds are entered into the Mel bank filter. In the proposed method, the number of Mel bank filters is equal to 173, the length of the time frame is equal to 0.1 s, the overlap of the frames in the Fourier transformation is 512 points, and the input sampling frequency is 44.1 KHz. After extracting the spectrogram with the above conditions using the tags in the database, the resulting matrix is divided into sub-matrices corresponding to different events. By choosing periods where the events do not overlap, the cross-correlation between each event is calculated with examples of events in other sounds in the database to determine the similarity between each event to itself and other events in sounds taken from the environment. The cross-correlation between an event with other events should show a significant difference in the effective frequency bands compared to the cross-correlation of an event with itself.

In order to assess each frequency band's effectiveness between every two events, a trimmed mean (between the 1st quartile and the 3rd quartile) of cross-correlation results between the Mel spectrogram of these events in all training samples was calculated. This routine was repeated between all events. Finally, the frequency bands where the trimmed mean of correlation data in an event with itself was higher than other events in at least 90% of cases were selected as the effective frequency bands. Figure 3 shows the frequency bands selected for each event in white. Finally, based on the selected frequency bands, a modified spectrogram is created and used during deep neural network training.

4   Dataset and Results

In this section, first experimental condition including dataset details and implementation conditions, are explained, and then, the result of deep neural network classifier with normal and modified spectrograms is compared against the latest references in this area, and its efficiency is demonstrated.

Fig. 3. Effective frequency bands in detecting each event marked in white.

4.1   Dataset

The dataset used in this work is URBAN-SED [22]. The URBAN-SED dataset is a public dataset built based on real sounds by adding urban environment and real conditions noise using the Scaper library written in Python. The dataset includes 10,000 different audio files of equal length. The events in this dataset include air conditioner, car horn, children playing, dog bark, drilling, engine idling, gun shot, jackhammer, siren and street music. The files are all single channel (mono), 44100 Hz, 16-bit, WAV format. For a more accurate comparison, the data is divided by the creator into three categories: validation set, training set, and test set, which contains 6000 data in the training category and 2000 data in the other two categories. Tags in this database are stored in JAMS format.

4.2 Results

AED methods based on spectrograms were chosen to validate the proposed method. All selected methods were retrained with the proposed modified spectrogram, and the results are compared to the network trained with the standard spectrogram. Since the spectrogram extraction parameters are usually different from the current assumptions, in all the comparisons, the effective frequency bands were recalculated based on new conditions, so the comparison conditions are completely the same. Johnson et al. [21] trained a ResNet deep neural network using spectrograms and reported their results on the URBAN SED database. In this work, the frequency of the input sound is first reduced to 22050 Hz, and then a Fourier transform of 2048 points is taken from it. In this way, the length of each sound frame is approximately 0.1 s. The overlap between the frames is 512 points, i.e. 25 ms, and the number of filters in the filter bank is 256. Hence, the length of the frame used for training, assuming that the length of each segment is 1 (one) second, was equal to $43 \times 256 \times 1$. This structure is retrained using the proposed modified spectrogram without data augmentation, and the results were compared with the original one.

Due to the proposed method structure, in selecting effective frequency bands for each audio event separately (different from other events), a separate network should be trained for each event. The original spectrogram of [21] has 256 frequency bands. After processing, a total of 1150 frequency bands were removed in 10 trained classifiers, which means removing 44.92% of the frequency bands and reducing the dimensions of the input image by almost half. The results were averaged on 10 events. Table 1 shows the results in terms of F-score values, where it can be seen that despite the reduction of spectrogram dimensions, the F-score value was improved compared to the normal spectrogram in ResNet, and there was a very small drop in CNN, which shows that removing frequency bands was done correctly. The improvement of the correct percentage in ResNet can be considered due to the removal of redundant samples and the reduction of additional information and noise in the input, which contributed to the stability of the system and the improvement of its efficiency.

Dinkel et al. [23] used a spectrogram including 64 filter banks. In the current research, the length of each sample is 40 ms, the Fourier transform has 2048 points, and the overlap between segments is assumed to be 1024 points or 20 ms. By examining the [23] spectrogram frequency bands, a total of 340 frequency bands (out of a total of 640 frequency bands) were removed, which means that 53.1% of the bands were unimportant. Due to the proposed method structure in selecting the effective frequency bands for each audio event separately, a binary classifier was trained for each event, and the final results were averaged. The

Table 1. F-score values of [21] AED system with normal and modified spectrograms.

Method	CNN	ResNet
Reported in [21]	0.587	0.589
Normal Spectrogram	0.523	0.518
Modified spectrogram	0.517	0.524

Table 2. F-score values of [23] AED system with normal and modified spectrograms.

Approach	F-scores
Base CNN [22]	56
SoftPool [25]	49.2
MaxPool [25]	46.3
AutoPool [25]	50.4
Multi-Branch [24]	61.6
Supervised SED [20]	64.7
duration robust CRNN (CDur) [23]	62.83
Our method	60.22

results of the F-score values are given in Table 2, which also show that despite the removal of almost half of the samples, the efficiency of the proposed method was reduced by only 2.6% compared to the method [66], which shows the efficiency of the proposed approach in selecting effective bands due to the halving of the input length.

5 Conclusion

In this article, a method based on cross-correlation was proposed to check the efficiency of different frequency bands of the spectrogram in AED. The proposed method separates the effective frequency bands in the detection of every audio event. Accordingly, the spectrogram can be modified based on the separated bands used as input for any deep neural network or any other classifier. Through discarding redundant features from the spectrogram, in addition to reducing the noise and error probability, the complexity of the deep neural network is also reduced, and the AED system becomes more optimized. The proposed method was tested on the URBAN SED database for a sample spectrogram, and a complete analysis of the frequency bands in each event was performed. With the fact that almost half of the frequency bands are practically useless, the proposed method was compared to state-of-the-art AED methods, and the performance of the user structures was investigated with and without the proposed method. The obtained results showed that the proposed method while reducing the dimensions of the spectrogram and the computational load, keeps accuracy in all cases.

The effective frequency bands in each event can be considered as the frequency pattern of that event, which can be used in future research to optimize the frequency methods of AED.

Acknowledgements. This article is partially a result of the project Safe Cities - "Inovação para Construir Cidades Seguras", with reference POCI-01-0247-FEDER-041435, co-funded by the European Regional Development Fund (ERDF), through the Operational Programme for Competitiveness and Internationalization (COMPETE 2020), under the PORTUGAL 2020 Partnership Agreement. The first author would like to thank "Fundação para a Ciência e Tecnologia" (FCT) for his Ph.D. grant with reference 2021.08660.BD.

References

1. Hajihashemi, V., Gharahbagh, A.A., Cruz, P.M., Ferreira, M.C., Machado, J.J.M., Tavares, J.M.R.S.: Binaural acoustic scene classification using wavelet scattering, parallel ensemble classifiers and nonlinear fusion. Sensors **22**(4), 1535 (2022)
2. Hajihashemi, V., Alavigharahbagh, A., Oliveira, H.S., Cruz, P.M., Tavares, J.M.R.S.: Novel time-frequency based scheme for detecting sound events from sound background in audio segments. In: Tavares, J.M.R.S., Papa, J.P., González Hidalgo, M. (eds.) CIARP 2021. LNCS, vol. 12702, pp. 402–416. Springer, Cham (2021). https://doi.org/10.1007/978-3-030-93420-0_38
3. Plenkers, K., Ritter, J.R.R., Schindler, M.: Low signal-to-noise event detection based on waveform stacking and cross-correlation: application to a stimulation experiment. J. Seismol. **17**(1), 27–49 (2013)
4. Plinge, A., Grzeszick, R., Fink, G.A.: A bag-of-features approach to acoustic event detection. In: 2014 IEEE International Conference on Acoustics, Speech and Signal Processing (ICASSP), pp. 3704–3708. IEEE (2014)
5. Lu, X., Tsao, Y., Matsuda, S., Hori, C.: Sparse representation based on a bag of spectral exemplars for acoustic event detection. In: 2014 IEEE International Conference on Acoustics, Speech and Signal Processing (ICASSP), pp. 6255–6259. IEEE (2014)
6. Espi, M., Fujimoto, M., Kinoshita, K., Nakatani, T.: Exploiting spectro-temporal locality in deep learning based acoustic event detection. EURASIP J. Audio Speech Music Process. **2015**(1), 1–12 (2015)
7. Pikrakis, A., Kopsinis, Y., Libra, M.L.I.: Dictionary learning assisted template matching for audio event detection (legato). Reconstruction **40**, 60 (2016)
8. Farina, A., Pieretti, N., Salutari, P., Tognari, E., Lombardi, A.: The application of the acoustic complexity indices (ACI) to ecoacoustic event detection and identification (EEDI) modeling. Biosemiotics **9**(2), 227–246 (2016)
9. Yang, L., Chen, X., Liu, Z., Sun, M.: Improving word representations with document labels. IEEE/ACM Trans. Audio Speech Lang. Process. **25**(4), 863–870 (2017)
10. Adavanne, S., Pertilä, P., Virtanen, T.: Sound event detection using spatial features and convolutional recurrent neural network. In: 2017 IEEE International Conference on Acoustics, Speech and Signal Processing (ICASSP), pp. 771–775. IEEE (2017)
11. Kim, H.-G., Kim, J.Y.: Environmental sound event detection in wireless acoustic sensor networks for home telemonitoring. China Commun. **14**(9), 1–10 (2017)

12. Lu, Z.: Sound event detection and localization based on CNN and LSTM. Detection Classification Acoust. Scenes Events Challenge, Technical report (2019)
13. Cordourier, H., Meyer, P.L., Huang, J., Del Hoyo Ontiveros, J., Lu, H.: GCC-PHAT cross-correlation audio features for simultaneous sound event localization and detection (SELD) on multiple rooms, pp. 55–58 (2019)
14. Cao, Y., Iqbal, T., Kong, Q., Galindo, M., Wang, W., Plumbley, M.: Two-stage sound event localization and detection using intensity vector and generalized cross-correlation. Technical report of Detection and Classification of Acoustic Scenes and Events 2019 (DCASE) Challenge (2019)
15. Noh, K., Jeong-Hwan, C., Dongyeop, J., Joon-Hyuk, C.: Three-stage approach for sound event localization and detection. Technical report of Detection and Classification of Acoustic Scenes and Events 2019 (DCASE) Challenge (2019)
16. Nguyen, T.N.T., Jones, D.L.: Gan, W.-S.: On the effectiveness of spatial and multi-channel features for multi-channel polyphonic sound event detection. In: DCASE, pp. 115–119 (2020)
17. Sampathkumar, A., Kowerko, D.: Sound event detection and localization using CRNN models. (2020)
18. Ick, C., McFee, B., Sound event detection in urban audio with single and multi-rate PCEN. In: ICASSP 2021-2021 IEEE International Conference on Acoustics, Speech and Signal Processing (ICASSP), pp. 880–884. IEEE (2021)
19. Nguyen, T.N.T., Watcharasupat, K.N., Nguyen, N.K., Jones, D.L., Gan, W.-S.: Salsa: spatial cue-augmented log-spectrogram features for polyphonic sound event localization and detection. IEEE/ACM Trans. Audio Speech Lang. Process. **30**, pp. 1749–1762 (2022)
20. Martín-Morató, I., Mesaros, A., Heittola, T., Virtanen, T., Cobos, M., Ferri, F.J.: Sound event envelope estimation in polyphonic mixtures. In: ICASSP 2019-2019 IEEE International Conference on Acoustics, Speech and Signal Processing (ICASSP), pp. 935–939. IEEE (2019)
21. Johnson, D.S., Lorenz, W., Taenzer, M., Mimilakis, S., Grollmisch, S., Abeßer, J., Lukashevich, H.: Desed-Fl and urban-Fl: federated learning datasets for sound event detection. In: 2021 29th European Signal Processing Conference (EUSIPCO), pp. 556–560. IEEE (2021)
22. Salamon, J., MacConnell, D., Cartwright, M., Li, P., Bello, J.P.: Scaper: a library for soundscape synthesis and augmentation. In: 2017 IEEE Workshop on Applications of Signal Processing to Audio and Acoustics (WASPAA), pp. 344–348. IEEE (2017)
23. Dinkel, H., Mengyue, W., Kai, Yu.: Towards duration robust weakly supervised sound event detection. IEEE/ACM Trans. Audio Speech Lang. Process. **29**, 887–900 (2021)
24. Huang, Y., Wang, X., Lin, L., Liu, H., Qian, Y.: Multi-branch learning for weakly-labeled sound event detection. In: ICASSP 2020-2020 IEEE International Conference on Acoustics, Speech and Signal Processing (ICASSP), pp. 641–645. IEEE (2020)
25. McFee, B., Salamon, J., Bello, J.P.: Adaptive pooling operators for weakly labeled sound event detection. IEEE/ACM Trans. Audio Speech Lang. Process. **26**(11), 2180–2193 (2018)

Age-Friendly Route Planner: Calculating Comfortable Routes for Senior Citizens

Andoni Aranguren, Eneko Osaba$^{(\boxtimes)}$, Silvia Urra-Uriarte,
and Patricia Molina-Costa

TECNALIA, Basque Research and Technology Alliance (BRTA), 48160 Derio, Spain
eneko.osaba@tecnalia.com

Abstract. The application of routing algorithms to real-world situations is a widely studied research topic. Despite this, routing algorithms and applications are usually developed for a general purpose, meaning that certain groups, such as ageing people, are often marginalized due to the broad approach of the designed algorithms. This situation may pose a problem in cities which are suffering a slow but progressive ageing of their populations. With this motivation in mind, this paper focuses on describing our implemented Age-Friendly Route Planner, whose goal is to improve the experience in the city for senior citizens. In order to measure the age-friendliness of a route, several variables have been deemed, such as the number of amenities along the route, the amount of *comfortable* elements found, or the avoidance of sloppy sections. In this paper, we describe one of the main features of the Age-Friendly Route Planner: the *preference-based routes*, and we also demonstrate how it can contribute to the creation of adapted friendly routes.

Keywords: Route Planning · Optimization · Smart Cities · Senior Citizens · Ageing People

1 Introduction

Nowadays, route planning is a hot topic for both urban planners and the research community. The reason for this popularity can be broken down into two factors. On the one hand, due to their complexity, it is a tough challenge to solve this kind of problems. Hence, the inherent scientific appeal of these problems is irrevocable. On the other hand, the business benefits of efficient logistics and the social advantages that this would bring make addressing these problems of great interest for companies and civil servants.

Evidence of this interest is the growing number of scientific publications that are added to the literature year after year [1–4]. It is also interesting the growing number of open-source frameworks for route planning that can be found in the community, which can be used to solve routing problems of different kind. Examples of this kind of frameworks are Open Trip Planner (OTP, [5]), Open-Source Routing Machine (OSRM, [6]), or GraphHopper.

A. Rocha et al. (Eds.): WorldCIST 2023, LNNS 802, pp. 192–202, 2024.
https://doi.org/10.1007/978-3-031-45651-0_20

Despite the large amount of research and developments made on the topic, routing algorithms and applications are usually developed for a general purpose, meaning that certain groups with mobility restrictions, such as ageing people, are often marginalised due to the broad approach of their designs. In most cases, routing algorithms aim to optimise efficiency factors, such as the traverse speed, distance, and public transport transfers. However, these factors can result in a route that is challenging for underrepresented groups with specific physical needs such as periodic resting, hydration to prevent heat strokes and incontinence. These groups, typically older people and people with physical disabilities/ physical limitations, require route planners with a different approach that integrates accessibility factors.

In line with this, many European cities are experiencing a slow but progressive ageing of their populations, arising a considerable spectrum of new concerns that should be taken into account. For this reason, and as a result of these concerns, policy makers and urban planners are in a constant search of novel initiatives and interventions for enhancing the participation in the city life of senior citizens.

In this context, Artificial Intelligence (AI) has emerged as a promising field of knowledge area for dealing with ageing people related concerns. For this reason, a significant number of municipalities and cities have adopted AI solutions into their daily activity, implementing various systems for constructing innovative functionalities around, for example, mobility. However, the potential of AI for the development of innovative age-friendly functionalities remains almost unexplored, for example the development of age-friendly route planners. Few efforts have been made in this direction, such as the work recently published in [7] describing a preliminary prototype is described for planning public transportation trips for senior citizens.

With this motivation in mind, the main objective of this research is to present our developed Age-Friendly Route Planner, which is fully devoted to providing senior citizens with the friendliest routes. The goal of these routes is to improve the experience in the city for these ageing users. In order to measure this friendliness, several variables are considered, such as the number of amenities along the route, the number of elements that improve the comfortability of the user, or the consideration of flat streets instead of sloppy sections.

Specifically, in this paper we detail one of the main functionalities of our Age-Friendly Route Planner: the *preference-based route planning*. Thanks to this functionality, adapted walking routes can be computed based on four weighted preferences inputted by the user and related to *i)* the duration, *ii)* the incline of the streets traversed, *iii)* the amount of amenities found throughout the route, and *iv)* the overall comfortability of the trip. The entire route planner has been implemented based on the well-known Open Trip Planner[1]

To properly develop this functionality, different real-world data have been used, and two ad-hoc data-processing engines have been implemented, namely, the Standardized Open Street Maps Enrichment Tool (SOET), and the Amenity

---

[1] https://www.opentripplanner.org/.

Projection Tool (AOT). These tools, along with the preference-based route planning functionality, and the overall structure of the Age-Friendly Route Planner are described in detail along this paper. In addition, we show some solution examples in the city of Santander, Spain, to demonstrate the applicability of the planner we have developed.

It should be pointed out here that the *preference-based route planning* functionality supposes a significant innovation for our Age-Friendly Route Planner regarding the vast majority of general-purpose route planners available in the literature, which do not compute this type of age-friendly routes. SOET and AOT also represent a remarkable contribution to this work, as they can be easily replicated in other route planners and Open Street Map (OSM, [8]) based applications.

The structure of this paper is as follows. In the following Sect. 2, we detail the overall structure of the Age-Friendly Route Planner. In Sect. 3, we describe the main data used by the planner to properly perform the functions that this paper focuses on. We also describe SOET and AOT in this section. In addition, in Sect. 4, we describe the preference-based route planning. In this section, we also introduce some examples of its applicability. Lastly, we finish this paper with conclusions and further work (Sect. 5).

2   The Age-Friendly Route Planner

After analysing a significant amount of the most popular open source route planners (such as GraphHopper, OptaTrip, Traccar, OSRM or MapoTempo, among many others), we found that most of them are mainly designed for vehicle routing. This fact highlights the need for a solution that is primarily aimed at citizens. With this in mind, and as mentioned in the introduction, OTP has been selected as the framework for being used in this work. This fact does not imply that the advantages of other alternatives should be underestimated, but the main features, flexibility, and benefits that OTP offers to developers led us to choose it as an excellent platform for achieving the main objectives established. On closer examination, several important reasons led us to choose OTP as the base framework, which are the following:

- It is fully open source, meaning that it can be fully customised to fulfill the research requirements.
- It works efficiently with widely known standards such as OSM or Geotiff (for defining city elevations).
- Being published in 2009, OTP is a platform with a long trajectory. Therefore it is very well documented and has a large and active community working on it. This facilitates the understanding of the framework.
- Both the API and the outcome JSON are fully customisable to the research requirements.

As for the main structure of the Age-Friendly Route Planner, it has a central module coined as *route planning module*, which is responsible for calculating routes using both the available data and the information entered by the user via API as input. In Fig. 1 we represent the overall architecture of the Age-Friendly Route Planner, considering also the data needed for its correct use and the ad-hoc tools implemented for gathering the correct data.

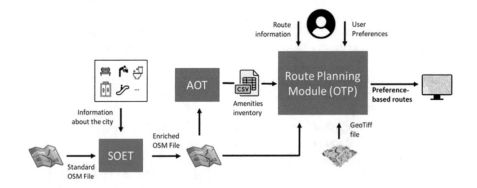

Fig. 1. Overall architecture of the Age-Friendly route planner.

Having said that, in order to properly contemplate all the requirements that the routing system should fulfill, the data sources described in the following section have been used. All these data sources are embedded in the OTP platform so that they can be taken into account in the correct planning of the routes.

3   Data Sources and Data Processing Engines

In order to generate the required routes appropriately, the Age-Friendly Route Planner must build the corresponding street network. To do this, we need the corresponding OSM map file of the city in question; in the case of this study, Santander, Spain. The OSM format is fully compatible with OTP, which automatically consumes the files and builds the corresponding road network.

In addition, this OSM file also takes into account important elements for the planner such as elevators, benches, fountains, toilets and automatic ramps. In line with this, it should be highlighted that the OSM files that can be openly obtained from open platforms are usually not as specific as we need them to be for our research. Open OSM files have proven to be very efficient for routing, but in terms of amenity related content, they are far from meeting the needs of this specific research. For this reason, we have developed an ad-hoc tool for enriching the Standardized OSM files. We have coined this tool as Standardized OSM Enrichment Tool, or SOET.

3.1 Standardized OSM Enhanced Tool - SOET

As explained earlier, for the Age-Friendly Route Planner, it is necessary to contemplate the amenities that are spread across the city, as this is a crucial factor for the success of the route planner. For this reason, the Santander's City Council, provided us with a series of files containing a list of amenities with the corresponding geospatial data. Among these amenities, we found benches, drinkable water sources, handrails and toilets. As the data was not provided in OSM format, it had to be pre-processed in order to be consumed. SOET was developed out of this motivation.

A data enrichment solution has been chosen over other options for this purpose, as data enrichment is in itself a crucial functionality for any project. Furthermore, as OSM is the basis for map creation in OTP, SOET provides a cascading effect for all applications based on the OSM file. Some examples are data visualisation in OTP and also the APT described later.

The required files for the Python-based SOET are the OSM file in which the data is stored and a *.csv* file containing the amenities to be loaded. In the following Fig. 2 we can see a clear example of an OSM file before enrichment (Fig. 2.a) and the result of applying SOET (Fig. 2.b). In this figure, we can see the newly added elements: Benches (pink dots), drinking fountains (blue dots) and garbage cans (red dots).

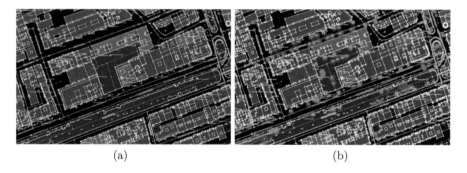

(a)   (b)

Fig. 2. A map excerpt of Santander before (a) and after (b) the SOET application. (Color figure online)

3.2 Amenity Projection Tool - APT

At this moment, four different types of amenities have been deemed in the route: public toilets, benches, handrails and drinking fountains. All these amenities have been extracted from the OSM maps. For this purpose, a new tool has been developed as part of this research. We have coined this tool as Amenity Projection Tool, or APT.

The APT was developed to correlate street segments (ways) and amenities (nodes). Thus, the AOT starts by reading the OSM file and then creates a bounding box that encloses each way. This bounding box is loose enough to enclose nearby nodes, and it is larger than the minimum bounding box by a user-defined maximum distance. So the bounding box is created to reduce the amount of correlation calculation which was $O(A_t * W_t)$, where A_t is the total amount of amenities and W_t is the total number of ways. It reduces it to the bare minimum by adding a $O(W_t)$ pre-process resulting in a $O(A_p * W_p + A_t) = O(A_p * W_p)$, where A_p and W_p are a subset of amenities and ways which are also less than or equal to A_t and W_t. This approach has been chosen to ensure that the algorithm is efficient. In Fig. 3.a we represent this situation graphically, considering that the green amenities are close enough to the street, in order to correlate these amenities with the road. After this first step, each amenity node (point) within the way bounding box is projected with an orthographic projection onto each path segment (lines). Several measures can be extracted from this projection, but only one is currently considered: the distance between the point and its projection. This distance indicates how far the amenity is from the segment. If it is equal to or less than the specified maximum distance parameter, then the amenity is added to its amenity type count. Figure 3.b represents this situation.

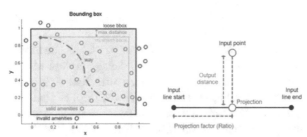

(a) APT criteria for consid-   (b) How APT relates ameni-
ering amenities   ties and ways

Fig. 3. Basic concepts of APT

After the execution of these two phases, all correlations among ways and amenities are stored in a .csv which contains the identifications of all the ways and the number of amenities of each type that are within the parameterised maximum distance from that way.

3.3   Elevation Data

In order to calculate friendly routes for senior citizens, the Age-Friendly route planner also considers the incline of the streets. This is a crucial aspect deeming that steep streets are usually preferred to be avoided and sometimes could create unwalkable routes for the older people. Consequently, a file containing the elevation of the city is compulsory. Luckily, OTP already allows the consumption of this information using the widely known GeoTIFF metadata standard. GeoTIFF permits the georeferenced of different information embedded into a *.tif* file. With such a file, OTP can assign to a certain elevation to its corresponding street. For this purpose, the *.tif* file has been obtained from the *SRTM 90m Digital Elevation Database* open platform[2].

4   Building Preference-Based Routes in the Age-Friendly Route Planner

In order to calculate routes based on user preferences, a functionality coined as *Square Optimization* has been implemented in the Age-Friendly Route Planner. This kind of optimization allows the user to define four different preferences for the calculation of walking routes (whose sum must equal 100%).

– *Slope*: this factor regards the incline of the route. The higher this factor, the flatter the routes calculated by the planner. The streets incline is calculated using the elevation data described in Sect. 3.3.
– *Duration*: this factor regards the duration of the route. The higher this factor, the shorter the routes calculated in terms of time.
– *Amenities found along the route*: this factor considers the amenities described in the previous Sect. 3.2. In this case: benches, toilets and drinking water fountains. The higher this factor, the more amenities will be found along the route. In other words, a high value of this factor implies that the route planner will prioritize going through streets that contain these amenities.
– *Comfortability factor*: the comfortability factor considers those elements that make the route more comfortable for the user. At the time of writing this paper, and because of a lack of additional data, only handrails have been included in this comfortability factor. In future stages of the Age-Friendly route planner, additional aspects such as shadows will be contemplated for this factor. Just like the amenities factor, the higher the comfortably factor, the more comfortable the routes will be.

In order to demonstrate the applicability of this kind of routes, a testing purpose webpage has been deployed based on OTP. This page is fully accessible to any interested reader[3]. In here the user is able to introduce their preferences using the interactive interface. We show in Fig. 4 two examples for this preference settings. Also, in this webpage, the user is able to choose routing options such as the origin and destination of the path.

---

[2] https://cgiarcsi.community/data/srtm-90m-digital-elevation-database-v4-1/.
[3] https://afrp.santander.urbanage.digital.tecnalia.dev/.

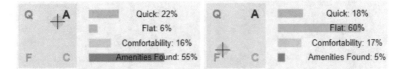

Fig. 4. Visual examples of the Square Optimization in the Age-Friendly route planner for walking routes.

Now we represent four different examples of walking routes, each one clearly prioritizing one of the deemed factors. As explained before, this demonstration is placed in the city of Santander, Spain; and in order to properly show the impact of the four factors, same origins and destinations are considered for each route. Thus, Fig. 5.a represents a route which prioritizes quickness. Secondly, Fig. 5.b depicts a path devoted to use low-incline routes. Meanwhile, Fig. 5.c shows a route prioritizing the appearance of amenities along the route. Lastly, Fig. 5.d represents a path which prioritizes comfortability, which means that it has handrails. Moreover, we depict in Table 1 a summary of the features of each route, in order to clearly visualize the main characteristics of each path, and how the choosing of different preferences is crucial for building adapted routes.

Table 1. Parameters and information about the routes calculated. *Incline*: sums the overall elevation-comfortability of the route (the less the better). *Duration*: duration of the route. *Amenities*: number of amenities found. *Comfortable*: amount of comfortable elements.

	Route I-Fig. 5.a	Route II-Fig. 5.b	Route III-Fig. 5.c	Route IV-Fig. 5.d
	Preferences of the route			
Slope Factor	14%	**74%**	4%	8%
Duration Factor	**72%**	8%	15%	3%
Amenity Factor	12%	2%	**66%**	22%
Comfortability Factor	2%	16%	15%	**67%**
	Information of the route			
Incline	487,3	**447,3**	504,5	514,5
Duration	**34 min**	38 min	37 min	38 min
Amenities	101	88	**161**	120
Comfortable Elements	40	46	60	**65**

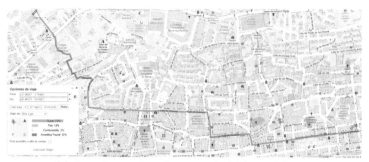

(a) A preference-based route prioritizing quickness.

(b) A preference-based route prioritizing low-slope streets.

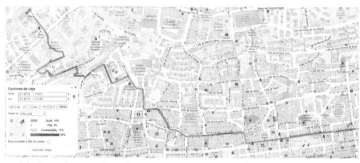

(c) A preference-based route prioritizing the finding of amenities.

(d) A preference-based route prioritizing the comfortability of the route

Fig. 5. Different examples demonstration the application of the preference-based walking routes functionality of the Age-Friendly Route Planner.

5   Conclusions

The application of routing algorithm to real-world situations has been a hot research topic in the last decades. As a result of this interest, the research carried out in this field is abundant. Despite this, routing algorithms and applications are usually developed for a general purpose, meaning that certain groups, such as ageing people, are often marginalized because of the broad approach of the designed algorithms. This situation may pose a problem in different parts of the world, such as Europe, in which many are experiencing a slow but progressive ageing on their populations, arising a considerable spectrum of new challenges and concerns that should be approached.

With this motivation in mind, this paper is focused in describing our own routing solution called Age-Friendly Route Planner. This planner is fully devoted to providing ageing citizens with the friendliest routes. The main objective of this route planner is to improve the experience in the city for senior people. To measure this friendliness, several variables have been taken into account, such as the number of amenities found along the route, the number of elements that improve the comfortability of the user along the path, the usage of urban infrastructures or avoiding sloppy sections.

Having shown and demonstrated one of the main functionalities of the Age-Friendly Route Planner, which is the *preference-based route planning*, several research lines have been planned as future work. As a short term, we will implement further features on our route planner, such as in public transportation routes. Or the consideration of people using wheelchair. As a long-term activity, we have planned to extend our Age-Friendly Route Planner to other European cities which might arise unique challenges.

Acknowledgments. This work has received funding from the European Union's H-2020 research and innovation programme under grant agreement No 101004590 (URBANAGE).

References

1. Precup, R.-E., et al.: Nature-inspired optimization algorithms for path planning and fuzzy tracking control of mobile robots. In: Osaba, E., Yang, X.S. (eds.) Applied Optimization and Swarm Intelligence. Springer Tracts in Nature-Inspired Computing, pp. 129–148. Springer, Singapore (2021). https://doi.org/10.1007/978-981-16-0662-5_7

2. Osaba, E., Villar-Rodriguez, E., Oregi, I.: A systematic literature review of quantum computing for routing problems. IEEE Access **10**, 55:805–55:817 (2022)

3. Precup, R.-E., et al.: Grey wolf optimizer-based approaches to path planning and fuzzy logic-based tracking control for mobile robots. Int. J. Comput. Commun. Control **15**(3) (2020)

4. Osaba, E., Villar-Rodriguez, E., Oregi, I., Moreno-Fernandez-de Leceta, A.: Hybrid quantum computing-Tabu search algorithm for partitioning problems: preliminary study on the traveling salesman problem. In: IEEE Congress on Evolutionary Computation (CEC), pp. 351–358. IEEE (2021)

5. Morgan, M., Young, M., Lovelace, R., Hama, L.: Opentripplanner for R. J. Open Source Softw. **4**(44), 1926 (2019)
6. Huber, S., Rust, C.: Calculate travel time and distance with openstreetmap data using the open source routing machine (OSRM). Stand Genomic Sci. **16**(2), 416–423 (2016)
7. Abdulrazak, B., Tahir, S., Maraoui, S., Provencher, V., Baillargeon, D.: Toward a trip planner adapted to older adults context: Mobilaînés project. In: International Conference on Smart Homes and Health Telematics, pp. 100–111 (2022)
8. Haklay, M., Weber, P.: Openstreetmap: user-generated street maps. IEEE Pervasive Comput. **7**(4), 12–18 (2008)

Information Systems and Technologies for Digital Cultural Heritage and Tourism

The VAST Collaborative Multimodal Annotation Platform: Annotating Values

Georgios Petasis[1](✉), Martin Ruskov[2], Anna Gradou[1], and Marko Kokol[3]

[1] Institute of Informatics and Telecommunications, National Centre for Scientific Research (N.C.S.R.) "Demokritos", GR-153 10, P.O.BOX 60228, Aghia Paraskevi, Athens, Greece
`{petasis,agradou}@iit.demokritos.gr`
[2] Department of Computer Science, Università degli Studi di Milano, Via Celoria 18, 20133 Milano, Italy
`martin.ruskov@unimi.it`
[3] Semantika Research, Semantika d.o.o., Zagrebška 40a, 2000 Maribor, Slovenia
`marko.kokol@semantika.eu`

Abstract. In this paper, we present the VAST Collaborative, Multimodal, Web Annotation Tool. It is a collaborative, web-based annotation tool built upon the Ellogon infrastructure, adapted to the content creation and annotation needs of digital cultural heritage. With the help of an annotation methodology and guidelines, the tool has been used to analyse and annotate intangible artifacts (mainly narratives) with moral values. This paper presents the tool and its capabilities, and an evaluation study for assessing its usability.

Keywords: annotation tools · inter-annotation reliability · collaborative annotation · web-based annotation · moral values

1 Introduction

It is widely spoken of the values inherited through literary heritage. However, being specific about how these values are expressed in intangible artifacts (e.g. narratives such as historical texts) is not straightforward. One widely adopted technique to externalise implicit content in text is qualitative content analysis, where experts annotate a text assigning labels that are not necessarily visible in the text itself.

In this paper, we present the VAST Collaborative, Multimodal, Web Annotation Tool. It is a collaborative, web-based annotation tool built upon the Ellogon infrastructure, adapted to the content creation and annotation needs of digital cultural heritage. Being based on a generic annotation platform in development for many years, offers a set of advantages, like the ability to support a wide range of annotation tasks and annotation schemata, robustness, as well as cross-domain features, like artifact/resource management, security, storage, etc. On the other hand, domains like the digital cultural heritage, may have specialised

A. Rocha et al. (Eds.): WorldCIST 2023, LNNS 802, pp. 205–216, 2024.
https://doi.org/10.1007/978-3-031-45651-0_21

requirements. In order to support content creation and annotation in the cultural heritage domain, the following methodology has been followed: a) collect requirements from cultural heritage professionals, with an emphasis on narratives and annotation with values; b) perform an analysis of state-of-art annotation tools and the percentage of requirements they support; c) adapt the selected tool to fulfil all requirements; d) perform content creation and annotation tasks, following defined methodologies and guidelines; and e) evaluate the performance of the tasks.

Innovative aspects of the VAST Tool include a) support for all modalities; b) ability to annotate long documents (either texts, audios, or videos); c) real-time annotation schema extension; and d) extensive support for annotation quality monitoring.

The structure of the is paper is as follows: Sect. 2 presents work related to annotation types and types for various modalities and domains, focusing on tools for cultural heritage, along with an overview of usability evaluation of annotation tools. Section 3 describes the requirements gathered from cultural heritage professionals, while Sect. 4 provides an overview of the VAST Tool and its main features. Section 5 presents the evaluation phase, while Sect. 6 concludes this paper, and presents some future directions.

2   Related Work

An annotation tool is a specialised application that aims to help annotators enrich multimedia artifacts, through the creation of additional metadata. These metadata (or "annotations") can be classified into two main categories, characterised by the granularity of their application: 1) "artifact properties" that are associated to an entire artifact, characterising it as a whole. Typical annotations of this category include labels on document or image level, e.g. classification categories associated with documents. 2) "artifact annotations", which typically associate labels with segments (parts) of an artifact. Typical annotation types of this category include a) *textual annotations*, labels associated with specific parts/segments of a text (e.g. words, sentences, paragraphs, etc.), b) *spatial annotations*, labels associated with areas of images and videos (e.g. points, lines, splines, key points and landmarks, 2D/3D boxes bounding objects, polygons, etc. [1]), and c) *temporal annotations*, which can associate labels with temporal segments that are determined by beginning and end timestamps, in audios and videos. Depending on the goal of data annotation, spatial and temporal annotation can be combined on the same artifact.

Historically, annotation tools started as desktop applications. The user needed to install special software, and the annotation process took place locally with the documents that were stored in a local machine. Nowadays, an increasingly number of on-line tools has been released, typically running within a Web browser and offering capabilities like collaborative annotation by multiple annotators. Regarding availability, options at one's disposal range from open source tools that can be adapted by developers, and freeware applications that can be used at no cost, to commercial applications.

Over the years, a plethora of annotation tools has been presented, mainly driven by applications of annotated data, such as machine learning. The vast majority of annotation tools has been driven by the needs of research areas such as natural language processing (NLP) and image analysis, since annotation tools are among the primary means for transferring human knowledge to artificial intelligence models through the assignment of labels to data [1]. Several surveys try to organise and compare annotation tools along dimensions that relate to features, tasks and modalities. A recent overview of text annotation tools is presented at [15], while a fairly recent extensive review and comparison of several annotation tools for manual text annotation can be found in [14]. Several image annotation tools are surveyed in [1,16], while surveys about audio and video annotation tools can be found in [6,10]. Finally, a recent survey regarding requirements and use of annotation tools can be found in [19].

In the cultural heritage domain, the function of annotation tools remains the same, aiming at digitising human knowledge from professionals, scholars, and quite often the crowd, that is involved in documentation, curation, restoration and enhancement of the cultural assets. While generic annotation tools can also support this domain, there are domains of application that require more specialised tools, such as the "ART3mis"[1] annotation tool for 3D objects [3] and the "Music Scholars Score Annotator"[2], which allows users to label digital musical scores [22]. A frequent requirement in this domain is the annotation of artifacts with structured knowledge, typically in the form of an ontology or a vocabulary encoded in Semantic Web technologies, such as "Culto" [9], and "CulHIAT" [20], along with the need to annotated 3D objects and scenes [2].

Usability is the discipline that provides tools and techniques to measure the quality of a piece of software. An important overarching perspective is given by the technology acceptance model which identifies a separation between perceived usefulness and perceived ease of use [11]. The System Usability Scale (SUS) is the most widely used quantitative measure of perceived ease of use [5,13]. It consists of 10 questions on a 5-point Likert-scale and produces an overall score in the range of 0–100 with higher numbers meaning better usability. It has been shown to be broadly equivalent to other popular measures, yet very efficient - achieving good statistical convergence over samples as small as 12 participants. A huge body of data indicates an average score of 68 with the end values of 0 and 100 actually obtainable by individual users. Based on this data, Sauro and Lewis have suggested that a score of at least 80 is representative for an above average user experience. Yet, they indicate that average scores vary by device, type of application, user personality traits, user experience, and application complexity. Comparisons across age and gender deliver mixed results, with a dominance of studies that fail to find significant differences. Geographic location does not appear to affect results [13]. Their studies also confirm that slight variations of the questions – to better address the task at hand – typically do not compromise the results.

---

[1] https://warmestproject.eu/tools-survey/art3mis/.
[2] https://trompamusic.github.io/music-scholars-annotator/.

3   Values in the VAST Project

VAST is a European H2020 Research project that aims to bring (moral) values to the forefront of advanced digitisation, and to investigate the transformation of core European Values, including freedom, democracy, equality, the rule of law, tolerance, dialogue, dignity, etc. across space and time. Through the analysis and the awareness raising on moral values, VAST wants to contribute to the public discourse about them and to understand how they are perceived.

Having as a starting point that morality is an individual construct, influenced and shaped by any aspect of a person's social life, VAST wants to study values in the context of social interactions that relate to arts (focusing on theatre through ancient Greek Drama of the 5th BCE century), folklore (focusing on folktales/fairytales of the 19th century), science (focusing on Scientific Revolution and natural-philosophy documents of the 17th century) and education.

VAST aims to research existing collections of intangible assets (expressed in natural language, from different places and from significant moments in European history) and trace and inter-link the values emerging from them. For these purposes, VAST has developed a collaborative semantic annotation platform, the "VAST Semantic Annotation Platform"[3]. It allows the collaborative, multimodal, analysis and annotation of artifacts (primary with values, but not limited to values). These services facilitate professionals like scholars analysing narratives, or museum curators who want to extend the available metadata on their collections.

3.1   VAST Semantic Annotation Platform Requirements

In order to collect a set of functional requirements related to the annotation tool used within the VAST Platform, an online survey has been conducted, involving mainly scholars, theatre and museum professionals (~80 professionals across Europe provided answers). Important features obtained through this survey are:

Artifact management: A comprehensive way of managing artifacts is required, supporting file types from multiple modalities (texts, images, audios, and videos), management of resources (such as importing/exporting, copying/ cloning, moving, sharing), and security (supporting both private and shared artifacts and annotations, network storage with automated backup procedures, provision of data privacy within the context of the annotation tool).

Multiple Annotation Methods: Multiple methods and capabilities in applying labels to artifacts are required. Users requested the ability for *document-level annotations*, *multi-label annotations*, and *annotation of relationships*. The ability to annotate with *multiple terminologies* and *ontologies* are also considered important by the users. *Pre-annotations* and *machine-learning assisted annotation* is not considered as important.

Regarding labelling capabilities, annotation of (overlapping) arbitrary text

---

[3] https://platform.vast-project.eu/ .

segments is required, along with bounding boxes in images, and temporal annotations in audios and videos.

Finally, the ability to support *multiple annotation schemes* and *many annotation tasks* has been characterised as important.

Annotation of Long Documents: Users emphasise the need to support the annotation of lengthy documents, and to visualise the document in its entirety, e.g. being able to visualise and annotate a whole theatrical play or a scientific manuscript. In addition, support for *various languages* has been requested.

Quality Control: The ability to inspect annotations, compare annotations/ documents/collections and acquiring various inter-rater/inter-annotation metrics has been characterised as important.

Partial Annotation: The ability to partially save documents has been requested.

Based on the aforementioned requirements, along with additional constraints (e.g. availability under an open source license, support for real-time collaborative annotation, etc.), a literature review has been conducted. The "Ellogon Annotation Platform" [15] has been selected as the most promising infrastructure for basing the VAST Semantic Annotation Platform on. The feature comparison dimensions can been seen on Table 1 (adapted from [15]).

4   The VAST Semantic Annotation Platform

The Annotation Tools that is included in the VAST Semantic Annotation Platform is based on the Ellogon Web Annotation Tool [15], and extends it along the following dimensions:

Artifact management:
- Added support for managing artifacts beyond texts, implementing support for image, audio, and video files. The management of artifacts has the same capabilities across modalities.
- Extended REST API, to support image, audio and video artifacts.

Multiple Annotation Methods:
- Added support for spatial annotations (bounding boxes), for image annotations.
- Added support for temporal annotations, when annotating audio and video artifacts.

Annotation of Long Documents:
- Added support for large audio and video artifacts.

With these extensions, the VAST Semantic Annotation Platform satisfies all the user requirements. The VAST Semantic Annotation Platform is publicly available as a) a publicly available cloud service[4], and b) open-source software

---

[4] https://platform.vast-project.eu/ .

Table 1. Feature comparison of existing annotation solutions.

	BRAT [17]	Clarin-EL [12]	Ellogon Annotation Tool [15]	GATE Teamware [4]	Label Studio [21]	WebAnno [7]
Open Source	Yes	Yes	Yes	Yes	Yes[a]	Yes
Collaborative Annotation (Real-time)	Yes (Yes)	Yes (Yes)	Yes (Yes)	Yes (No)	Yes (No)	Yes (No)
Role Management	Basic	Basic	Basic	Advanced	Advanced[b]	Advanced
Progress Monitoring	No	No	No	Yes	Yes	Yes
Annotation Statistics	No	No	Yes	Yes	Yes	Yes
Automatic Annotation	Yes	No	Yes	Yes	Yes	Yes
Inter-annotator Agreement	Plugin	No	Yes	No	Enterprise (Paid)	No
Annotation Comparison	Partial	No	Yes	No	Enterprise (Paid)	Partial
Long Document Annotation	No	Yes	Yes	No	No	No
Real-time Schema Extension	No	Yes	Yes	No	No	No

[a] Enterprise Features may not be included.
[b] Advanced in Enterprise (Paid).

under the Apache license[5]. More details about the Ellogon Web Annotation Tool can be found at [15].

In Fig. 1, we present a typical example of text annotation in the context of the VAST project. On the right panel, the user interface (UI) is adjusted to the used annotation schema. The user can add additional, custom labels and enrich the annotation schema by using the label creation button at the bottom of the screen. On the left panel, the text of "Cinderella" tale is displayed, along with some annotations. The user can read the text in its entirety and click on the colored segments to highlight them, in order to see annotation data and edit its details. An important feature of this application is the navigation through overlapping annotations (if there are any) by using the combo-box that exist at the bottom of the UI.

[5] The VAST Semantic Annotation Platform: https://github.com/vast-project/ellogon\penalty-\@M-annotation-tool.

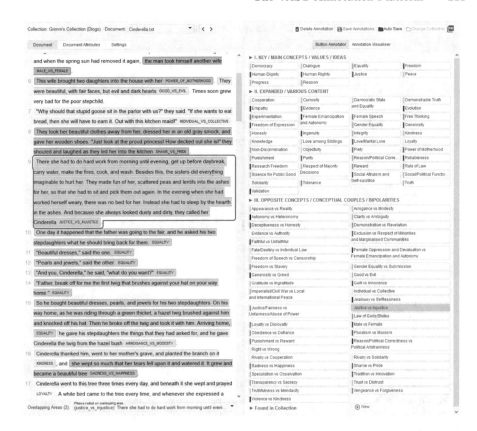

Fig. 1. Annotating a tale with values. The annotation scheme in use includes a set of annotation labels that represent concepts, moral values and ideas that are being tracked down the texts. Labels are organised under three categories: Key/Main Concepts/Values, Expanded Concepts, and Bi-polarities.

5   Evaluation

An evaluation with 27 users was performed in order to assess the usability of the annotation tool. Evaluators were invited to a in-person group session where each of them worked individually. They were instructed verbally about the process, asked to get acquainted with written materials and to express their written consent about their role in the evaluation. According to ethical and legal practice, the they were informed that their participation is anonymous and that they can withdraw from the activity at any point, in which case their collected data would be deleted. Then they were left to work independently with a researcher available for clarifications. For the purposes of the evaluation, the English translations of three one-page excerpts per pilot were selected, as shown in Table 2.

Table 2. Texts from which excerpts were selected for the evaluation

Pilot 1 - Ancient Greek Drama	
Text Title	**Text Author**
Antigone	Sophocles
Hecuba	Euripides
Peace	Aristophanes
Pilot 2 - Scientific Revolution	
Text Title	**Text Author**
On the Revolutions of the Celestial Spheres	Copernicus
Micrographia	Hooke
Mathematical Principles of Natural Philosophy	Newton
Pilot 3 - Fairytales	
Text Title	**Text Author**
Faithful Johannes	Grimm Brothers
Juniper Tree	Grimm Brothers
Little Snow White	Grimm Brothers

Evaluators were asked - whenever possible - to find and annotate five values: two main concepts, two extended concepts, and one conflict concept. A set of anonymous accounts were created and randomly distributed to evaluators. Each of them was able to choose the pilot to annotate without sharing any personal details. Upon completion of the tasks, evaluators were asked to complete an evaluation questionnaire. It contained three sections: demographics, System Usability Scale (SUS), and the following required open exploratory questions:

1. List one or more aspects that positively impressed you in using the VAST Annotation Tool.
2. List one or more aspects that negatively affected your annotation experience.
3. Do you have suggestions on how to improve the usability of the tool?
4. Do you have suggestions about possible functionalities that can be added to the tool?
5. Is there any question that you had and remained unanswered or difficult to answer when using the tool?

We note that the group of evaluators is balanced in terms of gender. It is mainly composed of young, yet educated people, consequence of recruitment in university setting. To probe the technical proficiency of evaluators, a list of 5 generic – yet broadly relevant to the task – technical skills were included in the demographics questions as a form of self-assessment. Overall, evaluators reported having a very good level of technical proficiency. The huge majority of them had no experience of annotation, with only 19% reporting repeated experience with this type of task.

The overall SUS score was 69.6 (standard deviation 20.0), which can be considered satisfactory given that we evaluate a piece of professional software [13]. Also in the responses to the open questions the overall sentiment towards the annotation tool was positive. Evaluators indicated that the tool was "clear and well organized", that "the interface is user-friendly", and that "there were clear instructions for each step of annotation". They explicitly listed the highlighting paradigm, colour-coding and ease of value selection as positive features. One participant summarised this with "It's very important that this tool is similar to Word tools". When asked to elaborate on their annotation experience, participants offered opinions that for evaluation purposes can be seen as falling in three groups: divergent perceptions of features, suggestions for improvement, and complexity.

Divergent Perception of Features. In some situations, we note that some evaluators highlighted criticisms that some others stressed as positive aspects. A typical example of this is the interaction to define the matches between text selections and values. Some evaluators, including ones without prior technical experience, found the process to be very intuitive. Others reported getting confused about the order of selection. One particular evaluator without annotation experience perceived it more intuitive to first select values and then the text. They wrote "when you select the attribute if you accidentally still have the text from before it changes that attribute" and made two particular suggestions: asked for the possibility to match starting from the value and suggested that after a match the program should automatically deselect it. However, others asked for the possibility to allow for multiple values for the same selection. Yet another evaluator just commented that "visualization of overlapping highlights is not straightforward to understand." This topic is further discussed in the section "Suggestions for improvement".

Suggestions for Improvement. Some suggestions for improvement were more conceptual, while others were practical. Two evaluators expressed the need to add personal notes, one of them illustrating the suggestion by referring to the comment feature of Microsoft Word. A theme that emerged as a recurring challenge is the visualisation of overlapping annotations. One evaluator proposed as a possible solution to add filtering to visualise only a sublist of values of current interest. Another hypothesised that there could be a way to use colouring to show overlapping selections. Other ideas were adapting selections to disregard differences in punctuation, autosave option, indication of used or unused annotations.

Complexity. Some evaluators commented about the learning curve to work with the software. One explained it like this "At first, it was hard to find the texts. It's not an intuitive software and it takes quite some time to master it". Another evaluator provided emotional feedback by saying "It was difficult in the beginning but then I enjoyed it." Other evaluators commented on the difficulty of the texts. One wrote "It was easy to use the tool. The text was harder considering it was a translation of the original Greek text." This corresponds to the difficulty of working with translations, highlighted in the participants section of the survey.

6 Conclusions and Future Work

The VAST Annotation Tool was designed as an integrative part of the VAST Platform and as an instrument addressing the specific task of annotation of historical documents. Results from this first evaluation suggest that it already matches the expectations from a typical software aimed to be used in a specific professional context.

With this in mind, the answers to the exploratory questions suggest that the annotation process could benefit from an in-depth analysis of the cognitive load, employed by annotators, possibly understanding what constitutes intrinsic and extraneous cognitive load in the annotation activity and look for ways to reduce the extraneous part [18].

The evaluation presented here covers only the annotation of text. Further studies are needed to assess other types of annotation. Also, aspects of quality control, could also be considered subject to further evaluations. In particular, it could be of interest to explore and evaluate the possibilities to compare across documents and collections or across annotators. Some work regarding the first of these directions has already been done by subjecting the resulting data to further qualitative analysis [8]. However, this research is still in progress and needs to be expanded and results need to be critically interpreted by professionals in the humanities.

Acknowledgments. The research leading to these results has received funding from the European Union's Horizon 2020 research and innovation programme, in the context of VAST project, under grant agreement No 101004949. This paper reflects only the view of the authors and the European Commission is not responsible for any use that may be made of the information it contains.

References

1. Aljabri, M., AlAmir, M., AlGhamdi, M., Abdel-Mottaleb, M., Collado-Mesa, F.: Towards a better understanding of annotation tools for medical imaging: a survey. Multimedia Tools Appli. **81**(18), 25877–25911 (2022). https://doi.org/10.1007/s11042-022-12100-1
2. Apollonio, F.I., Gaiani, M., Bertacchi, S.: Managing cultural heritage with integrated services platform. The International Archives of the Photogrammetry, Remote Sensing and Spatial Information Sciences (2019)
3. Arampatzakis, V., et al.: Art3mis: ray-based textual annotation on 3d cultural objects. In: CAA 2021 International Conference "Digital Crossroads" (2021)
4. Bontcheva, K., et al.: Gate teamware: a web-based, collaborative text annotation framework. Lang. Resour. Eval. **47**(4), 1007–1029 (2013)
5. Brooke, J.: SUS: a retrospective. J. Usability Stud. **8**(2), 29–40 (2013)
6. Cassidy, S., Schmidt, T.: Tools for Multimodal Annotation, pp. 209–227. Springer Netherlands, Dordrecht (2017). https://doi.org/10.1007/978-94-024-0881-2_7

7. de Castilho, R.E., Biemann, C., Gurevych, I., Yimam, S.M.: Webanno: a flexible, web-based annotation tool for clarin. In: Proceedings of the CLARIN Annual Conference (CAC) 2014 (Oct 2014)
8. Ferrara, A., Montanelli, S., Ruskov, M.: Detecting the semantic shift of values in cultural heritage document collections (short paper). In: Damiano, R., Ferilli, S., Striani, M., Silvello, G. (eds.) Proceedings of the 1st Workshop on Artificial Intelligence for Cultural Heritage, pp. 35–43. No. 3286 in CEUR Workshop Proceedings, Aachen (2022). https://ceur-ws.org/Vol-3286/04_paper.pdf
9. Garozzo, R., Murabito, F., Santagati, C., Pino, C., Spampinato, C.: Culto: An ontology-based annotation tool for data curation in cultural heritage. ISPRS - International Archives of the Photogrammetry, Remote Sensing and Spatial Information Sciences 42, 267–274 (2017)
10. Gaur, E., Saxena, V., Singh, S.K.: Video annotation tools: A review. In: 2018 International Conference on Advances in Computing, Communication Control and Networking (ICACCCN), pp. 911–914 (2018). https://doi.org/10.1109/ICACCCN.2018.8748669
11. Hornbæk, K., Hertzum, M.: Technology acceptance and user experience: A review of the experiential component in hci, vol. 24(5) (Oct 2017). https://doi.org/10.1145/3127358, https://doi.org/10.1145/3127358
12. Katakis, I.M., Petasis, G., Karkaletsis, V.: CLARIN-EL web-based annotation tool. In: Proceedings of the Tenth International Conference on Language Resources and Evaluation (LREC 2016), pp. 4505–4512. European Language Resources Association (ELRA), Portorož, Slovenia (May 2016). https://aclanthology.org/L16-1713
13. Lewis, J.R.: The system usability scale: past, present, and future. Inter. J. Hum.-Comput. Interact. 34(7), 577–590 (2018). https://doi.org/10.1080/10447318.2018.1455307
14. Neves, M., Ševa, J.: An extensive review of tools for manual annotation of documents. Briefings Bioinform. 22(1), 146–163 (2019). https://doi.org/10.1093/bib/bbz130
15. Ntogramatzis, A.F., Gradou, A., Petasis, G., Kokol, M.: The ellogon web annotation tool: Annotating moral values and arguments. In: Proceedings of the Thirteenth Language Resources and Evaluation Conference, pp. 3442–3450. European Language Resources Association, Marseille, France (Jun 2022), https://aclanthology.org/2022.lrec-1.368
16. Pande, B., Padamwar, K., Bhattacharya, S., Roshan, S., Bhamare, M.: A review of image annotation tools for object detection. In: 2022 International Conference on Applied Artificial Intelligence and Computing (ICAAIC), pp. 976–982 (2022). https://doi.org/10.1109/ICAAIC53929.2022.9792665
17. Stenetorp, P., Pyysalo, S., Topić, G., Ohta, T., Ananiadou, S., Tsujii, J.: Brat: a web-based tool for nlp-assisted text annotation. In: Proceedings of the Demonstrations at the 13th Conference of the European Chapter of the Association for Computational Linguistics, pp. 102–107. Association for Computational Linguistics (2012)
18. Sweller, J., van Merriënboer, J.J.G., Paas, F.: Cognitive architecture and instructional design: 20 Years Later. Educ. Psycho. Rev. 31(2), 261–292 (2019). https://doi.org/10.1007/s10648-019-09465-5, http://link.springer.com/10.1007/s10648-019-09465-5
19. Tan, L.: A survey of nlp annotation platforms (2020). https://github.com/alvations/annotate-questionnaire

20. Theodosiou, Z., Georgiou, O., Tsapatsoulis, N., Kounoudes, A., Milis, M.: Annotation of cultural heritage documents based on XML dictionaries and data clustering. In: Ioannides, M., Fellner, D., Georgopoulos, A., Hadjimitsis, D.G. (eds.) EuroMed 2010. LNCS, vol. 6436, pp. 306–317. Springer, Heidelberg (2010). https://doi.org/10.1007/978-3-642-16873-4_23
21. Tkachenko, M., Malyuk, M., Holmanyuk, A., Liubimov, N.: Label Studio: Data labeling software (2020-2022). https://github.com/heartexlabs/label-studio, open source software
22. Tomašević, D., Wells, S., Ren, I.Y., Volk, A., Pesek, M.: Exploring annotations for musical pattern discovery gathered with digital annotation tools. J. Math. Music **15**(2), 194–207 (2021). https://doi.org/10.1080/17459737.2021.1943026

Knowledge Representation Technologies for Narratives, Digital Humanities and Cultural Heritage

Gian Piero Zarri(✉) 🆔

STIH Laboratory, Sorbonne University, 75005 Paris, France
gianpzarri@gmail.com, zarri@noos.fr

Abstract. This paper describes the conceptual tools that, in an NKRL context (NKRL = Narrative Knowledge Representation Language), allow us to obtain a (computer-usable) description of full *narratives* as logically-structured associations of the constituting (and duly formalized) *elementary events*. These conceptual tools must then be able to formalize those *connectivity phenomena* – denoted, at surface linguistic level, by logico-semantic coherence links like causality, goal, co-ordination, subordination, indirect speech etc. – which assure the conceptual unity of the whole narrative. The unification-based solutions adopted in this context like, e.g., *completive construction* and *binding occurrences*, allow us to take into account the connectivity phenomena by reifying the formal representations used to model the constitutive elementary events.

Keywords: Elementary events · connectivity phenomena · reification · completive construction · binding occurrences · inference rules

1 Introduction

NKRL, the "Narrative Knowledge Representation Language", is both a conceptual modeling tool [1] and a (fully implemented) computer science environment [1: Appendix A] created, thanks mainly to several European projects, for dealing with *narrative information* [2, 3] in an (digitalized) innovative way. Informally, a *narrative* is a general unifying framework used for relating real-life or fictional stories involving concrete or imaginary characters. It materializes as work of speech, writing, song, film, television, video games, photography, theatre, iconographic items, etc. An important characteristic of any possible kind of narratives (fictional, non-fictional, textual, visual etc.) concerns their well-known *ubiquity property* see, in this context, the first paragraph of a famous Roland Barthes' paper [4: 1]. This property is also reinforced by the *extreme ease with which full-fledged narratives can be created*: the presence of a verb (more precisely, of a *generalized predicate*, even a noun or an adjective, see "Jane's *amble* in the park") and of few arguments/modifiers is sufficient to give birth to a narrative. The interest of having at our disposal a tool like NKRL, able to represent/manage full narratives keeping the loss of information as little as possible, should then be apparent[1].

There is already a large literature about NKRL in general and its possible benefits, see [1] for example. The present paper will focalize then on *specific aspects* of this language that, however, are of a *general interest* in a Digital Humanities context – e.g., from a Computational Linguistics point of view. These aspects concern how the NKRL's tools can be used to produce the *formal representation* of those *connectivity phenomena* that are denoted, at surface linguistic level, by *logico-semantic coherence links* like causality, goal, co-ordination, subordination, indirect speech etc. The two connectivity phenomena taken into account here are i) the need to refer to an event *as an argument of another event* (see, e.g., an event X where someone speaks about Y, where Y is itself a specific event or a logically coherent set of events), and ii) the need for *linking together within wider conceptual* frameworks – like complex events, scripts, scenarios, narratives, storyboards, episodes, etc. – events/situations that can still be regarded as *independent entities* (as an event X being linked to another event or set of events Y by causality, goal, coordination, alternative etc. relationships). As we will see, the formal expression of the first connectivity mechanism is called *completive construction* in an NKRL context, and that of the second *binding occurrences*.

In the following, we will present first, Sect. 2, a quick recall of some fundamental NKRL's principles; this recall is necessary in order to understand the examples of the following Sections. Section 3 is the central component of the paper. The modalities for dealing with the connectivity phenomena in an NKRL context will be illustrated through examples of interest in a Digital Humanities context. Section 4 will mention briefly other important aspects of NKRL; Sect. 5 is a short Conclusion.

2   Basic Notions About NKRL

NKRL innovates with respect to the current ontological paradigms by adding an *ontology of elementary events* to the usual *ontology of concepts*.

The ontology of concepts is called HClass (hierarchy of classes) in an NKRL context and includes presently (April 2023) more than 7,500 standard concepts – standard means here that the properties used to define a given concept are simply expressed as *binary (i.e., linking only two arguments) relationships* of the property/value type. From a purely formal point of view HClass, see [1: 43–55, 123–137], is not fundamentally different, then, from the ontologies that can be built up by using the original frame version of Protégé [8].

Before describing the structure of the ontology of elementary events, an important methodological premise must be made. According, in fact, with the vision shared by

---

[1] The interest of NKRL from an actual point of view is also evidenced from the fact that it has been used mainly within a (very concrete) *non-fictional narratives context*, i.e., for applications in domains like actuality news, defence/security, accidents in the oil/gas industry, sentiment analysis, IoT etc. Recently, applications of the NKRL's techniques in a *very specific fictional narrative domain* have concerned the so-called *iconographic narratives*, i.e., the *stories* told according to a *strict visual modality* that are supported, e.g., by cultural heritage items like paintings, drawings, frescoes, mosaics, sculptures, murals etc., see [5, 6, 7].

several eminent narratologists see, e.g., [3] in this context, a narrative is seen in NKRL as a *coherent set* (i.e., its components are logically and chronologically related), possibly of cardinality $= 1$, of spatio-temporally constrained *elementary events* describing *some specific activities, states, experiences, behaviors* etc. of the (human and non-human) entities involved in the global narrative. From an operational point of view, the representations of the elementary events are obtained in NKRL as *instances of formal n-ary structures called "templates" that correspond to general categories of elementary events* like move a physical object, be present in a place, having a specific attitude towards someone/something, produce a service, asking/receiving an advice, etc. The different NKRL *n*-ary templates *correspond to the nodes of the ontology of elementary events*, which is then denoted as HTemp (hierarchy of templates). Templates, in opposition to the *static/basic* notions (like human being, book, color, artefact, level of temperature ...) denoted by the binary HClass concepts, take into account the *dynamic/structured* component of the narrative information. NKRL templates can be represented schematically according to Eq. 1 below:

$$(L_i(P_j(R_1, a_1) (R_2, a_2) \ldots (R_n, a_n)))$$ (1)

In Eq. 1, L_i is the *symbolic label* denoting the particular *n*-ary structure corresponding to a specific template; symbolic labels are of a fundamental importance in the context of the association of (formalized) elementary events, see next Section. P_j is a *conceptual predicate* denoting the semantic *type/class* of the template. R_k is a generic *functional role* [9] used to specify the logico-semantic function of its *filler a_k* with respect to the predicate. a_k is then a *predicate argument* introduced by the role R_k.

When a template following the general syntax of Eq. 1 and denoted as Move:TransferMaterialThingsToSomeone in NKRL (see also Table 1 below) is *instantiated* to provide the representation of a simple elementary event like "Bill gives a book to Mary", the predicate P_j (MOVE) will introduce its three arguments a_k, JOHN_, MARY_ and BOOK_1 (*individuals*, i.e., instances of HClass concepts) through the three functional relationships (R_k roles) SUBJ(ect), BEN(e)F(iciary) and OBJECT. The global *n*-ary construction is *reified* through the symbolic label L_i *and managed then as a coherent block*. According to the NKRL's jargon, the instances of templates are called *predicative occurrences* and correspond to the representation of *specific elementary events*, see the examples in the following Section. To avoid the ambiguities of natural language and any possible combinatorial explosion problem, see [1: 56–61], both the conceptual predicate of Eq. 1 and the associated functional roles are *primitives*. Predicates P_j pertain then to the set {BEHAVE, EXIST, EXPERIENCE, MOVE, OWN, PRODUCE, RECEIVE}, and the functional roles R_k to the set {SUBJ(ect), OBJ(ect), SOURCE, BEN(e)F(iciary), MODAL(ity), TOPIC, CONTEXT}. Roles in NKRL coincide then, in practice, with those included in the – well-known and widely used in a Computational Linguistics (CL) framework – EAGLES (Expert Advisory Group on Language Engineering Standards) system [10].

The global HTemp hierarchy is structured into *seven branches*, where each branch includes only the templates created (according to the general syntax of Eq. 1) around one of the *seven predicates* (P_j) admitted by the NKRL language. Table 1 reproduces the template Move:TransferMaterialThingsToSomeone used to produce the predicative occurrence formalizing the elementary event "Bill gives a book to Mary" of the above example. The constituents (as SOURCE, MODAL, (*var2*) etc. in Table 1) included in square brackets are *optional*. HTemp includes now (April 2023) about 150 templates, very easy to specialize and customize, see [1: 137–177].

As we can see from Table 1, the arguments of the predicate (the a_k terms in Eq. 1) are actually represented by variables (*var$_i$*) with *associated constraints*. These are expressed as *HClass concepts or combinations of concepts*. When creating a predicative occurrence from a template, the constraints linked to the variables are used to specify the *legal sets of HClass terms (concepts or individuals) that can be substituted for these variables within the occurrence*. In the occurrence corresponding to the above example, we must verify, e.g., that JOHN_ and MARY_ are actually HClass instances of individual_person, a specific term of human_being_or_social_body, see the constraints on the SUBJ and BENF functional roles of the template of Table 1[2].

Table 1. A template of the Move: branch of HTemp.

---

name: Move:TransferMaterialThingToSomeone
father: Move: TransferToSomeone
position: 4.21
NL description: Transfer a Material Thing (e.g., a Product, a Letter…) to Someone

MOVE:
SUBJ: *var1*: [*var2*]
OBJ: *var3*
[SOURCE: *var4*: [*var5*]]
BENF: *var6*: [*var7*]
[MODAL: *var8*]
[TOPIC: *var9*]
[CONTEXT: *var10*]
{ [modulators], ≠abs }

var1: human_being_or_social_body
var3: artefact_
var4: human_being_or_social_body
var6: human_being_or_social_body
var8: activity_related_property, process_, service_
var9 : sortal_concept
var10 : situation_, symbolic_label
var2, *var5*, *var7*: location_

---

[2] *Determiners* (or *attributes*) are constants/operators that can be added to templates or predicative occurrences to introduce *further details* about their formal representation. In particular, determiners/attributes of the *location* type – represented by *lists of instances* of the HClass location_ concept and of its specialization terms – can be associated through the colon operator, ":", with the *arguments of the predicate* (i.e., the fillers) introduced by some functional roles of a template, see Table 1. Another category of determiners/attributes associated, in this case, to a *full, well-formed template or predicative occurrence* to particularize its meaning are constants of the *modulator* type. Modulators are classed into three categories, *temporal* (begin, end, obs(erve)), *deontic* (oblig(ation), fac(ulty), interd(iction), perm(ission)) and *modal modulators* (for, against, wish, negv, ment(al), etc.) – see Subsect. 3.2 below for a short discussion about negv, the modal modulator denoting a *negated event*.

3   Linking Elementary Events

In NKRL, the *connectivity phenomena* are dealt with making use of *Higher Order Logic (HOL) structures* – according to HOL, a predicate can take one or more other predicates as arguments – obtained from the *reification* of generic (i.e., not only predicative, see below) occurrences. Concretely, the reification is based on the use of the *symbolic labels* denoted by the L_i terms in Eq. 1 above. "*Reification*" is intended here as the possibility of creating new *first-class entities* out of already existing entities and to *say something* about them without making explicit reference to the original ones.

3.1 Completive Construction

A first example of HOL connection between elementary events is represented by the *completive construction*. This consists in using as *filler* of a functional role in a predicative occurrence pc_i the *symbolic label* L_j of another (*generic*) occurrence c_j. Note that the c_j (*indirectly*) used as fillers can correspond not only to predicative occurrences pc_i, but also to those *binding occurrences* bc_i we will introduce in the next sub-section. Constraints proper to the completive construction are:

- *Only* the OBJ, MODAL, TOPIC and CONTEXT functional roles of pc_i can accept as filler the symbolic label L_j of a c_j, and *only one of these four roles* can be utilized in the context of a *specific instantiation* of the completive construction mechanism.
- L_j must denote a *single* symbolic label, i.e., any L_j represented under the form of an association of labels *cannot be used* in a completive construction framework.
- For implementation reasons, this *single* label L_j is prefixed, in the *external* NKRL format, by a *sharp*, "#", code. The general format of a completive construction filler corresponds then to #symbolic_label, see the example below. Symbolic_label is a *standard HClass concept* whose instances are all the *actual labels* used to denote (predicative and binding) occurrences within a specific NKRL application.

As an example, we reproduce in Table 2 a fragment of a scenario that deals with a well-known episode in the life of Jeanne d'Arc known as "La Marche sur Reims".

In this fragment, during a face-to-face conversation that takes place in Saint Benoît sur Loire, Jeanne d'Arc sends to Dauphin Charles, see the predicative occurrence ex.c1, the message represented by the occurrence ex.c3, i.e., that he must move *necessarily* to Reims to be crowned. The modulator oblig(ation) is one of the *four deontic modulators* introduced in Note 2 above, and has been used in ex.c3 to denote the *absolute necessity*, according to Jeanne d'Arc, of moving to Reims before the end of June. The coding of Table 2 is also interesting because of the use, in the context of the *same completive structure*, of predicative occurrences corresponding to *two different types* of Move: templates. The first, Move:StructuredInformation is used *to represent the transmission of a complex information whose content is described by one or more occurrences*. The second, Move:APersonOrSocialBodyMovesAutonomously, is used to encode the physical displacement of a person/social body in the form of a SUBJ that *moves itself as an OBJ*. The *initial location*, Giens in Table 2, is linked to the filler of the SUBJ role, whereas the *final one*, Reims, is associated with the OBJ role's filler; possible *intermediary steps* can be denoted by an ordered sequence of locations included, before the final destination, in the list denoting the arrival location.

Table 2. An example of completive construction.

---

ex.c1:   MOVE:
          SUBJ : JEANNE_D'ARC: (SAINT_BENOIT_SUR_LOIRE)
          OBJ: #ex.c3
          BENF: DAUPHIN_CHARLES
          MODAL: one_to_one_conversation
          date-1: 21/6/1429
          date-2:

Move:StructuredInformation (4.42)

*On 21/6/1429, in Saint Benoît sur Loire, Jeanne d'Arc transmits to Dauphin Charles during a one-to-on
conversation what is described in the predicative occurrence* ex.c3.

ex.c3:   MOVE:
          SUBJ: DAUPHIN_CHARLES: (GIENS_)
          OBJ: DAUPHIN_CHARLES: (REIMS_)
          CONTEXT: (SPECIF coronation_ DAUPHIN_CHARLES): (REIMS_)
          { oblig }
          date-1: 21/6/1429, 30/6/1429
          date-2:

Move:APersonorSocialBodyMovesAutonomously (4.31)

Before the end of June, Dauphin Charles must necessarily, deontic modulator oblig(ation), *move to Reims
in the framework of his coronation.*

3.2 Binding Occurrences

A second, more general way of linking together NKRL elementary events consists in
making use of *binding occurrences*. These are lists labelled with specific *binding operators* Bn_i whose arguments arg_i correspond to labels L_j of (*predicative or binding*) c_j
occurrences. The general expression of a binding occurrence bc_i is then:

$$(Lb_k(Bn_iL_1L_2\ldots L_n)), \tag{2}$$

where Lb_k is now the symbolic label identifying the whole (*autonomous*) binding structure. Unlike templates and predicative occurrences, binding occurrences are then characterized by the absence of any predicate or functional role. The eight Bn_i operators are:
ALTERN(ative), COORD(ination), ENUM(eration), CAUSE (the strict causality operator), REFER(ence, the weak causality operator), GOAL (the strict intentionality operator), MOTIV(ation, the weak intentionality operator), COND(ition). More details and a
formal definition of these operators can be found, e.g., in [1: 91–98] and [9]. The binding
occurrences bc_i must necessarily conform to the following mandatory restrictions to be
considered as *well formed*:

- Each term (argument) L_j that, in a binding list, is associated with one of the Bn_i operators denotes exactly a *single* predicative or binding occurrence c_j *defined externally
  to the list*. Therefore, the arguments L_j are always *single terms*.
- In the occurrences of the ALTERN, COORD and ENUM type, *no restriction is
  imposed on the cardinality of the list*, i.e., on the number of terms (arguments) L_j.
- In the occurrences labelled with CAUSE, REFER, GOAL, MOTIV, COND *only two
  arguments L_m and L_n are admitted*. These arguments can denote in general *either*

predicative or binding occurrences, with the exception of the COND occurrences where the first argument, L_m, *must correspond to a predicative occurrence pc_i*.

The binding structures are particularly important in an NKRL context given that, among other things, the *top-level occurrence introducing the full representation of any kind of NKRL-encoded narrative structure necessarily has the form of a binding occurrence*. A – reduced, because of space limitations, see a more complete version in [7] – example of complete NKRL narrative is given in Table 3, which refers to the so-called *hidden painting problem*. This concerns the identification of the woman represented in the portrait, visible only in *x*-rays, surely painted by Leonardo Da Vinci and that lies beneath Mona Lisa on the same poplar panel [11]. In Table 3, we explain in particular that the CAUSE of what was declared in occurrence gio2.c6 – i.e., the assertion that WOMAN_52, Mona Lisa, *is not* WOMAN_53, the hidden painting woman – comes from in the publication in 1988 (see gio2.c7, an instance of the Move:StructuredInformation template) of a paper by Lillian Feldman Schwartz in "The Visual Computer" journal.

CAUSE is one of the *binding operators* introduced above. The constraint concerning the number of arguments, two, of a CAUSE list *is respected*, see the binding occurrence gio2.c5, given that, in this last occurrence, gio2.c7 and #gio2.c8 are considered as a *unique occurrence* according to the modalities of the completive construction introduced in 3.1 above. The predicative occurrence gio2.c6 represents a typical example of the *NKRL modelling of the notion of negation under the form of "negated events"*. This consists in adding to a complete and well-formed predicative occurrence a specific *modal modulator*, negv (negated event, see Note 2 above) to point out that *the corresponding elementary event did not take place*. In our example, WOMAN_53, the woman represented in the hidden painting, has not been recognized as WOMAN_52 – i.e., as Mona Lisa, according to the label (WOMAN_52) conventionally assigned in NKRL to Mona Lisa. WOMAN_53 is probably Isabella d'Aragona, the wife of the Duke of Milan Ludovico il Moro, one of the Leonardo's employers [11]. The association of the temporal modulator obs(erve) with today_ has the standard *as of today* meaning; today_ is a HClass term pertaining, through intermediate terms, to the time_period branch of HClass.

The second argument, gio2.c7 #gio2.c8, of the CAUSE clause (i.e., the *reason* of the failure to identify) refers us to the publication of a Lillian Feldman Schwartz's paper where she had listed all the *incoherencies* – like the strong dissimilarities existing among the mouths, eyes, noses, chins, hairlines etc. of the two women – that prevent any possible identification. In the NKRL encoding, all these incoherencies are denoted under the form of a set of *predicative occurrences* grouped within a *binding occurrence* of the COORD type labelled as gio2.c8 in Table 3 – only one of these predicative occurrences, gio2.c9, is actually presented in Table 3. We can remark that the binding occurrence gio2.c8 is introduced in gio2.c7 under the form of a *completive construction*, see 3.1. This particular form of *completive construction*, introduced by the template Move:StructuredInformation, is particularly important in an NKRL context because it represents the *standard form* used to denote any sort of *transmission of messages* [1: 165–167].

Eventually, thanks to Table 3 we can note the importance, in NKRL, of the use of the SPECIF(ication) operator for building up *structured arguments* (also called *expansions*)

Table 3. Binding occurrences and NKRL representation of a full narrative.

---

gio2.c5: (CAUSE gio2.c6 gio2.c7 #gio2.c8)

The elementary event modelized by gio2.c6 *has been caused by what is collectively described in the completive construction involving occurrences* gio2.c7 *and* gio2.c8.

gio2.c6: BEHAVE:
        SUBJ: (SPECIF WOMAN_53 (SPECIF identified_with WOMAN_52))
        { obs, negv }
        date-1: today_
        date-2:

Behave:HumanProperty (1.1)

We can remark (temporal modulator obs(serve)) *today that the elementary event represented by* gio2.c6 *is a negated event (modal modulator* negv), *i.e.,* WOMAN_53 *is not* WOMAN_52.

gio2.c7: MOVE:
        SUBJ: LILLIAN_FELDMANN_SCHWARTZ
        OBJ: #gio2.c8
        MODAL: (SPECIF SCIENTIFIC_PAPER_2 (SPECIF published_on
                                  THE_VISUAL_COMPUTER_JOURNAL))
        date-1: 1/1/1988, 31/1/1988
        date-2:

Move:StructuredInformation (4.42)

Mrs. Schwartz has circulated what denoted in gio2.c8 *by means of a paper in "The Visual Computer" journal.*

gio2.c8: (COORD1 gio2.c9 ...)

The binding occurrence gio2.c8 *includes a list of predicative occurrences denoting each one a difference between* WOMAN_52 *and* WOMAN_53.

gio2.c9: OWN
        SUBJ: (SPECIF eye_ (SPECIF cardinality_ 2) WOMAN_53))
        OBJ: property_
        TOPIC: (SPECIF different_from (SPECIF eye_ (SPECIF cardinality_2) WOMAN_52))
        MODAL: x_ray_analysis
        { obs }
        date-1: today
        date-2:

Own:CompoundProperty (5.42)

Only the dissimilarity between the eyes of WOMAN_52 *and* WOMAN_53 *has been taken into account in this Table. In reality, also the mouth, nose, chin and hairline of the two women are incompatible.*

---

of the NKRL's predicates, see [1: 68–70]. Structured arguments concern in general the possibility of representing the predicate arguments a_k in Eq. 1 as a *set of recursive lists*. These are introduced by the four AECS operators: the *disjunctive operator* ALTERN(ative) = A, the *distributive operator* ENUM(eration) = E, the *collective operator* COORD(ination) = C and the *attributive operator* SPECIF(ication) = S. This last is used *to introduce additional details about the first argument of the SPECIF list*. For example, in gio2.c7, Lillian Feldman Schwartz's paper is qualified as *published*, and a second SPECIF list is used to specify *where*. The interweaving of the AECS operators is controlled by a so-called *priority rule* [1: 69], mnemonically formulated as (ALTERN (ENUM (COORD (SPECIF)))), which forbids, e.g., the use of COORD lists within the scope of lists SPECIF, while the inverse is wholly admissible.

4   Additional Issues and Remarks About NKRL

Querying/reasoning are particularly important topics in an NKRL context. They range from the *direct questioning* of a knowledge base of NKRL formal structures making use of *search patterns* – basically, partial instantiation of standard templates – to the execution of *high-level inference procedures*. These include, e.g., the *transformation rules* that try to *adapt*, from a semantic point of view, a search pattern that *failed* (that was unable to find a unification within the knowledge base) to the real contents of this base making use of a sort of *analogical reasoning*. Another important category of high-level procedures is represented by the *hypothesis rules*. These allow us to build up a *causal explanation* for an elementary event (a predicative occurrence) retrieved by direct query within a NKRL KB. By following a sequence of partially pre-defined *reasoning steps*, the hypothesis rules can generate search patterns able to retrieve information supporting the explanation hypothesis. The readers interested in the querying/inferencing topics in NKRL can refer, e.g., to [1: 183–243].

To move on to a quick analysis of the state of the art in the context of the *formalization/digitalization of narratives*, we can note that the first really advanced proposals for formally representing *narratives* correspond to the solutions, the *correlation theory*, suggested in the fifties-sixties by Silvio Ceccato [12] in a Mechanical Translation context. Correlations were syntactic/semantic *n-ary (triadic) structures* arranged around a *central correlator* having a function similar to that of a functional role in NKRL; they could be embedded to represent situations like "Mary sings because she loves Peter". Narratives were then represented as *recursive networks of triadic structures*. NKRL has been strongly inspired by the expressive *n*-ary approach proposed by Silvio Ceccato.

From a *concrete, operational point of view*, NKRL must also acknowledge its debt to all those systems derived from the *adaption*, in the last sixties, of the Charles Fillmore's Case Grammar theory [13] to an Artificial Intelligence/Knowledge Representation context, see [9] in this context. A well-known representation model of this type is the Conceptual Dependency theory of Roger Schank [14]. In this, the underlying meaning (*conceptualization*) of a given narrative was expressed as the association of semantic predicates chosen from a set of twelve formal *primitive actions* (like INGEST, MOVE, ATRANS etc.) with seven *role relationships* in the Case Grammar style. A system in this style is also represented by John Sowa's Conceptual Graphs, CGs, theory [15]. A conceptual graph is a finite, connected, bipartite graph that makes use of two kinds of nodes, *concepts* and *conceptual relations* – these last corresponding largely to NKRL's *functional roles*. This is not the sole similarity existing between CGs and NKRL: contexts in CGs are dealt with, e.g., making use of *second order (nested graphs) extensions* that bear some resemblance to NKRL's constructs like completive construction and binding occurrences. Important differences also exist, the most important concerning John Sowa's choice of leaving *completely free*, for generality's sake, the selection of those *predicates* that, in CGs as in NKRL, *represent the focal element of the formal representation of an elementary event* – they correspond, on the contrary, to *primitives* in NKRL and are rigorously limited to the seven ones introduced in Sect. 2 above. Other general and well-known knowledge representation systems that share, at least partly, the Fillmore-derived *strict symbolic approach*, are CYC [16] and Topic Maps [17].

With respect now to the Semantic Web (SW) tools and languages, we can note that they follow a *formal/logical-oriented paradigm*, see [18], which is *orthogonal, in a sense,* to the paradigm proper to NKRL and to the previously mentioned systems, more focused on *linguistic and psychological interests.* Moreover, *from an operational point of view*, the strict binary approach adopted in a SW context, is *hardly compatible* with the *n*-ary approaches adopted by NKRL and the related systems. Even if, in a specific NKRL context, a (partial) *meeting point* with the SW technologies could be represented *by the strict binary approach adopted to build up the HClass ontology.*

5   Conclusion

In this paper, we have described the conceptual tools that, in an NKRL context, allow us to obtain a (computer-usable) description of full *narratives* as logically-structured associations of the constituting (and duly formalized) *elementary events*. These conceptual tools are then able to supply a *formal description* of those *connectivity phenomena* – denoted, at surface linguistic level, by logico-semantic coherence links like causality, goal, co-ordination, subordination, indirect speech etc. – which assure *the conceptual unity of the whole narrative.* The unification-based solutions adopted by NKRL in this context, like *completive construction* and *binding occurrences*, take into account the connectivity phenomena *by reifying the formal representations used to model the constitutive elementary events.* Some examples are provided in this context.

References

1. Zarri, G.P.: Representation and Management of Narrative Information. Theoretical Principles and Implementation. Springer, London (2009)
2. Bal, M.: Narratology: Introduction to the Theory of Narrative, 2nd edn. University Press, Toronto (1997)
3. Jahn, M.: Narratology: A Guide to the Theory of Narrative, version 2.0. English Department of the University of Cologne, Cologne (2017)
4. Barthes, R.: Introduction à l'analyse structurale des récits. Communications – Recherches sémiologiques, l'analyse structurale du récit 8, 1–27 (1966). English version: Barthes, R., and Duisit, L.: An Introduction to the Structural Analysis of Narrative. New Literary History, vol. 6 (2), 237–272 (1975)
5. Zarri, G.P.: Use of a knowledge patterns-based tool for dealing with the "narrative meaning" of complex iconographic cultural heritage items. In: Proceedings of the 1st International Workshop on Visual Pattern Extraction and Recognition for Cultural Heritage Understanding (VIPERC 2019), CEUR Publications, vol. 2320, pp. 25–38. CEUR-WS.org, Aachen (2019)
6. Amelio, A., Zarri, G.P.: Conceptual encoding and advanced management of leonardo da vinci's mona lisa: preliminary results. Inform. MDPI **10**(10), 321 (2019)
7. Amelio, A., Zarri, G.P.: A knowledge representation framework for managing leonardo da vinci's mona lisa: case study of the hidden painting. In: Proceedings of the 1st International Virtual Conference on Visual Pattern Extraction and Recognition for Cultural Heritage Understanding (VIPERC 2022), CEUR Publications, vol. 3266. CEUR-WS.org, Aachen (2022)

8. Noy, N.F., Fergerson, R.W., Musen, M.A.: The knowledge model of protégé-2000: combining interoperability and flexibility. In: Knowledge Acquisition, Modeling, and Management, Proceedings of EKAW 2000, LNCS vol. 1937, pp. 17–32. Springer, Berlin (2000). https://doi.org/10.1007/3-540-39967-4_2

9. Zarri, G.P.: Functional and semantic roles in a high-level knowledge representation language. Artif. Intell. Rev. **51**(4), 537–575 (2019)

10. The EAGLES Lexicon Interest Group: EAGLES LE3–4244, Preliminary Recommendations on Lexical Semantic Encoding – Final Report. ILC-CNR, Pisa (1999).   http://www.ilc.cnr.it/EAGLES96/EAGLESLE.PDF (Accessed 19 November 2022)

11. Amelio, A.: Exploring Leonardo Da Vinci's mona lisa by visual computing: a review. In: Proceedings of the 1st International Workshop on Visual Pattern Extraction and Recognition for Cultural Heritage Understanding (VIPERC 2019), CEUR Publications, vol. 2320, pp. 74–85. CEUR-WS.org, Aachen (2019)

12. Ceccato, S.: Automatic translation of languages. Inform. Storage Retrieval **2**(3), 105–158 (1964)

13. Fillmore, C.J.: The Case for Case. In: Bach, E., Harms, R.T. (eds.) Universals in Linguistic Theory, pp. 1–88. Holt, Rinehart and Winston, New York (1968)

14. Schank, R.C.: Identification of conceptualizations underlying natural language. In: Schank, R.C., Colby, K.M. (eds.) Computer Models of Thought and Language, pp. 187–247. W.H. Freeman and Co., San Francisco (1973)

15. Sowa, J.F.: Knowledge Representation: Logical, Philosophical, and Computational Foundations. Brooks Cole Publishing Co., Pacific Grove, CA (1999)

16. Lenat, D.B., Guha, R.V., Pittman, K., Pratt, D., Shepherd, M.: CYC: Toward programs with common sense. Commun. ACM **33**(8), 30–49 (1990)

17. Pepper, S.: Topic Maps. In: Bates, M.J. (ed.) Encyclopaedia of Library and Information Sciences, 3rd edn., pp. 5247–5260. Taylor & Francis, Abingdon (2010)

18. Baader, F., Calvanese, D., McGuinness, L., Nardi, D., Patel-Schneider, P.F.: The Description Logic Handbook: Theory, Implementation. Applications. Cambridge University Press, Cambridge (2003)

A Dashboard to Enable New Opportunities for Rural Development by Overcoming the Dominant Segmentation of European Pilgrimage Routes

Martín López-Nores[✉], Sergio Arcay-Mallo, Roi Martínez-Portela, José Juan Pazos-Arias, Alberto Gil-Solla, and Raúl Estévez-Gómez

AtlanTTic Research Center for Information and Communication Technologies, Department of Telematics Engineering, University of Vigo, Vigo, Spain
mlnores@det.uvigo.es

Abstract. The numerous pilgrimage routes that cover the European continent entail an untapped potential to offer slow travel experiences with which to promote the development of rural areas that many people travel through, but very few take time to explore. The rurAllure Horizon 2020 project is investigating ways in which the points of cultural and historical interest in the vicinity of the routes could be connected and put on the map for pilgrims, tourists and locals. Most often, this requires investing in new hospitality services, as well as in the restoration and promotion of specific assets. This paper explains how the IT platform created in rurAllure can help in the decision-making that precedes the investments, by revealing the key locations and needs that, properly connected and catered for, would unlock leisure opportunities that cannot be realised otherwise.

Keywords: Pilgrimage routes · tourism · cultural heritage · rural development · trip planning · accessibility · reachability · service availability

1 Introduction

Every year, thousands of people set out to travel over several days or weeks along European pilgrimage routes like Camino de Santiago, Via Francigena or the St. Olav Ways. Those who undertake a religious or spiritual pilgrimage find themselves along with increasing numbers of people who travel for leisure or sport, as a retreat from modern life, or for cultural enlightenment.

Some pilgrimage routes have become notable economic and political assets [7], while others are experiencing significant growth thanks to the support provided by regional, national and European institutions. Notwithstanding, it has been observed [1] that the impact of the pilgrimage routes and related modes of *slow tourism* is mostly perceived in locations found directly on the official

© The Author(s), under exclusive license to Springer Nature Switzerland AG 2024
A. Rocha et al. (Eds.): WorldCIST 2023, LNNS 802, pp. 228–236, 2024.
https://doi.org/10.1007/978-3-031-45651-0_23

paths, rarely permeating into the surrounding rural areas. For example, the scientific literature indicates that the impact of traditional pilgrimage travels on the Way of St. James is noticeable within an area of 200–300 metres from the official paths, almost as if defining a tunnel through which the pilgrims move [4]. This way, entire regions of a predominantly rural nature become passive witnesses of the flows of pilgrims and tourist, in spite of having remarkable interest from the perspectives of historical, natural and/or ethnographic heritage.

The Horizon 2020 project rurAllure (www.rurallure.eu) was motivated by a shared belief that (i) the rural areas surrounding the pilgrimage routes of Europe can add much content and value to the experiences of pilgrims and tourists, and (ii) it is possible to generate impact well beyond the abovementioned geographical outreach. Since the project started in January 2021, we have been developing an IT platform with specialised content management facilities, that will help pilgrims and tourists plan personalised trips by considering points of interest in rural areas. We see this as a cornerstone to put thousands of heritage sites, museums and small businesses on the map, so that the pilgrimage routes act as backbones for territorial development.

This paper documents the experience from the work conducted in 2021 and 2022 on selected segments of the pilgrimage routes involved in the four rurAllure pilots:

– Camino de Santiago in Spain and Portugal.
– Via Francigena, Via Romea Strata and Via Romea Germanica from different Northern countries towards Rome, in Italy.
– The St. Olav Ways in Norway.
– The Way of Mary in Slovakia, Hungary and Romania.

It has been repeatedly noticed that, at least, certain profiles of pilgrims and tourists could be interested in taking detours from the official paths, lured by the heritage of nearby sites and by the activities offered by local providers. However, we have found that the detours would not generally be feasible unless new hospitality services were created. Therefore, we explain how the project's IT platform can help in the decision-making that precedes the investments, by revealing the key locations and needs that, properly connected and catered for, would unlock leisure opportunities that cannot be realised otherwise.

The paper is organised as follows. Next, Sect. 2 presents an overview of the scientific literature about the impact of forms of slow tourism that develop along territorial lines. In Sect. 3 we describe the main features of the rurAllure IT platform as a whole, and the trip planning service in particular. Then, Sect. 4 summarises the observations gathered so far that made it necessary to create a tool to inform regional decision-makers of the potential opportunities. Finally, Sect. 5 presents our conclusions and future work.

2   Background

Many authors have analysed the economic effects generated by forms of slow tourism that develops along territorial lines, highlighting the economic potential of this sector for the direct, indirect and induced impacts [2,3,10].

In this line, it has been shown that slow tourism activated along a line can encourage a more equitable and balanced dissemination of the benefits [6, 11] and that a virtuous supply chain can be established, with which to involve not only the territories directly crossed, but also the surrounding ones, due to the fact that most of the slow travellers' consumption is made up of goods and services that generate more jobs (direct and indirect).

One of the clearest examples of a supply chain is certainly that linked to agricultural production and the transformation of food. Moscarelli [7] calculated the so-called "employment multiplier", i.e. the number of jobs generated directly, indirectly and induced, with an increase of one million euros in the visitor's final demand. The result is that each euro spent by a slow traveller generates up to 18% more jobs than the expenditure of another type of tourist. For example, the costs for food represent 61% of the expense of a slow traveller, while for a mainstream tourist they represent only 26%. Moreover, the slow traveller shows a preference for local produce and this generates work locally. The mainstream tourist then spends a large part on transport (about 23% of the total cost of the trip), a sector that has a low multiplier effect on the local scale. This expense, on the contrary, is almost nil in the case of the pilgrim, excluding the cost to get to the pilgrimage route and start the journey on foot or by bike.

Thus, even a quantitatively modest number of slow travellers —in comparison with mainstream tourists— can have a qualitatively high impact in territorial balance and rural development. These observations lead to advocating slow tourism projects as priority axes on which to concentrate public investments.

A very relevant point for rurAllure to appeal to slow tourists relates to their perceptions about information technology. More than a decade ago, Zago [13] observed that in slow tourism the information technologies played a different role than in mainstream tourism. Then, both offer and demand of slow tourism could show attitudes not much favourable to being permanently connected. However, in the 2020s the situation may have changed radically, as IT has now evolved in a manner that not only do not disturb slow travellers, but improve their experience instead [12]. Such is the underlying approach of the rurAllure IT platform and, specifically, the use of geographically-annotated multimedia narratives to help gain better understanding of the territory, its history and its heritage.

On the supply side, it is interesting to note that many of the services created or kept alive by the presence of pilgrims also represent services for residents, which is particularly important for the smaller and less densely populated areas that face the risks of demographic decline and gradual disappearance of services. A recent study on socio-economic impact [5], based on fieldwork and statistical data analysis on a local scale, found that the Galician municipalities crossed by the French Way —though affected by the global decline— had systematically better demographic and economic trends than other municipalities with comparable variables, which is once again a motivator for public funding.

The perception of technology reported in [8] is also relevant to the supply side of slow tourists. Years ago, vendors from rural areas saw IT as unwanted, inaccessible or unaffordable, it is now increasingly perceived by those concerned with direct impacts as a necessary tool to gain visibility among the target public. Problems are still reported, though, in relation with lack of skills and fragmentation of the IT service market, as most of the small businesses cannot afford being present on many competing platforms.

3   The RurAllure IT Platform and the Trip Planning Service

As shown in Fig. 1, the rurAllure IT platform comprises various mobile/web frontends and backend services, originally aimed at three types of users:

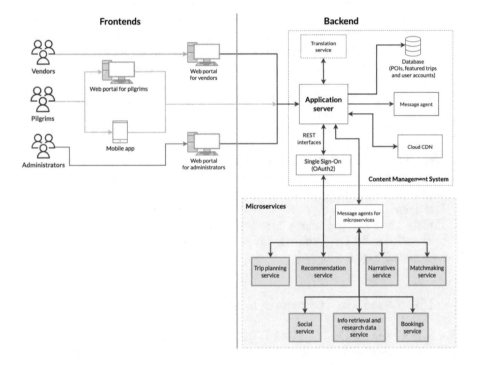

Fig. 1. The overall architecture of the rurAllure IT platform.

- **Pilgrims/tourists:** users who are on a trip along a pilgrimage route or planning to undertake such a trip.
- **Vendors:** users interested in offering activity/hospitality points of interest (POIs) or certain services to pilgrims/tourists.
- **Administrators:** users with the credentials to supervise the information offered for a given pilgrimage route. This includes creating POIs, approving vendors and their offerings, creating featured pilgrimage plans, uploading bundles of multimedia contents (narratives) for the travellers to discover the heritage of the places they visit, etc.

The backend services include a content management system, a trip planner, a recommender system to help discover potentially interesting POIs and services, services to retrieve information from external sources and to serve open data for research purposes, and tools to handle focused communications among travellers and vendors.

In particular, the Trip Planning service was designed to provide personalized experiences to pilgrims/tourists, providing them with a tailor-made itinerary from some basic parameters (starting point, destination, travelling mode, thematic preferences, . . .) and letting them make changes at will. The following procedure is executed for each request:

1. Firstly, the best route to reach the desired pilgrimage route from the starting point indicated by the user, and to get off the route towards the destination (in case those points are not directly on the official paths) is located, taking the official paths of the pilgrimage routes as a reference backbone.
2. A list of reachable POIs within successive ranges of distance from the official paths is computed.
3. The reachable POIs are filtered and sorted for the requesting user by the Recommendation service.
4. The Trip Planning service explores possibilities to take detours from the official paths through the reachable POIs, taking into consideration aspects of accessibility, reachability and service availability. For instance, it takes care not to surpass the desired average number of hours for the successive journeys and to include stops to rest, have meals and stay for the night at suitable locations.
5. Finally, the resulting choices are displayed to the user, presenting the official paths with solid lines and the proposed detours with dashed lines as in the example of Fig. 2.

4   The Need for a Fourth Frontend

The beta version of the rurAllure IT platform was made available in December 2021, and thereafter tested in the context of the four project pilots. Soon it was found that, up to step 3 of the procedure explained in Sect. 3, there were usually many relevant possibilities to visit POIs in the vicinity of the official paths. However, the filtering implemented in step 4 led to a drastic reduction, due to the specific constraints raised by travelling on foot or by bicycle, to the relief of the territory and/or to the social-demographic features of the rural territories. The following are common reasons why some alternatives had to be discarded, notwithstanding the potential high interest of certain POIs:

– Steep slopes or poorly-maintained paths, leading to negative accessibility checks.
– Sparse populations with insufficient hospitality services (if any at all!).
– POIs that could not be reached on suitable times, given the starting and ending locations for the daily journey and a suitable place to stop for lunch.

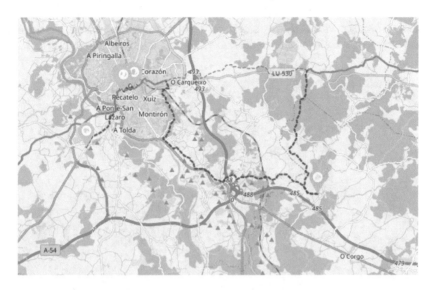

Fig. 2. A proposed trip with a detour from the official paths of Camino de Santiago.

- Lack of infrastructure, transport and/or emergency services over lengthy detours.
- Dubious or unchecked data on the OpenStreetMap sources.

As a result, the Trip Planning service would only come up with proposed itineraries along the official paths or including POIs in their immediate vicinity, rarely venturing further than 1.2 kilometres. This may serve to put hundreds of POIs on the map for pilgrims/tourists, that would go unnoticed otherwise, but falls short to achieve the desired widespread impact and to furnish greater opportunities for territorial development. Especially because of the current availability of hospitality services, it is generally not possible to overcome the predominant segmentation of the pilgrimage routes: once a person has chosen a particular route, the starting and ending locations of the successive daily journeys is largely predetermined —the same stages and the same locations for practically everyone! At the most, the traveller may decide to stay for more than one night in any of those locations and explore the surroundings, but the impact remains concentrated.

Faced with these observations, we have started the design of a fourth frontend for the rurAllure IT platform: a dashboard for regional decision-makers. This will provide visualisations of the mesh of alternative itineraries that could be laid through a given territory, revealing the key locations and needs that, properly connected and catered for, would lead to satisfying the accessibility, reachability and service availability checks that are currently unfulfilled. It will also display estimations of the potential interest of the different POIs scattered over the region for the different pilgrim/tourist profiles, provided by the Recommendation service. A sample visualisation based on isochrones (areas accessible from the official paths of a route within a certain time threshold) is shown in Fig. 3.

Fig. 3. Visualisations of POIs reachable within different ranges of distance from the official paths of Camino de Santiago.

With the aid of this dashboard, we want to support such decisions as the following ones:

- To create new inns at certain locations.
- To provide training and resources to the local people so that they can host visitors at their homes.
- To invest in the enhancement of certain POIs with dining/accommodation facilities or other services.
- To focus promotion efforts on specific locations on the official paths.
- To recover and restore specific assets, relevant from the perspective of cultural or natural heritage.
- To facilitate round-trip transport to distant POIs (with previous reservation).

We see this as a crucial tool to systematically attain positive effects observed as a result of specific actions implemented in the rurAllure pilots during 2022.

5   Conclusions and Future Work

The work conducted in the rurAllure project in 2021 and 2022 has shown that it is difficult to come up with feasible itineraries to lure pilgrims and slow tourists into the rural territories that surround the numerous pilgrimage routes of Europe. The pilgrimage phenomenon is growing steadily, but the experiences remain homogeneous and repetitive. The predominant segmentation of the routes, as documented in many guides and followed by thousands by the book, becomes engraved in the territory due to the concentration of services along a line —actually, on specific locations thereon, which act as isolated poles of direct economic impact.

The original idea of rurAllure (providing personalised trips with detours into the rural surroundings of the routes, motivated by visits to selected POIs) is feasible within limited ranges. However, we believe that the full potential embedded in the historical, natural and ethnographic heritage can only be achieved by systematically revealing the key locations and needs that could receive public and/or private investments in order to unlock possibilities that are nowadays discarded because of accessibility, reachability and service availability concerns. To the best of our knowledge, the dashboard proposed in this paper will be the first tool to pursue such a goal.

In parallel with the development of the dashboard, which will be the focus of research during 2023, the rurAllure team is working to improve the Trip Planning service so that it considers as a reference backbone not only the official paths of the pilgrimage routes, but also a set of featured trips curated manually via the administrators' frontend. A number of such trips have been created in the project pilots during 2022, embedding substantial knowledge of the traversed territories and going through greater numbers of POIs than the planner would consider. The new version of the service that takes into account the featured trips will be tested during 2023.

Acknowledgements. This work has been funded by the European Union's Horizon 2020 Research and Innovation programme under grant agreement no 101004887.

References

1. Allgemeiner Deutscher Fahrrad-Club. Travelbike Bicycle Travel Analysis - Summary report. Berlin, Germany. Retrieved from https://www.adfc.de/%20leadmin/%20user_upload/Expertenbereich/Touristik_und_%20Hotellerie/Radreiseanalyse/Downloads/Datenblatt_%20Radreiseanalyse2019_en_-_neu.pdf (2019)
2. Downward, P., Lumsdon, L., Weston, R.: Visitor expenditure: the case of cycle recreation and tourism. J. Sport Tourism **14**(1), 25–42 (2009)
3. Dunkelbergt, D., Püschel, R., (Eds.) (2009) Grundlagenuntersuchung Fahrradtourismus in Deutschland. Berlin, Germany: Bundesministerium für Wirtscha und Technologie. Retrieved from https://digital.zlb.de/viewer/resolver?urn=urn:nbn:de:kobv:109-opus-259062
4. Fernández-Fernández, M., Lazovski, O., Real Neri, G.: Tourism impacts in an emerging destiny through the local entrepreneurship perception: The Fisterras case. J. Tourism Heritage Res. **3**(2), 269–285 (2020)
5. Fernández, M., Riveiro, D.: A exclusión territorial como unha forma de manifestación dos procesos de exclusión social. SÉMATA, Ciencias Sociais e Humanidades **30**, 145–165 (2018)
6. Lois González, R. C., López, L.: The singularity of the Camino de Santiago as a contemporary tourism case. In P. Pileri & R. Moscarelli (Eds.), Cycling & walking for regional development. How slowness regenerates marginal areas (pp. 221–234). Cham, Germany:Springer (2021)
7. Moscarelli, R.: Slow tourism, public funding and economic development. A critical review on the case of the Way of St. James in Galicia. Revista Galega de Economía, **30**(3), 7522 (2021)
8. Med Pearls project. Research study on Slow Tourism international trends and innovations. http://www.slow-tourism.net (2021)
9. Moira, P., Mylonopoulos, D., Kondoudaki, A.: The application of slow movement to tourism: is slow tourism a new paradigm? J. Tourism Leisure Stud. **2**(2) (2017)
10. Piket, P., Eijgelaar, E., Peeters, P.: European cycle tourism: a tool for sustainable regional rural development. Appl. Stud. Agribus. Comm. **7**(2–3), 115–119 (2013)
11. Pileri, P., Giacomel, A., Giudici, D.: (2015) VENTO. La rivoluzione leggera a colpi di pedali e paesaggio. Mantova, Italia: Corraini (2015)
12. Pileri, P., Moscarelli, R: Cycling and walking for regional development. How slowness regenerates marginal areas. Cham, Germany: Springer (2021)
13. Zago M., (ed.) Guidelines for the Slow Tourism, Research report of the cross-border cooperation Italy-Slovenia 2007-13 (2011)

Information Systems and Technologies for the Steel Sector

New Measurement Techniques Describing the Sinter Process

Emanuel Kashi Thienpont[1]([✉]), Thorsten Hauck[1], Tolga Erolglu[1], Tobias Kleinert[2], and Kerstin Walter[3]

[1] VDEH Betriebsforschungsinstitut gmbH, Sohnstr. 69, 40237 Düsseldorf, Germany
lncs@springer.com
[2] RWTH Aachen University, Turmstr. 46, 52062 Aachen, Germany
[3] DK Recycling und Roheisen GmbH, Werthauser Str. 182, 47053 Duisburg, Germany

Abstract. Sinter with high and consistent quality, produced with low costs and emissions is very important for iron production. Transport and storage degrade sinter quality, generating fines and segregation effects. Conventional sinter quality monitoring is insufficient as it is slow and expensive. Consequently, sinter quality and the impact of different sinter quality on daily BF operation is extremely non-transparent. In this work, a new approach will be implemented to strengthen the data base that describes the sinter quality at a sinter plant. New on-line measurements will be established, combined, and analyzed with Big Data technologies. This break-through in continuous quality monitoring should help to get continuous quality indices for sinter and will enable combined optimization of sinter plant and BF.

Keywords: Sinter Plant · Hot Crusher · Acoustic Measurement Technique · Sinter Quality Indices · Data Mining

1 Introduction

Steel making using the integrated blast furnace (BF) process route is still the dominant steel production method covering 70% of worldwide steel production and iron ore sintering is one of the most important iron sources within this process, which represent 70–80% of charged iron-bearing materials into modern BF [1]. Inside this route, sinter plants offer unique capability to recycle residues and fines from iron ores for the steel production.

Along the iron and steel industry, iron ore deposits are becoming increasingly scarce and are therefore exposed to fluctuations in quality and chemical composition due to the varying raw material structure of the different mining areas as well as from other used raw materials. Due to the intensive material usage of sinter plants, this has a strong impact on product quality and energy consumption. The current standard methods for monitoring the sinter performance of iron ore are the tumbler test and screening analysis. Both are very laboratory and time intensive and are unable to detect rapid changes in raw material or process changes [2]. At the same time, the political and economic pressure to decrease energy consumption and emissions is gaining momentum. Sinter plants contribute the largest share of pollutant emissions in the ironmaking plants and are the second largest emitter after municipal solid waste incineration [3].

Hence, the steel research must be intensified and propose innovations which improve sinter plants. Despite intensive activities in the areas of the steel research in the past, there has been a lack of research dealing with the sinter process. Work focusing on the permeability of the sinter cake are available in [4, 5] and prediction of the burn-through point (BTP) in [6] but cannot replace the manual quality monitoring.

Maintenance and sinter quality management have not yet evolved to an industry 4.0 standard but are still under manual control. The current industrial practice to monitor sinter quality is still primarily based on manual sampling. The sampling method is slow and costly due to its manual laboratory analysis, particularly in view of cold strength and particle size distribution (PSD). The sinter plant is generating a large variety of operational data during the sinter process which is not yet used to evaluate the sinter quality. This is an untapped opportunity which can enhance the frequency of quality controls and long-term also reduce costs. The major focus of this work is the optimization of continuous monitoring of sinter quality indices by developing new measurement techniques leveraging big data technologies and machine learning (ML) approaches to predict the Sinter quality indices. The present paper will introduce three new measurement techniques, an optical-, acoustical measurement as well as the integration and further research of the power consumption. On the one hand, the new high-frequency measurement results should help to generate additional information regarding the sinter process and allow quantification of short time events during the sinter production. On the other hand, expand the data basis for machine learning (ML) methods that can be used to predict sinter quality indices. The enhanced information density of sinter quality information should also allow better knowledge about the charged sinter before burdening at the BF.

1.1 Sinter Plant

Sinter plants agglomerate fine ferrous compounds, mostly iron ore fines, but also internal waste such as flue dust, mill scale and/or dusts from electrostatic precipitators at high temperature to form a product that can be (re-)used in a blast furnace. The product obtained, called "sinter", are small, porous, and irregularly shaped lumps of iron ore mixed with small amounts of other minerals.

The process of sintering to improve the physical and chemical properties of iron ore for use in blast furnaces is well documented in the literature and will be not the scope of this study. But to understand the process, a short overview will be given.

Figure 1 gives a schematic overview about a common sinter plant and the sinter process flow. The raw materials stored in bins such as: Iron ore fines, other iron-bearing wastes, flux (limestone/lime), and coke breeze (fuel) are homogenized first in a drum mixer. Water is added to the raw mix to achieve the desired moisture level where materials form a cohesive mass with enhanced formation of micro-pellets and to improve in such a way the sinter bed permeability [7, 10, 4]. The resulting mix is continuously charged together with hearth layer material onto the moving sinter grates. The coke breeze in the sinter mixture is ignited at the surface of the bed by passing the ignition hood. Air

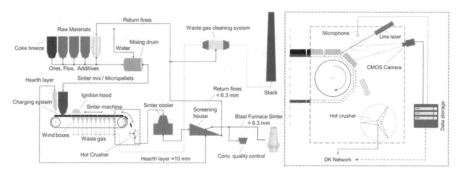

Fig. 1. Schematic view of Sinter plant [7–9] (left,) Schematic illustration of new measurements at the sinter discharge (right)

is sucked continuously through the sinter cake by suction boxes and a burning zone is formed, which is linearly distributed over the height and along the sinter belt in the sinter cake. After ignition the sinter process begins, initiated by the heat generated in the burning zone. The vertical sinter speed is controlled by strand speed and gas flow rate to ensure that the sinter process is finished prior to the end of the strand where the sinter cake is being discharged [10]. During the sinter process, the raw materials fuse into a single porous mass. After discharge from the sinter strand the hot sinter cake is crushed by the hot crusher and cooled in the following sinter cooler. The cold sinter is screened, and the undersized fraction (return fines < 6.3 mm) is delivered back to the raw material feed system. The oversized fraction is transported to the BF. The main quality requirements for sintered products are high resistance to mechanical stress under blast furnace conditions, good reducibility, and uniformity of particle size distribution (PSD), and high iron content [7, 11–14].

2 New Measurements Setup at the Sinter Discharge (Optical-, Acoustical Measurement, Power Consumption)

The optical assessment of the fracture behavior of the sinter cake at the discharge position of the sinter strand is realized by a combination of CMOS camera and a line laser setup at the sinter plant of DK Recycling und Roheisen GmbH (DK). The CMOS camera unit "DALSA GENIE NANO-5G-C4040" has a resolution of 4112 × 3008 Pixel. To reduce fading the camera is equipped with a UV bandpass filter with a transmission wavelength of 550 nm and bandwidth of 300 nm and anti-reflection coating. The class 1M line laser, "LH520–10-24(20 × 80)-PL-C5000-WP" has a wavelength of 520 nm. The line laser is aligned parallel to the sinter cake surface and pointed to reach the front surface of the sinter cake (breaking edge). The CMOS camera is aligned in such a way that it can capture the fracture edge of the sinter cake over its entire width as shown in Fig. 1. (right). The CMOS / laser measuring device should be able to detect the fracture edge of the sinter cake at the discharge before the sinter cake falls into the hot crusher. The shape of the laser should allow to reflect the geometry of the breaking edge of the Sinter to draw conclusions about the fracture behavior. The CMOS camera records images with a framerate of 2 s.

The acoustical measuring is realized by a piezoelectric pressure sensor (microphone) "KistlerZ20167-01e" with a pressure range of 0–10 bar and a sensitivity of 0–1 bar, resonance frequency of 35 kHz and a response threshold of 5 μbar. The pressure recording is carried out with a resolution of 100 kHz (12MB/min) and is equipped with a 100 Hz high-pass filter. The sensor is installed inside the sinter box roof directly above the hot crusher, Fig. 1 (right). At this position, the sensor can detect the crushing noise of the sinter inside the hot crusher. The raw signal from the acoustic sensor as well as the raw images from the optical measurement are stored on an industrial computer (IPC). The IPC is connected to the DK network infrastructure and accessible via a virtual machine for further data processing.

The measurement of the power consumption of the hot crusher is not a new measurement technique, it is recorded by default from the data acquisition of the sinter plant. Here it will be considered to get additional information which could help to describe the sinter quality. The assumption is that hard sinter with high sinter strength should lead to higher power peaks than loose sinter with low sinter strength.

The data transfer concept is shown in Fig. 2. The operating data from the database of the DK sinter plant and signals from the new measuring methods are first pre-collected in a sub database. From there the collected data is transferred via MQTT protocol to a database in the BFI for detail analysis.

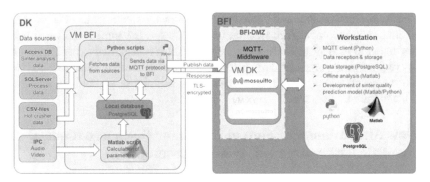

Fig. 2. Data flow chart DK-BFI

2.1 Optical Measurement

The image processing takes place in three main parts, the image preparation, filtration, and calculation of key performance indicators (KPIs). The first step of the developed MATLAB® filter is to crop the relevant area from the raw image (see Fig. 3 blue rectangle). In the next sub step the cropped image is converted to HSV color values which gives direct information about the Hue, Saturation and (color) Value. After transformation the green pixels are filtered by the characteristic green value range. Groups of individual green pixels are merged by morphological closure, i.e., individual black pixels surrounded by green pixels are converted into green pixels. The output at this stage is a binary image that can be represented as a simple binary matrix.

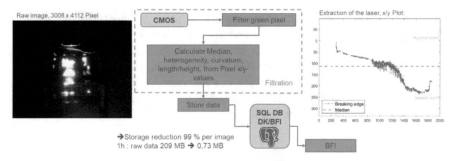

Fig. 3. Basic flowchart of the image filter (Color figure online)

In the next step the coordinates of all white pixels (former green pixel) are located and geometrical characteristic KPIs are calculated like median from x/y direction, heterogeneity, curvature, and the width from the outermost pixel as well as the height from the highest and lowest pixel. Additional all coherent pixel cluster with a size of more than 50 pixels were counted and measured after width and height. All these key values are transmitted to a PostgreSQL database. The last step is the deletion of the raw image and compressing the cropped image for backup reasons. With this image filtrations procedure, the raw image storage size can be reduced by 99% (from 428 kb to 4.28 kb) per image. The image processing algorithm is repeated on each recorded image (2 s, frame rate of CMOS camera). The steps are summarized in Fig. 3. The filtered pixels are depicted as blue curve which was extracted from the raw image Fig. 3 (left side). The breaking edge of the sinter cake is not recognizable without the support of the line laser. Due to the line laser the breaking edge becomes visible but is still not sufficiently evaluable. After applying the filter algorithm mentioned before, the green laser line can be isolated as shown Fig. 4. And the geometric KPIs can be calculated. Based on these values the image data of the breaking edge will be compressed to quantitative information and can support the estimation of sinter quality indices.

2.2   Acoustical Measurement

The processing of the acoustic signal results in two main steps. The raw signal is first compressed almost lossless and then filtered in the second step. The compression is performed by a frequency decomposition of the raw signal in specific frequency domains. The subsequent filtration is used to remove the background noise and to emphasize the sinter crusher noise. The overall processing route is depicted in Fig. 4.

The raw signal of the acoustic measurement is recorded in 5-min segments, each with a sampling rate of 100 kHz. In Fig. 4 (left hand) the raw signal is depicted after Fast Fourier Transformation (FFT) for a random single 5-min block. The crushing noise peaks from the sinter in the hot crusher can be observed after 0.5, 1.8, 3 and 4.2 min in the image but also smaller following peaks after the main peak at 3–3.5 and 4.2–4.5 min. These vertical peaks reach mainly from 0.10–25 kHz but are also detectable in higher frequencies over the hole frequency bandwidth up to 50 kHz. The horizontal peaks that occur throughout the sample at constant frequencies result from sources such as fans, strand drives, and other constant noise sources from the sinter plant. To extract only the vertical peaks of the crushing noise a two-step filter was developed as shown in Fig. 4. The two-step filter consists of a 19-frequency band decomposition and thereby leads to a compression of the raw signal from 99,95%. Thus, the signal can be transferred faster and stored to a PostgreSQL database. The tow step filtration has the advantage, that the raw signal is only compressed nearly lossless, and no frequency information is lost. The compressed signal can be easily stored and can remain as backup.

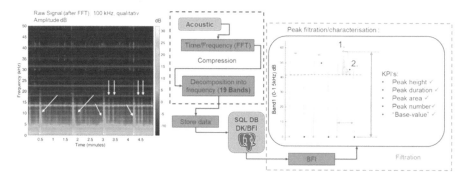

Fig. 4. Summary of the processing of acoustic data

The two-step filter will be introduced in detail in this section. In the first step the raw data is applied to a FFT to get time, frequency, and power information of the signal. In the next sub step the signals are separated into 50 frequency bins, á 1 kHz. After uniform frequency separation, the bins are combined to 19 specific frequency domains by creation of median values for each domain (consisting of several 1 kHz bins). The boundaries of the 19 specific frequency domains have been chosen to exclude most constant frequency zones as far as possible. The resulting (compressed) signal is stored on a PostgreSQL database. The signal filtering algorithm starts with a sliding maximum filter with a sliding window k applied to each of the 19 frequency bands to detect all maximum values on the noisy signal. In the next sub step, a variable threshold value is defined for each band by a moving minimum value with specific offset. In the next sub step, all values in each band that are below the threshold value are set to zero. As a result, only the peak values remain while the lower noises are removed from the signal. In Fig. 4 right hand, three filtered peaks are shown after passing the filter algorithm showing the 0–1.5 kHz band. In the last step the resulting peaks are characterized by maximum power value (dB), peak duration, area, and numbers of subpeaks. All these KPIs are stored and can be considered to describe the sinter state at the end of the sinter process.

2.3 Power Consumption of the Hot Crusher

The peak characteristic for the power consumption is determined in a similar procedure as shown for the acoustic signal. After locating the local maximum, a sliding cut off value based on the minimum values is defined. To follow fluctuations during the time a moving minimum value was selected, analogue to the acoustic filter. The peak beginnings and ends are determined by the intersection points of the cut-off value and the moving maximum of the raw signal. In the last step all values between start and end values were set to zero. In this way, the duration, location, and maximum height of the peaks can be determined. Interrelated peaks are merged by this approach and can later be assigned to single sinter drops or grates. Subpeaks are also located by ignoring the maximum values of the main peak. Finally, for each sinter drop all mentioned features can be determined and are stored on a PostgreSQL database.

2.4 Validation of New Measurement Methods

At this stage no validation values were present to verify the new gained KPIs and their correlation to sinter quality indices. It is proposed to conduct sampling campaigns at the DK sinter plant that allow adequate resolution of the quality data (screen analysis and tumbler test). The existing daily quality data does not have sufficient resolution to reflect short term effects. For this reason, the values generated in this study can only be compared with other process data, but not directly with quality indices. This will only be possible after the sampling campaigns with increased sampling intervals for short time effects. First trends of the KPIs showed a good sensitivity after short shutdowns of the sinter plant which indicates a good reaction to short term effects as shown in Fig. 5. Displayed for the hot crusher-, and acoustic peak values. The average run time indicate the duration for the passing of one strand length. The values are median values summarizing 10 events (stoppages).

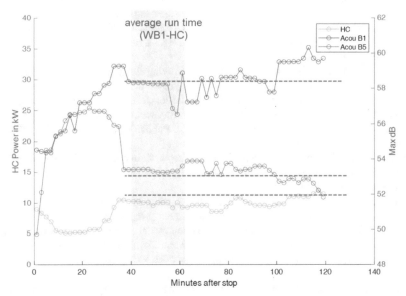

Fig. 5. Trend of hot crusher (HC) and acoustic peaks values (Acou B1 0–1.5 kHz, Acou B5 13–17.35.kHz) after 20 min interruption of sinter production

3 Conclusion and Outline

Three novel measurement methods which could help to characterize sinter quality indices were presented. These combine three different measurement techniques and approaches: An optical-, acoustical measurement and the analysis of the hot crusher power data. These make it possible to obtain continuous KPIs in high resolution sampling rates form 1 Hz up to 100 kHz directly after finished sintering at the end of the sinter machine. These should allow short time conclusion about the sinter process. First comparisons with operating data and trends of the KPIs showed already good agreement and sensitivity to the actual sinter process and events, e.g., after short shutdowns of the sinter plant. In the next step, sampling campaigns will be carried out at the sinter plant from DK to validate the new values with increased sampling intervals regarding the relation to PSD and cold strength information. Later, these could be used as incoming predictor variables for ML methods that allow to predict the sinter quality indicators in advance. Other standalone approaches to detect short time fluctuations in sinter quality or productions indicator are also possible.

Funding. Thanks are given to the financial supports from the "Minimise sinter degradation between sinter plant and blast furnace exploiting embedded real-time analytics" (MinSiDeg) Project (GA 847285), EU-RFCS-2018

References

1. Hesham M. Ahmed, E. A. Mousa, M. Larsson, N. N. Viswanathan,: Recent trends in ironmaking blast furnace technology to mitigate CO2 emissions: top charging materials. In: Cavaliere, P. (ed.) Ironmaking and Steelmaking Processes: Greenhouse Emissions, Control, and Reduction, pp. 101–124. Springer International Publishing, Cham (2016). https://doi.org/10.1007/978-3-319-39529-6_6
2. Fan, X., Li, Y., Chen, X.: Prediction of iron ore sintering characters on the basis of regression analysis and artificial neural network, Bd. 16, pp. 769–776. Elsevier, Energy Procedia (2012)
3. Kan, T., Evans, T., Strezov, V.: Risk assessment and control of emissions from ironmaking. In: Cavaliere, P. (ed.) Ironmaking and Steelmaking Processes: Greenhouse Emissions, Control, and Reduction, pp. 321–339. Springer International Publishing, Cham (2016). https://doi.org/10.1007/978-3-319-39529-6_19
4. Ray, T., et al.: Quality and productivity enhancement of sintering process using correlation-based online permeability by infrared-based flow measurement technique. Ironmaking Steelmaking. **48**(7), 788–795 (2021)
5. Vannocci, M., Colla, V., Pulito, P., Saccone, M., Zagaria, M., Dimastromatteo, V.: Advanced monitoring system of sinter plant. Ironmaking Steelmaking **42**, 424–432 (2015)
6. Du, S., Wu, M., Chen, L., Pedrycz, W.: Prediction model of burn-through point with fuzzy time series for iron ore sintering process. Engi. Appli. Artifi. Intell. **102** (2021)
7. Babich, A., Senk, D., Gudenau, H.W.: Ironmaking, p. 51. Verlag Stahleisen GmbH, Düsseldorf (2016)
8. Cheng, Z., et al.: Kuching, Sinter strength evaluation using process parameters under different conditions in iron ore sintering process. Presented at the 18th Conference Process Integration, Modelling and Optimisation for Energy, pp. 36–70. Malaysia Elsevier (2015)
9. Łechtanska, P., Wielgosinki, G.: The use of ammonium sulfate as an inhibitor of dioxin synthesis. Ecol. Chem. Eng. **21**(1), 59–70 (2014)

10. Wang, S.-H., Li, H.-F.: A hybrid ensemble model based on ELM and improved AdaBoost.RT algorithm for predicting the iron ore sintering characters. Comoutational Intell. Neurosc., 11 (2019)
11. Pankratz, C.: Bestimmung von Kriterien zur Beurteilung der Sinterqualität. Leoben : Lehrstuhl für Eisen- und Stahlmetallurgie an der Montanuniversität Leoben, pp. 1–20 (Oktober 2014)
12. Fernández-González, D., et al.: Iron ore sintering: quality indices. Mineral Process. Extract. Metall. Rev. **38**(4), 254–264 (2017)
13. Mochón, J., et al.: Iron ore sintering Part 2 Quality indices and produtivity. Dyna **81**(183), 168–177 (2014)
14. Cores, A., et al.: The influence of different iron ores mixtures composition on the quality of sinter. ISIJ Inter. **50**(8), 1089–1098 (2010)

A Data Model for the Steel Production

Christoph Nölle[(✉)]

VDEh-Betriebsforschungsinstitut (BFI), Sohnstraße 69, 40237 Düsseldorf, Germany
christoph.noelle@bfi.de

Abstract. A data model for the steel production is presented, based loosely on ETSI's SAREF4INMA standard. It models the production resources, such as steel billets, slabs or coils, production equipment, such as a continuous casting machine or a hot rolling mill, and is capable of representing the tracking uncertainty characteristic of the production of steel long products. A property graph representation of the involved entities is used, where not only each entity is typed, but also the links between entities. The model has been developed as part of a digital twin software platform in the CAPRI project and serves as documentation for the application programming interfaces. A decision support system for through-process quality control is being realised as the main application based on the platform.

Keywords: Data model · Semantics · Digital twins · Steel · Long product tracking

1 Introduction

Large amounts of production data are collected by sensors in the process industry and stored in various automation systems, such as the Level2 or MES. A common approach to exploit this rich source of information about the production processes by means of innovative software applications is the deployment of an industrial internet of things (IIoT) platform, which provides consolidated access to data from the different automation systems. The concept of Digital Twins is often used to organise access to the data by means of application programming interfaces, simulation services and user interfaces.

In order to enable efficient application development and the introduction of cognitive capabilities into the software solutions, it is of utmost importance to convey precise semantic information about the data provided. A data model that defines the meaning of all attributes is a common tool for this purpose. In this paper we present one such model for a steel mill producing steel long products. The model is sufficiently generic, however, that it should be useful in similar application scenarios, as well, with small adaptations. The model is based on the ETSI standard SAREF for industry and manufacturing standard (SAREF4INMA) [1], with adaptations, and uses the NGSI-LD metamodel [2], consisting of the three concepts *Entity*, *Property* and *Relationship*.

A core problem for the introduction of Digital Twins in the production of steel long products is the difficulty to track individual (semi-)products through the complete processing chain and hence to convert the time-based process data to length-based product data. A key requirement for our model is hence that it be capable of expressing tracking information, including tracking uncertainty. On the other hand, detailed modeling of individual processes, as addressed for instance by the Process Specification Language (PSL) [3], is not in the scope of the model.

The data model presented in this paper is of a high-level nature, many details of relevance to specific applications are left out. In the literature, several examples of data modeling strategies for specific applications can be found, some of which could be valuable additions to the present high-level model. This includes, for instance the SCRO ontology for steel cold-rolling processes [4], which has been introduced in the context of condition-based maintenance of rolls. Other more generic models for predictive maintenance applications may be of interest, too, such as the ones proposed in [5] or [6]. Another example is the modeling approach presented in [7], where a multi-scale model for indexing high-resolution measurement data is proposed, optimized for serving data via a tile server, similarly to those used by online maps. It aims to enable very performant visualizations of statistical data aggregations over large sets of coils in a flat steel production scenario. Finally, Bao et al. propose an ontology for the converter steelmaking process [8].

2 Data Model

2.1 High-Level Structure

The basis for our data model is ETSI's SAREF4INMA standard (SAREF for industry and manufacturing) [1], part of the SAREF (Smart Applications REFerence) ontologies. The high-level structure of classes is shown in Fig. 1.

One main class is *ProductionResource*, which represents among others steel products (subclass Item), such as a billet being processed, or batches of material, such as liquid steel in a ladle (subclass MaterialBatch). The other main class is *ProductionEquipment*, which represents both single machines and larger aggregates, such as a hot rolling mill, which consists of multiple individual machines. Most of the classes represented in Fig. 1 are actually subclasses of *Entity* (not explicitly shown to avoid cluttering the figure), the only exceptions being *Transformation*, which is a *Relationship*, as shown explicitly, and *ChemicalComposition*, which models a complex *Property*. Transformations can be applied to links between entities in a property graph, in our example the relationship is given by the *processedBy* attribute in the *ProductionResource* class, whose target is a *ProductionEquipment*. This is used to model the transformation that an item is subjected to in a machine. Multiple transformations are supported, since *processedBy* is a list. If an item ceases to exist after a specific transformation, then the latter will be the last element of that item's *processedBy* list. The newly created resource or resources will be equipped with a list of *processedBy* transformations as well, whose first entry models the specific transformation considered. Furthermore, a link will be created from the source resource to the target resource(s), by means of the attribute *produces*, which indicates that one item or resource is made from the other. The inverse relationship is hence labelled *madeFrom*. An example is shown in Fig. 2, where a batch of liquid steel undergoes solidification in the continuous casting process and is transformed into a set of billets, one of which is shown in the figure.

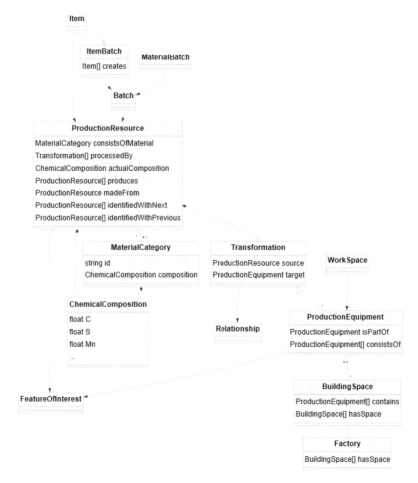

Fig. 1. High-level class structure of the data model.

The steel grade of a production resource is indicated by means of a link *consistsOf-Material* to a material category entity, which also contains information about the target chemical composition. On the other hand, resources may contain information on their actual chemical composition in the attribute *actualComposition*, which can be used to indicate measured compositions. Note that, depending on the requirements of the application, the steel grade model used here may be too simplistic, since it does not allow for the provision of allowed content ranges. The model is thought of as representing the target values for individual elements, if ranges are required they could be added as additional fields or even replace the target values.

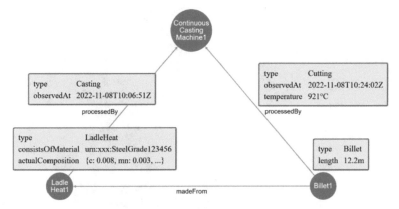

Fig. 2. Model for a billet made from a batch of liquid steel in the casting machine.

2.2 Production Resource Classes and Product Tracking

Typical *ProductionResource* elements are semi-products, such as billets, blooms or slabs, final products, such as bars or coils, and batches of material, such as liquid steel in a ladle (a heat). The corresponding classes are shown in Fig. 3.

A typical problem in the steel production, particularly relevant for long products, is the difficulty of tracking individual products through the process chain. In the use case that motivated this work an innovative tracking system has been developed [9], but a significant uncertainty remains, for instance due to missed items or wrong identifications. It is therefore not feasible to simply model one physical (semi-)product as a single entity and assign the data from different processing steps to it, but it is rather necessary to create one entity per process and create links between the instances to indicate that they represent the same physical item. This way it is possible to model both missed identifications, represented by two entities without connecting links, and duplicates, represented by two or more connecting links from a single entity.

Figure 4 shows an example of a billet tracked successfully between the continuous casting machine and the hot rolling mill, represented by the two *Billet* instances shown at the bottom and referring to the respective plants. The link between the two instances is formed by the *identifiedWithPrevious* relationship (the reverse relationship is not shown).

2.3 Production Equipment Classes and Transformations

The *ProductionEquipment* hierarchy is shown in Fig. 5, consisting mainly of a set of plant types. The exact plant types required and the level of granularity of the equipment model are of course strongly dependent on the use cases considered and on the factory equipment present, therefore the model in Fig. 5 should be seen as an example. A plant that consists of multiple others is modeled in terms of the *consistsOf* relationship and/or its inverse, called *isPartOf*. An example of such relationships is shown in Fig. 4.

Production resources and the equipment are connected by means of *Transformation* relationships. Some examples of such transformations which are important in our steel production use case are shown in Fig. 6. They include temperature transformations

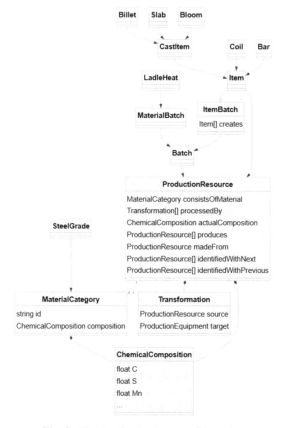

Fig. 3. The *ProductionResource* hierarchy.

(heating and cooling), shape transformations (rolling), cutting and surface transformations, such as descaling and coating. For concrete applications the individual subclasses will have additional properties not shown here, for instance *TemperatureTransformation* could have two properties *temperatureStart* and *temperatureEnd*.

3   Applications

In our demonstration case the data model presented here serves as documentation for the application programming interfaces (APIs) of a digital twin software platform [9]. For this purpose, an OpenAPI schema [10] can be generated from the json-schema files encoding the model specification.

Furthermore, the graph structure of the underlying data is exploited in the form of a GraphQL API, which allows app developers to specify their data requirements in terms of custom templates and thus to retrieve complex data graphs in a single query. The GraphQL API is autogenerated from the data model specification via a GraphQL schema [11], which provides query types for all classes in the data model. A sample GraphQL

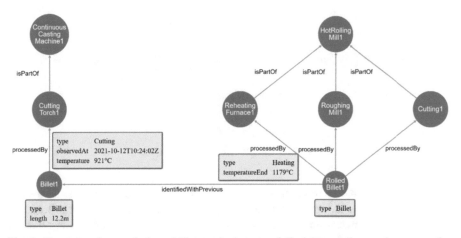

Fig. 4. Example of a graph for a billet tracked successfully between the continuous casting machine and the hot rolling mill. A reference *identifiedWithPrevious* has been created between the hot rolled billet and the cast billet.

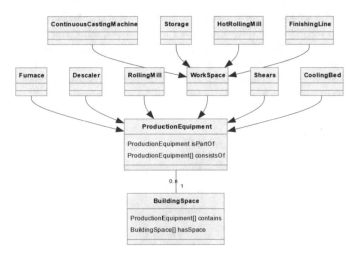

Fig. 5. The *ProductionEquipment* hierarchy.

query for subgraphs of the form shown in Fig. 4 is shown in Fig. 7. A set of applications is being developed in the project that access data via the APIs and that make use of the model. They include a soft sensor for the solidification process in the continuous casting machine, which uses the the product class *LadleHeat* as input data, carrying all relevant information from the secondary metallurgical processses, and creates new instances of type *Billet*, one per newly cut billet from the casting strands. For the hot rolling mill there are two soft sensors, one of them models the temperature evolution of billets and the other one models the build-up of mill scale on the surface of the billets, respectively of the bars cut from the billets at the end of the hot rolling mill. These two models use the *Billet* class as input data, along with process data coming from the hot rolling mill,

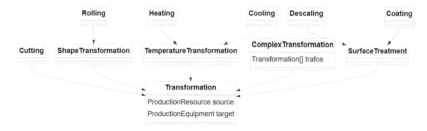

Fig. 6. Examples of *Transformations*.

and they amend the data representation of the billets by means of their model results. As explained above, the tracking uncertainty from casting machine to hot rolling mill implies that the Billet instances used by the hot rolling mill soft sensors are distinct from those generated by the solidification sensor at the casting machine, but get linked to them by means of the *identifiedWithPrevious* relationship, which is supplied by the internal billet tracking system.

Finally, a risk estimation for surface defects is under development, which aims to take into account multiple processing steps, from secondary metallurgy and continuous casting via the hot rolling and finishing lines to the final quality assessment, and hence must evaluate the tracking information provided for the different products. Its input are instances of the *LadleHeat*, *Billet* and *Bar* types, as well as custom quality assessment results for the training of the estimator. A decision support system enabling the through-process quality control will be realised based on the risk estimator.

```
1 · {
2 ·     Billets {
3           id
4 ·         identifiedWithPrevious {
5               id
6               length
7           }
8 ·         processedBy {
9               id
10 ·             isPartOf {
11                  id
12              }
13          }
14      }
15  }
```

Fig. 7. A sample GraphQL query for subgraphs similar to the one shown in Fig. 4.

4   Conclusions

A somewhat generic data model for the steel production has been presented, which builds on two ETSI standards, NGSI-LD and SAREF4INMA. The model has been used for API documentation and generation in a Digital Twin software platform for a steel plant, modeling processes such as secondary metallurgy, continuous casting and hot rolling of long products. The model is well suited to express the tracking uncertainty characteristic of steel long products. A set of cognitive applications have been realised in terms of the APIs which improve the observability of the steel production and aim at helping detect process anomalies early in the production chain.

The data model specification has been published on Github [12].

Acknowledgements. This work has been supported by the CAPRI project, which has received funding from the European Union's Horizon 2020 research and innovation programme under grant agreement No. 870062.

References

1. ETSI. ETSI TS 103 410–5 SmartM2M; Extension to SAREF; Part 5: Industry and Manufacturing Domains (2020). https://saref.etsi.org/saref4inma, (Accessed 11 Aug 2022)
2. ETSI. ETSI GS CIM 009 Context Information Management (CIM); NGSI-LD API (2021) https://www.etsi.org/deliver/etsi_gs/CIM/001_099/009/01.04.02_60/gs_CIM009v01 0402p.pdf, (Accessed 11 Aug 2022)
3. NIST. The Process Specification Language (PSL) Overview and Version 1.0 (2000). https://www.nist.gov/publications/process-specification-language-psl-overview-and-version-10-specification, (Accessed 4 Jan 2023)
4. Beden, S., Cao, Q., Beckmann, A.: SCRO: a domain ontology for describing steel cold rolling processes towards industry 4.0. Information **12**(8), 304 (2021). https://doi.org/10.3390/info12 080304
5. Cao, Q., Giustozzi, F., Zanni-Merk, C., de Bertrand de Beuvron, F., Reich, C.: Smart condition monitoring for industry 4.0 manufacturing processes: an ontology-based approach. Cybern. Syst. **50**, 82–96 (2019). https://doi.org/10.1080/01969722.2019.1565118
6. May, G., Cho, S., Majidirad, A., Kiritsis, D.: A semantic model in the context of maintenance: a predictive maintenance case study. Appl. Sci. **12**(12), 6065 (2022). https://doi.org/10.3390/app12126065
7. Brandenburger, J., Colla, V., Nastasi, G., Ferro, F., Schirm, C., Melcher, J.: Big data solution for quality monitoring and improvement on flat steel production. IFAC-PapersOnLine **49**(20), 55–60 (2016). https://doi.org/10.1016/j.ifacol.2016.10.096
8. Bao, Q., Wang, J., Cheng, J.: Research on ontology modeling of steel manufacturing process based on big data analysis. In: MATEC Web of Conferences, vol. **45**, 04005 (2016). https://doi.org/10.1051/matecconf/20164504005
9. Nölle, C., Arteaga, A., Egia, J., Salis, A., De Luca, G., Holzknecht, N.: digital twin-enabled application architecture for the process industry. In: Proceedings of the 3rd International Conference on Innovative Intelligent Industrial Production and Logistics – ETCIIM, pp. 255–266 (2022). https://doi.org/10.5220/0011561800003329
10. OpenAPI specification. https://spec.openapis.org/, (Accessed 8 Nov 2022)
11. GraphQL specification. https://spec.graphql.org/, (Accessed 8 Nov 2022)
12. CAPRI steel data model. https://github.com/bfi-de/capri-steel-model and https://zenodo.org/record/7440292, (Accessed 4 Jan 2023)

Human Impact in Complex Classification of Steel Coils

Joaquín Ordieres-Meré[1][(✉)], Antonio Bello-García[2], Ahmad Rajabi[3], and Jens Brandenburger[3]

[1] Universidad Politécnica de Madrid, 28006 Madrid, Spain
`j.ordieres@upm.es`
[2] Universidad de Oviedo, 33209 Gijon, Asturias, Spain
[3] VDEh-Betriebsforschungsinstitut GmbH, Sohnstrasse 69, 40237 Dusseldorf, Germany
`https://biba.etsii.upm.es/jom/`

Abstract. This work explores different artificial intelligence based alternatives to create automatic classification system based on salience maps of coil surface when heavy unbalanced datasets are considered and where the labels have been assigned by human operators, considering different complex rules. After testing the possibilities of classifier setup process, additional effort was spend create synthetic features based on the characteristics of the salience maps and such features have been used to verify the need for additional check of scores coming from the human operators. Although it is a preliminary result, it provide evidences of the significant impact that confusing scores can have on the classifier final performance.

Keywords: surface defects · steel coils · salience maps · Artificial intelligence · Human effect on sample labeling

1 Introduction

Industry 4.0 has gained a significant amount of traction since becoming known. Smart and connected technologies are being embedded in organizations, assets and even people in the case of wearable devices, taking advantage of emerging capabilities from artificial intelligence (AI) to quantum computing (QC), to foster classical manufacturing and additive manufacturing processes with the help of the Internet of Things(IoT) [5,13,14].

There is a constant growth in the area of Artificial Intelligence driven by new available modeling techniques, platforms and solutions. That ecosystem quickly evolves toward better algorithms and wider practical applications in Industry. Although certification of these tools is a must to be achieved, its progressive implementation is making the costs to decrease.

AI has several advantages for the manufacturing industry when applied with the right approach. Some of the most relevant advantages for the manufacturing sector today:

A. Rocha et al. (Eds.): WorldCIST 2023, LNNS 802, pp. 256–265, 2024.
https://doi.org/10.1007/978-3-031-45651-0_26

- **Error reduction**. After being trained, intelligent algorithms can perform very well tasks that are susceptible to errors in processes executed by humans, such as data-driven technical quality assessment [4]. Since algorithms are not susceptible to external factors, they should be unlikely to suffer these factors' consequences.
- **Cost reduction**. More attention to the asset management, integrating production scheduling with maintenance and operation tracking will provide a significant increase of overall equipment effectiveness (OEE) optimization [1].
- **Revenue Growth**. With fewer errors and employees focused on more critical processes, decision-makers will have more time to think about the core business and leave other AI tasks [2].

Although there are many articles reporting successful AI application to manufacturing industry [7], much less contributions have shown the potentially disturbing effects in limited datasets that noncalibrated human decisions can have in the overall performance of AI models. This is the main goal that this paper wants to analyze. First, in Sect. 2 a use case related to the quality classification of flat steel products having difficulties in interpretation will be introduced. Section 3 explores the effectiveness of different AI based models and then Sect. 4 discusses the influence that operator's decision being used in the labeling process means. Finally, conclusions and future research lines are presented in Sect. 5.

2   Use Case for Analysis

The selected use case is related to a downstream flat steel production process, where a continuous annealing with zinc coated is produced (Hot Dip Galvanizing process). The quality process aims to identify surface defects by using a specialized computer vision system [3, 9, 10].

Single defects, usually from 5 cm - 50 cm (or rare up to 2 m), depending on their severity and (if no rework is allowed) reason for blocking and downgrading. The following defect is classified into subgroups, considering the position on the strip and the appearance of the defect. Surface defects may have different origins. Starting from the caster and hot strip mill, the origin of slivers is continued, followed by the scale from the hot strip mill. The root causes of zinc pimples are various and can be found in HSM, Cold Roll mill and also in the Galva process itself. Zinc dross is a defect caused by dross from the zinc Pott in the Galva process. Even if all of those defects may have different origins and dimensions, the defect severity increases or pops up with the Galva process. The outcome of the visual inspection system with controlled illumination is processed to decide about the acceptability of the coils.

For storing the images and because of the coil size a tiling process is implemented in such a way that all the surface is resampled into 128*256 chunks and the defects found in each tile are cummulated. Figure 1 shows examples of the different classes.

(a) Example of OK status (b) Example of NoK status (c) Human accepted coil

Fig. 1. Digital representation of coils with the adopted resolution

The goal is to create an AI based model able to interpret the complex rules being applied by the human operator to decide whether the existing defects are acceptable or not. To learn such complex rules a data-driven strategy was adopted, and to this end a dataset was created to learn from experience (by reviewing a significant set of classification images already assessed by human operators, for the same defect type) is a consistent alternative approach. In this case, the dataset covers 4649 images, where 4405 are OK and 244 are not OK.

```
num_classes=2
inChannel = 1
x, y = 128, 256
input_img = Input(shape = (x, y, inChannel))
def clasifier(input_img):
    conv1 = Conv2D(8, (3, 3), activation='relu', padding='same')(input_img) #128 x 256 x 32
    b1 = BatchNormalization()(conv1)
    pool1 = MaxPooling2D(pool_size=(2, 2))(b1) #64 x 128 x 32
    conv2 = Conv2D(16, (3, 3), activation='relu', padding='same')(pool1) #64 x 128 x 48
    b2 = BatchNormalization()(conv2)
    pool2 = MaxPooling2D(pool_size=(2, 2))(b2) #32 x 64 x 48
    conv3 = Conv2D(32, (3, 3), activation='relu', padding='same')(pool2) #32 x 64 x 64
    b3 = BatchNormalization()(conv3)
    pool3 = MaxPooling2D(pool_size=(2, 2))(b3) #16 x 32 x 64
    conv4 = Conv2D(64, (3, 3), activation='relu', padding='same')(pool3) #32 x 64 x 64
    b4 = BatchNormalization()(conv4)
    pool4 = MaxPooling2D(pool_size=(2, 2))(b4) #16 x 32 x 64
    flat = Flatten()(pool4)
    den = Dense(256, activation='relu')(flat)
    den2= Dense(32, activation='relu')(den)
    out = Dense(2, activation='softmax')(den2)
    return out
```

Fig. 2. Base architecture adopted as a reference

A significant constraint to be handled is the heavily imbalanced datasets (4405 vs 244) which can affect the learning process, as the initial sample distribution makes one class much more probable than the other one. Different deep learning techniques have been tested, including Convolutional Neural Networks (CNNs), but also in which different predefined configurations have been used to extract relevant feature sets, such as VGG16. Other alternatives, such as encoders and dimension compressors, that enable classifiers to be used in much lower dimensions were also tested.

Actually, the existing rule-based classifier is able to identify 643 as OK and 200 as NoK. For all the remaining 3806 coils, the automatic classification system was not able to decide, and they have been sent to human operators for a final decision. Of them, 3762 have been labeled as OK by the human deciders, and 44 have been discarded. An example of the coils that are accepted is presented in Fig. 1c, where the input is the set of image pixels and the output is the binary classification. This model requires 2.1 million parameters.

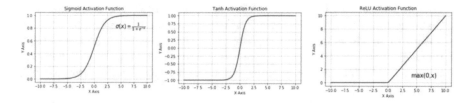

Fig. 3. Different activation functions tested.

3   AI Models in Quality Control

To establish a base case, to compare against, a classical CNN was used, with an architecture such as presented at Fig. 2.

An equal-size training set from the classes was created with 300 images in total, randomly selected. The results show the limited success found, mainly due to the limited size of the training set, and the interest in identifying the NotOK coils:

– Success ratio in class OK   40%.
– Success ratio in class NotOK   67%.

Different activation functions were tested to assess relevance without specific impact (see Fig. 3). As is well known from the different applications carried out for CNNs, overfitting occurs after a few epochs, which prevents many epochs from being used [8].

To try to determine the right strategy fighting against the imbalanced classes, currently there is a wide variety of solutions for this problem, although they can be grouped into:

– Adjustment of model parameters to penalize failures to detect minority class during training.
– Modify the data set to balance the classes. Within this group, it is possible to distinguish between oversampling techniques, which consist of increasing the number of samples of the minority class, and subsampling, which consist of reducing the number of samples of the majority class.

For the case under study, the first solution did not significantly improve the results, so the second group of solutions will be studied in more depth. SMOTE technology [6] was implemented by studying both oversampling and undersampling to create a balanced training data set. Starting with oversampling, the dataset is separated into train, validation, and test:

- **Train**: is the set of images with which the model is trained, following the procedure discussed in the theoretical foundation. It consists of 3009 images of steel coils from the category defined as inspector_OK (insp_OK), and 170 images from the NoK category. After applying SMOTE, the training set is made up of 6018 images in total, half inspector_OK and half NoK.
- **Validation**: allows for evaluating the predictive capacity of the model before moving on to the test stage, as well as detect overfitting. In this case, it is made up of 37 images from the inspector_OK category and 37 from the NoK category.
- **Test**: It is used to evaluate the already trained models. In this case, three test sets will be used to study how the model behaves for each of the three categories. The first will be made up of the 643 images cataloged as OK, the second will be made up of the remaining 716 images of the inspector_OK category, and the third the remaining 37 images of the NoK class.

Table 1 shows the results obtained when training and testing the different CNN models tested, where three different sequences for the number of features in the convolution layers and three different sizes were used in the flatten layer. Indeed, different sizes of kernel for convolution were tested, and different sizes of maxpooling window was implemented. The last three architectures were extending the number of casacade blocks of con2d+batchnorm+masxpooling used, that produced more complex features. It should be noted that each of the models has been trained and tested ten times and the mean and standard deviation of the results have been calculated, to be able to compare in a consistent way.

The numerical values in green represented in Table 1 the mean percentage of correctly classified coils, while the values in yellow are the standard deviations, and allow us to estimate the dispersion or variability between the ten runs carried out for each model. In principle, to choose the best model, priority has been given to the detection of NOK coils, for security reasons, considering the importance of minimizing the value of the standard deviation. For this reason, among the first six models, CNN3 was chosen as the basis for continuing the experiments. However, increasing the size of the kernel of the convolution layers improved the NOK plate detection ability from 51.89% to 57.57%, in addition to decreasing the standard deviation of the results. Therefore, among the first nine models, CNN7 was chosen as the basis for further experimentation. In the third stage of testing, the architecture of the models was modified, and it was shown that in this case, reducing the number of layers improves the results. The CNN10 model is considered to have the best performance, as it correctly detects more than 60% of the NOK coils and reduces the standard deviation of its results somewhat more. Therefore, this will be taken as a reference to compare with the rest of the machine learning models.

Table 1. CNN over Data-set treated with Smote. Training and Validation accuracy depending on Epochs.

Models	Average(% Success)			Standard Deviation		
	OK	*Insp.OK*	*NoK*	*OK*	*Insp.OK*	*NoK*
CNN1	71.12	70.41	47.57	16.66	19.06	24.33
CNN2	78.96	78.30	42.97	9.72	10.10	14.87
CNN3	73.17	73.13	51.89	20.96	21.58	22.35
CNN4	75.27	76.45	40.54	17.34	16.98	22.74
CNN5	75.77	75.81	43.24	21.63	21.99	25.55
CNN6	76.58	75.17	50.51	18.57	20.78	27.61
CNN7	66.70	66.26	57.57	18.69	16.36	17.18
CNN8	71.15	70.60	52.43	16.55	15.69	20.41
CNN9	83.11	81.79	37.03	9.83	11.20	20.51
CNN10	68.79	64.53	60.54	14.66	17.72	17.06
CNN11	79.30	76.87	47.03	16.51	18.52	20.59
CNN12	68.27	68.42	55.68	20.58	21.89	24.86
CNN13	81.99	82.07	38.65	18.13	16.58	19.75

The other approach, the subsampling [6], was also explored, in view of the results obtained so far. The same experiments as in the previous approach (CNNx) were carried out, but using subsampling to solve the class imbalance problem and balance the training set. The random undersampling method allows random undersampling of the majority class until the classes are balanced. With this proposal, in addition to solving the problem of data imbalance, the computational load is reduced, and therefore the training time. However, it also has certain drawbacks since for the models to generalize correctly, it is usually necessary to use a large amount of data, and with this technique useful images can be eliminated, which could worsen the results. This time the total set of images has been divided into:

- **Train**: 150 images from the inspector_OK category and 150 NoK images.
- **Validation**: 47 images from the inspector_OK category and 47 NoK images.
- **Test**: again, we have chosen to define three test sets to study how the model behaves with each of the classes. The first one is formed by the 643 OK images, the second by 3565 inspector_OK images, and the third by 47 NoK images.

Table 2 shows the results obtained by the different models for each of the categories defined in the dataset. At first glance, we see that all the models work better than those studied previously, so for this application, the subsampling technique works better than SMOTE and oversampling. In this case, among the first six models, CNN2 was chosen as the best, with a precision of 73.19% in the detection of NoK coils and a very low standard deviation value, indicating

Table 2. CNN over Data-set treated with Undersampling Smote. Training and Validation accuracy depending on Epochs.

Models	Average(% Success)			Standard Deviation		
	OK	Insp.OK	NoK	OK	Insp.OK	NoK
CNN1	69.88	66.51	71.70	9.05	10.08	9.13
CNN2	69.21	66.65	73.19	8.62	7.67	4.87
CNN3	73.11	70.11	70.43	6.10	7.72	7.74
CNN4	68.62	66.68	69.36	5.77	5.98	6.94
CNN5	71.70	69.46	68.72	11.05	10.66	11.10
CNN6	74.65	72.81	65.74	8.42	9.29	10.53
CNN7	68.91	64.40	72.34	10.19	11.69	12.94
CNN8	73.92	70.59	67.23	7.12	7.95	11.97
CNN9	68.71	66.49	74.04	9.88	9.85	9.51
CNN10	67.40	66.86	69.57	12.18	11.25	9.57
CNN11	73.33	71.24	70.43	6.72	7.38	5.82
CNN12	71.70	69.50	65.74	10.42	9.75	10.09
CNN13	65.65	61.71	68.09	13.42	15.10	23.62

low variability in results. Starting from the latter as the reference model and continuing with the experiments to compare with the rest of the machine learning models.

Another technique used was the autoencoders, where the goal was to compress the input to a reduced space that contains its main characteristics, and a decoder, whose job is to reconstruct the original input starting from said compressed information. The idea is to study whether this can be used to improve the results obtained by conventional convolutional classifiers. The strategy to follow in this case has been to design several auto-encoders to see which of them best reconstructed the input images. After choosing it and assuming that its reduced space was therefore the one that best represented the initial image, the idea is to take advantage of the encoder part as a feature extractor and add the relevant layers to build an algorithm capable of classifying coils correctly. First, the reconstruction capacity of six auto-encoders with different architectures was studied. To do this, two different patterns are defined that will facilitate the subsequent description of the autoencoders. Once these patterns have been defined, the architecture of the autoencoders can be described from them (in all the cases the encoder and decoder were symmetric) describing the encoder (two, three, and four type 1, and the same for the type 2 pattern).

The number of feature maps of the convolution layers, as well as other parameters, has been chosen considering the results obtained in tests with conventional convolutional classifiers. And by its training, the learned lesson can be formulated as: *The greater the number of layers, the blurrier, and the more convolutions, the better the reconstruction.*

Another technique under exploratory analysis was the extraction of engineering features based on the pre-trained VGG model [11] to identify objects on Imagenet (more than 14 million images from more than 1000 different classes), where the only transformation was to accommodate the input images in the VGG model ($224 \times 244x3$). A transfer learning strategy with eleven different configurations was tested, producing the results presented in Table 3, where a different number of layers were defined as nontrainable, and a different number of neurons was adopted in dense layers.

Table 3. Results obtained from the VGG based classifiers.

Models	Average(% Success)			Standard Deviation		
	OK	*Insp.OK*	*NoK*	*OK*	*Insp.OK*	*NoK*
VGG16_1	20.06	20.01	80	39.89	39.98	40
VGG16_2	60	60	40	48.99	48.99	48.99
VGG16_3	71.76	68.11	65.53	15.35	17.34	32.95
VGG16_4	70.89	69.22	73.19	9.56	11.82	6.54
VGG16_5	71.94	71.64	74.04	8.36	10.4	7.04
VGG16_6	67.22	67.77	76.6	9.32	10.81	9.13
VGG16_7	69.55	69.38	77.87	4.84	5.85	5.65
VGG16_8	79.04	79.16	59.57	11.94	12.41	31.59
VGG16_9	82.02	81.39	61.28	3.9	3.61	8.86
VGG16_10	66.78	65.54	78.72	8.03	8.42	12.41
VGG16_11	72.1	71.14	77.02	5.76	5.86	7.89

Although the best model was VGG16_7, it should be noted that, depending on the set of images that randomly configured the training set, the network learned or not, that is, the method was not robust enough.

4 Human Influence

Since the quality criteria is decided by human operators, a potentially relevant factor is the human bias or the multi-criteria developed by the different operators. To clarify such an effect, a heuristic feature extraction based on saliency map technology [12] was implemented where a Gaussian function was placed over every single defective pixel (see 4). After integrate all the damage functions contours can be derived.

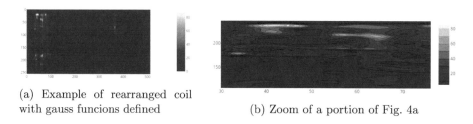

(a) Example of rearranged coil with gauss funcions defined

(b) Zoom of a portion of Fig. 4a

Fig. 4. Digital representation of the defaults exhibited.

For each coil, defects over 40,25 and 10 levels of damage are obtained and the total area and number of clusters for each of the levels. In Fig. 5 there is no clear segmentation between OK and NoK, neither by the extension of the damaged area nor by the number of clusters. This analysis shows that the operator decision is not fully consistent as they do not produce a separated area between the OK and NoK coils. Therefore, either more congruence is needed the operators' criteria or additional characteristics need to be consided for making a robust decision system more congruent with the adopted decisions. As a side effect, to keep the labeling as indicated in Fig. 5, evidences the main reason why regardless of the DL technology being adopted, there is an upper bound to the accuracy reached.

(a) Defect area per coil vs the coil label.   (b) Clusters of defect intensity.

Fig. 5. Synthetic feature map of coils.

5   Conclusions

This application shows how digitalization technology can help solve complex intrinsic problems by significantly improving success rates. Both deep learning and deep transfer learning have been tested to identify the quality of the provided solution, where ten runs have been elaborated to identify intrinsic variability due to the method itself. Obviously, there is a limit in the separation capabilities (about 70%), where additional effort is required to clarify the impact of the operators. Another reason to suppose such external influences is the significant behavior found during the sub-sampling instead of the Smote strategy. The reason why learning behavior increases with a lower number of randomly selected samples is that increasing the training set introduces a higher level of noise. Different strategies need to be decided in order to assess different criteria from the human operators, or additional factors need to be considered. To this end, we developed a dynamic workflow enabling the reassessment of synthetic maps, with shape rotation and different coil resequences. In this way, the impact of the human operator can be independently assessed. More work needs to be done in the coming weeks to allow a better assessment of human-based quality decisions.

Acknowledgements. The authors want to thank the EU RFCS research program for the support of this research through the grant with ID 847202 (AutoSurveillance project) and the grant ID 101034037 (DeepQuality project). It also received funding from the Agencia Estatal de Investigación (program MCIN/AEI/10.13039/501100011033), through the grant RTI2018-094614-B.I00 (Smashing).

References

1. Branca, T.A., et al.: The challenge of digitalization in the steel sector. Metals **10**(2), 288 (2020)
2. Brock, J.K.-U., Von Wangenheim, F.: Demystifying AI: what digital transformation leaders can teach you about realistic artificial intelligence. California Manag. Rev. **61**(4), 110–134 (2019)
3. Chakraborty, A., et al.: Investigation of a surface defect and its elimination in automotive grade galvannealed steels. Eng. Failure Analy. **66**, 455–467 (2016). https://doi.org/10.1016/j.engfailanal.2016.05.010, https://www.sciencedirect.com/science/article/pii/S1350630716302837, issn: 1350-6307
4. Chiarini, A.: Industry 4.0, quality management and TQM world. a systematic literature review and a proposed agenda for further research. TQM J. (2020)
5. Li, F., et al.: Ensemble machine learning systems for the estimation of steel quality control. In: 2018 IEEE International Conference on Big Data (Big Data), pp. 2245–2252. IEEE (2018)
6. Mishra, S.: Handling imbalanced data: SMOTE vs. random undersampling'. Int. Res. J. Eng. Technol. **4**(8), 317–320 (2017)
7. Pellegrini, G., et al.: Successful use case applications of artificial intelligence in the steel industry. In: Iron & Steel Technology (AIST), Digital Transformations, pp. 44–53 (2019)
8. Perin, G., Buhan, I., Picek, S.: Learning when to stop: a mutual information approach to fight overfitting in profiled sidechannel analysis. Cryptology ePrint Archive (2020)
9. Rose, A., Wandera, C., Favor, E.: Parameters influencing the hot dip galvanizing processes of sheet metal. Am. J. Mater. Synth. Process. **6**(1), 1 (2021)
10. Saravanan, P., Srikanth, S.: Surface defects and their control in hot dip galvanized and galvannealed sheets. Inter. J. Adv. Res. Chem. Sci. (Ijarcs) **5**(11), 11–23 (2018)
11. Simonyan, K., Zisserman, A.: Very deep convolutional networks for large-scale image recognition. arXiv preprint arXiv:1409.1556 (2014)
12. Song, G., Song, K., Yan, Y.: Saliency detection for strip steel surface defects using multiple constraints and improved texture features. Opt. Lasers Eng. **128**, 106000 (2020). https://doi.org/10.1016/j.optlaseng.2019.106000
13. Villalba-Diez, J., González-Marcos, A., Ordieres-Meré, J.: Quantum cyber-physical systems. Sci. Rep. **12**(1), 1–11 (2022)
14. Villalba-Diez, J., et al.: Quantum deep learning for steel industry computer vision quality control. IFAC-PapersOnLine **55**(2), 337–342 (2022)

Enabling Software Engineering Practices Via Latest Development's Trends

An ML-Based Quality Features Extraction (QFE) Framework for Android Apps

Raheela Chand[1]([✉]), Saif Ur Rehman Khan[1], Shahid Hussain[2],
and Wen-Li Wang[2]

[1] Department of Computer Science, COMSATS University Islamabad (CUI),
Islamabad, Pakistan
`raheela.chand@yahoo.com`
[2] Department of Computer Science and Software Engineering, School of Engineering,
Penn State University, Behrend, Erie, USA

Abstract. Context: The generic quality attributes fail to comprehend
the current state-of-the-art challenges and constraints of mobile apps.
Objectives: The goal of this study is to fill the gap in the systematic
procedures to identify and extract specific quality features relevant to
Android apps. Method: To accomplish the objective, we have proposed
an ML-based Quality Features Extraction (QFE) framework for Android
apps. QFE analyzes, parses, and gains insights from use reviews utilizing
Natural Language Processing (NLP), Sentimental Analysis, Topic Modelling, and Lexical Semantics. Results: This study was tested on three
different datasets and QFE successfully discovered 23 unique Android-
specific quality features. Moreover, a comparative study with related
studies was conducted and the analysis delineates that QFE provides a
more reliable, efficient, and easy-to-use approach. Contribution: Briefly,
(i) an ML-based empirical framework is proposed for discovering quality
features for Android apps; (ii) the popular Topic Modelling technique
is enhanced by RBLSALT, that is to automate the manual process of
labeling topics in Topic Modelling; and finally, (iii) the pseudo-code and
Python implemented notebook of the framework is also given to provide
ease in the applicability of QFE. Conclusion and Future Work: Future
work, is planned to evaluate the framework by comparing it with different techniques of feature extraction and to propose a specific features-
oriented comprehensive quality model based on Android apps.

Keywords: Quality Framework · Quality Features · Topic Modelling ·
Feature Extraction · Android Games

1 Introduction

From testing emerging aspects of Android apps viewpoint, user expectations are
very different than those of other software applications i.e., desktop applications

A. Rocha et al. (Eds.): WorldCIST 2023, LNNS 802, pp. 269–278, 2024.
https://doi.org/10.1007/978-3-031-45651-0_27

[1]. The state-of-the-art mobile app concerns are related to energy consumption, user trust, excessive ads, and so on. However, the conventional evaluation is limited to a set of well-known standard quality attributes (i.e. Performance, Maintainability, Efficiency) [2]. Similarly, the literature review also lacks sufficient rigorous representations and a few authors have attempted to identify the specific quality attributes of Android apps. For instance, Inukollu et al. [3] discovered an Android-specific quality feature - an excessive amount of ads and Xiang et al. [4] presented software aging.

This paucity of information has encouraged practitioners of the Android industry to employ traditional quality models knowing the fact that Android development is different than general software development. Consequently, the practice of relying on generalized quality attributes has profane various quality dimensions of Android apps [5]. For example, El-Dahshan et al. [6] discussed that anti-patterns in Android apps are the basic cause of low quality and are not instituted in the traditional software quality models. In Android apps, especially usability has so many sub-categories that can never be generalized.

Thus, this research paper tried to provide a rich profusion of the quality features related to Android apps with a practical structure to support and guide researchers and practitioners to assess and extend the evaluating matrices specific to Android apps in the shape of ML-based QFE (Quality Features Extraction) framework for Android apps. In addition to that, a Rule-based Lexical Semantics Automated Labelling Technique (RBLSALT) is introduced as an enhancement to the popular Topic Modeling technique. Furthermore, the framework has attained promising results when evaluated by a case study on Google Play Store's top trending Android games and a comparison with the till-date literature discovered features for Android apps. In a nutshell, the main contributions of this research paper are:

1. Develop a novel practical framework QFE to help in discovering significant features for Android apps.
2. Review the literature focused on quality features for Android apps.
3. Identify and validate a list of QFE-generated current state-of-the-art quality breaching features for Android apps.
4. Introduce an RBLSALT technique to automate the manual labeling process of Topic Modelling.
5. Devise a pseudo-code and a Python-implemented notebook demonstrating the practical steps of the proposed framework highlighting the usability of QFE.

The rest of the paper is organized as follows. Section 2 describes the related work, and Sect. 3 presents the proposed framework. In contrast, Sect. 4 defines the evaluation criteria, while the pseudo-code to implement the proposed framework is described in Sect. 5. In Sect. 6, the results and observations are discussed, while Sect. 7 discloses the threats to validity. Section 8 provides the implications for the research community. Finally, Sect. 9 concludes this paper.

2   Related Work

In this section, the literature review of relevant studies useful in the identification of features specific to Android apps is summarized. Inukollu et al. [3] mentioned that an excessive amount of ads during app usage is disturbing for users. Chuang et al. [7] reported that malware in Android apps overheats the phone, drains the battery, and jeopardizes the user's privacy. Shan et al. [8] considered resume and restart errors as the most crucial factor in Android apps. El-Dahshan et al. [6] discussed anti-patterns in Android apps. Xiang et al. [4] presented a novel quality attribute called software aging. Finally, Moreira et al. [9] discussed energy consumption problems in Android apps. While in ML-based studies Jianfeng Cui et al. [10] focused on conducting android vulnerability prediction analysis and suggested considering object-oriented metrics and RF algorithms in the software development process for android-based intelligent IoT systems. In contrast, Dehkordi et al. [11] highlighted the absence of a repository and introduced a repository of 100 successful and 100 unsuccessful apps of Android from Google PlayStore. Thus, from the previously discussed work, it is observed that there is a lack of systematic procedures to identify and extract quality features of Android apps.

3   Proposed Framework

This section presents the proposed framework which is depicted in Fig. 1. QFE contains six main steps with some additional sub-steps in each step. The complete step-by-step explanation of how the proposed framework discovers quality features is illustrated as follows:

3.1   Step 1: Pre-Processing User Reviews

Pre-processing is the first main step of the QFE framework. This step has four sub-steps which are discussed as follows:

Unwanted Character Removing. This step aims to remove all special characters, emojis, and brackets from the whole document of the user reviews utilizing regular expressions (aka Regex).

Stop Words Removing. It aims to remove articles and pronouns containing extra information. To achieve this, the split function has been utilized.

Noise Removing. It objects to removing noisy words except for nouns, adverbs, adjectives, and verbs in the documents. To achieve this, POS tagging and TextBob functions are been utilized.

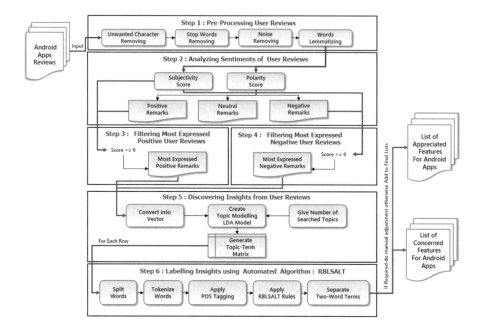

Fig. 1. The Proposed Quality Feature Extraction (QFE) Framework

Words Lemmatizing. It is the final sub-step of Pre-Processing step. It aims to analyze the words and return the base or dictionary form of those words i.e., play from playing or played. Doing so will increases processing speed and unique search strings.

3.2   Step 2: Analyzing Sentiments of User Reviews

The principal purpose of this step is to scrutinize fake reviews if encountered. The negative effects are reduced by calculating the normalized polarity and subjectivity score of each user review. Where 0 is the least value and 1 is the highest value.

3.3   Step 3 and Step 4: Filtering Most Expressed Positive and Negative Reviews

This step aims to identify the reviews that should provide meaningful content useful for feature extraction. The minimum threshold of the framework is a 0.6 sentiment score. The reason for choosing 0.6 is to use the least possible threshold. The greater threshold will reduce the number of reviews as input to the model(Topic Modeling with LDA). Likewise, more decreasing the 0.6 will include tangential statements to the focus of the study. Moreover, the benchmark is tested and verified after experimenting with real datasets. Therefore, 60 percent balances the uniqueness as well as an ample amount of reviews for input.

3.4   Step 5: Discovering Insights from User Reviews

This step intends to identify the significant features for the analysis of quality in Android apps. For this purpose, the data analysis technique Topic Modelling with LDA has been adopted in this work. To implement the Topic Modeling, Yin et al., [12] study has been followed. Briefly, the steps include converting the user reviews into a vector form, selecting the number of terms and topics to be identified by the LDA model, feeding the LDA model, and finally generating a topic-term matrix.

3.5   Step 6: Labelling Insights Using Automated Algorithm: RBLSALT

This step aims to introduce the Rule-based Lexical Semantics Automated Labelling Technique (RBLSALT) for labeling the topic-term matrix. It is an algorithm based on the data mining technique and motivated by the idea of Dubey et al. [13]. Although various other advanced rule-based POS (Parts of Speech) tagging patterns were also available in academia [14,15]. However, the high complexity of procedures enables them very difficult to apply to a data. Dubey et al. provided patterns of tags for extracting two-word features from reviews based on nouns, verbs, and adjectives. Similarly, RBLSALT generated a list of two-word terms by analyzing nouns, verbs, and adjectives based on association rules. It searches for nouns with associated verbs and adjectives. If the algorithm finds the pair, it concatenates them, otherwise, makes them independent terms.

>   Rule 1:   Search for nouns with associated verbs and adjectives.
>   Rule 2:   If the algorithm finds a pair, concatenate them.
>   Rule 3:   Otherwise, treat them as independent terms.

A flowchart of the rules is provided in Figs. 2, 3 and 4 and the detailed pseudocode of working and rules is given in Sect. 6.

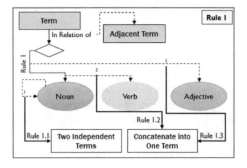

Fig. 2. The Flowchart of Rule 1

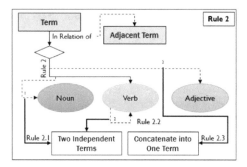

Fig. 3. The Flowchart of Rule 2

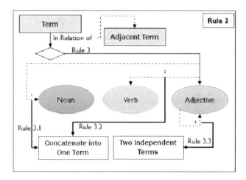

Fig. 4. The Flowchart of Rule 3

4   Evaluation Criteria

This section provides the evaluation criteria used in the study. To evaluate the proposed framework, a two-step evaluation criteria is used, including a case study and a comparison with related studies. The case study is based on three datasets of Google Play Store app reviews [16]. However, the practical implementation was performed in Anaconda Navigator (Python version - 3.8.3) on the Windows operating system (version - 11). The final lists of terms are provided in Table 1 and Table II. Nevertheless, for detailed results, the Python notebook can be downloaded using the link [16].

The second criterion of evaluation is a comparison with related studies. This step aims to analyze the competence of QFE with the existing related approach. It is observed from the literature review that Teymourzadeh et al. [17], Inukollu et al. [3], Chuang et al. [7], Shan et al. [8], El-Dahshan et al. [6] and Xiang et al. [4] none of them provided a systematic guideline for features extraction. On the contrary, this paper presents a framework. Moreover, Jian et al. [1] suggested generic attributes. On the other hand, this study provides more relatable to the current state-of-the-art Android challenges. Finally, in terms of ML-based work, the literature indicated sterile grounds, however, QFE has successfully incor-

Table 1. List of QFE generated concerning features for Android apps

SR	Generated Features
1	Energy Consumption
2	Account Problems
3	Resume and Restart Issues
4	Customer Service Issues
5	Scam App Features
6	Extensive Waiting/ Response Time
7	In-app Purchase Issues
8	Help/Support Issues
9	Update Issues
10	Excessive Money-oriented Features
11	Functional Malfunctions
12	Phone Infesting Features
13	Game Glitches

Table 2. List of QFE generated appreciated features for Android apps

SR	Generated Features
1	Enjoyable Graphics
2	Adorable Characters
3	Free Play
4	Pleasing Music
5	Easy and Interesting Challenges
6	Help Items
7	Easy to Win Stars
8	Great Download
9	Exciting Game Updates
10	Amazing In-game Support

porated ML-based techniques in the Android domain. Besides, the majority of the reported studies focused on the development and functional requirements of Android apps. On second thought, this research has provided a user perspective on the Android industry.

5   Results and Observations

This section summarized the consequences of the considered studies. The final quality features were empirically generated through QFE and a total of 23 extracted features are provided (please refer to Table 1 and Table 2). These quality features are from user perspectives for Android apps. The developers should consider these factors as quality checks to create a quality-oriented Android app. Moreover, In the existing literature, the researchers have identified some Android-specific quality features. However, the identification was entirely based on observation. The QFE framework can discover a wide range of attributes according to the case study. Nevertheless, the proposed QFE framework has applied the popular Topic Modelling technique. However, there were still manual steps needed to label the topics. This motivated us to provide an improved version of Topic Modelling to completely automate the steps. The RBLSALT procedure in the QFE was developed for this purpose. Procedure 5 presents this automated method. The QFE is indeed a novel and practical framework with the ability to algorithmically discover quality features for Android apps.

6   Pseudo-code to Implement QFE

In this section, the pseudo-codes of each step to perform QFE are described.

Algorithm 1. QFE Framework Pseudocode

Require: Set *Reviews* for analysis
  while *Reviews* ≥ 0 **do**
      Remove unwanted characters and lemmatize
      Analyze sentiments and filter reviews
      Discover insights using LDA topic modeling
  end while

Algorithm 2. RBLSALT Pseudocode

Require: Set *TopicTermMatrix* for processing
  while *rows* $\neq 0$ **do**
      Process terms using POS tagging and RBLSALT rules
  end while
Require: Set *Terms* for adjacency rules
Ensure: *AdjacentTerm* $\leq$ *Term*
  while *Terms* $\neq 0$ **do**
    if *Term* is a Noun **then**
      Find adjacent term
      if *AdjacentTerm* is a Noun **then**
        Two independent terms
      else if *AdjacentTerm* is a Verb or Adjective **then**
        Merge terms
      end if
    else if *Term* is a Verb **then**
      Similar processing as for Noun
    else if *Term* is an Adjective **then**
      Similar processing as for Noun
    end if
  end while

---

7   Research Opportunities

This section explains some implications for the research community:

1. Three test datasets were used in the case study. However, all three of the datasets contained reviews of Google Play Store users. Similar to Google Play Store, Amazon is also available, and researchers can test the framework on different platforms and further check the efficiency of the framework.
2. Especially, in rules lots of research could be conducted for improvement. For example; several more significant rules could be discovered.

3. Furthermore, while analyzing the keywords for labeling in Step 6 of the case study, it was observed that most of the users were complaining about various malware in Android apps. Researchers can target each quality feature listed in this study and suggest strategies to resolve the issues.

Hence, the above were a few examples, but not limited to only mentioned dimensions, in which this study can provide opportunities for the Android industry research community. Thus, this study is open to the research community. Research it, adopt it, adapt it, and discover more opportunities and improvements.

8 Limitations and Threat to Validity

This section presents some factors that can affect the efficiency of the QFE results and the study's limitations. For example, this study lacks incorporating filters to verify the user reviews. Thus, the authenticity of input can be a thread. Another rule is, grammar checks are also not implemented on review input. Due to possible spelling or grammar errors, there are chances of losing some valuable information undetected in the QFE framework. Finally, RBLSALT is only designed to extract two-words term. There is a high chance that a three-word term can better label the topics. Thus, this side of the edge is not explored in the current study.

9 Conclusion

This paper proposed a Quality Feature Extraction (QFE) framework that provides the most instrumental and effective methodology for feature extraction compared to the related work in the Android app's context. The attained promising results provide evidence of the QFE's competence and reliability. Moreover, this paper also proposed an improvement RBLSALT, which is a rule-based algorithm introduced to automate the Topic Modelling manual procedures. Furthermore, the pseudo-code as well as the Python-implemented notebook is provided for future research purposes. For future work, we plan to continue the investigation of additional quality features for Android apps in which developers' perspectives would also be incorporated. Moreover, we intend to propose an enhanced quality model based on specific quality features for the Android apps' context.

References

1. Jain, P., Sharma, A., Aggarwal, P.K.: Key attributes for a quality mobile application, pp. 50–54, January 2020
2. Koushki, M.M., AbuAlhaol, I., Raju, A.D., Zhou, Y., Giagone, R.S., Shengqiang, H.: On building machine learning pipelines for android malware detection: a procedural survey of practices, challenges, and opportunities. Cybersecurity 5, 1–37 (2022)

3. Inukollu, V.N., Keshamoni, D.D., Kang, T., Inukollu, M.: Factors influencing quality of mobile apps: role of mobile app development life cycle. Int. J. Softw. Eng. Appl. **5**, 1–34 (2014)
4. Jianwena, X., et al.: Software aging and rejuvenation in android new models and metrics. Softw. Qual. J. **28**, 85–106 (2019)
5. Maia, V., Rocha, A.R., Gonçalves, T.G.: Identification of quality characteristics in mobile applications, pp. 1–119, January 2020
6. El-Dahshan, K.A., Elsayed, E.K., Ghannam, N.E.: Comparative study for detecting mobile application's anti-patterns, pp. 1–8, April 2019
7. Chuang, H.-Y., Wang, S.-D.: Machine learning based hybrid behavior models for android malware analysis, pp. 201–206, August 2015
8. Shan, Z., Azim, T., Neamtiu, I.: Finding resume and restart errors in android applications. ACM SIGPLAN Not. **51**, 864–880 (2016)
9. Moreira, J.S., Alves, E.L.G., Andrade, W.L.: A systematic mapping on energy efficiency testing in android applications, pp. 1–10, December 2021
10. Cui, J., Wang, L., Zhao, X., Zhang, H.: Towards a predictive analysis of android vulnerability using statistical codes and machine learning for IoT applications. Comput. Commun. **155**, 125–131 (2020)
11. Dehkordi, M.R., Seifzadeh, H., Beydoun, G., Nadimi-Shahraki, M.H.: Success prediction of android applications in a novel repository using neural networks. Complex Intell. Syst. **6**, 573–590 (2020)
12. Bin, Y., Yuan, C.-H.: Detecting latent topics and trends in blended learning using LDA topic modeling. Educ. Inf. Technol. **973**, 1–24 (2022)
13. Dubey, G., Rana, A., Ranjan, J.: A research study of sentiment analysis and various techniques of sentiment classification. Int. J. Data Anal. Tech. Strat. **8**, 122–142 (2016)
14. Govilkar, S., Bakal, J.W., Rathod, S.: Part of speech tagger for the Marathi language. Int. J. Comput. Appl. **119**, 29–32 (2015)
15. Alebachew, C., Yitagesu, B.: Part of speech tagging: a systematic review of deep learning and machine learning approaches. J. Big Data **9**, 1–25 (2022)
16. Chand, R., Rehman Khan, S.U., Hussain, S., Wang, W.: TTAG+R: a dataset of google play store's top trending android games and user reviews, pp. 580–586 (2022)
17. Teymourzadeh, M., Gandomani, T.J.: Introducing a particular quality model in mobile application development: the mobile application developers' perspective. J. Softw. **12**, 339–347 (2017)

AI Incorporating NLP – to Boldly Go, Where No Algorithms Have Gone Before

Atif Farid Mohammad[1]([✉]), Neela Alagappan[2], and M. Basit[2]

[1] University of North Carolina at Charlotte, Charlotte, NC 28223, USA
atif.farid.mohammad@uncc.edu
[2] California State University, Long Beach, CA 90840, USA
nalagapp@uncc.edu, m.basit@csulb.edu

Abstract. This document is about understanding Natural Language Processing (NLP) - an area of research and application intended to explore how machines or computers can be used to understand and manipulate natural language text or speech conveyed by humans. In other words, it is discipline that focuses on understanding and interpreting raw or structured data which are in form of speech or text into meaningful interpretation. In recent times, NLP is trending technology and hot spot for researchers and technology experts. With exponentially growing data every day, NLP has already emerged as powerful technique in the history of mankind. NLP breaks down language into shorter tokens and attempts to understand the relationship between tokens. The foundations of NLP lie in several disciplines, namely, computer and information sciences, linguistics, mathematics, artificial intelligence and robotics, electrical and electronic engineering psychology and electrical and electronic engineering. This paper discusses about Techniques, Uses, Mathematics behind NLP and trending research on NLP with established real time examples.

Keywords: Artificial Intelligence · Natural Language Processing · Machine Translation · Tokens · Causes · Algorithms

1 Natural Language Processing (NLP)

1.1 Introduction

Natural language processing shortly NLP is a critically important and rapidly developing area of computer science. Any modern technology practitioner or data scientist needs a unified understanding of both machine learning algorithms and linguistic fundamentals. Humans can communicate through some of language which can either be verbal or sign language. With advancement in technology, humans have already started to communicate through texts. The next advancement in technology driven world is to make humans interact with computers. To achieve this, computers should be made to understand natural language used by humans on daily basis. Natural language processing is all about making computers to learn, process and manipulate natural language. NLP allows computers to communicate with people, using a human language. In recent times, NLP is one of the hot spot research areas. Numerous publications and thesis are being submitted online with application of NLP ranging from NLP in simple website to manned missions. The application areas of NLP is vast and extends from machine translation, spamming filters, sentiment analysis, social media text analysis, chat bots, auto pilot vehicles. The core part of NLP is the process of building computer programs that slowly evolve to understand and interpret natural language. NLP system may begin at the word level to determine the structure and nature such as part-of-speech or meaning of the word and eventually move to sentence level. This progress is made to understand word order, grammar and meaning of the entire sentence. NLP pipeline is a complex process. Hence, it has process categorization starting from sentence segmentation followed by word tokenization, predicting parts of speech for each token, lemmatization, identifying stop words, Dependency Parsing, Finding Noun Phrases, Named Entity Recognition (NER) and Coreference Resolution.

1.2   Motivation

Natural Language Processing involves complex process which allows computers to communicate with humans. NLP breaks down language into shorter, more basics pieces called tokens. The primary step of this process varies from high level to low level depending on the application that employs NLP technique. The high-level technique categorizes the following features of NLP. Content categorization being the first among it. This includes identifying alerts, duplication detection, search, and indexing of input from humans to machine both in speech and textual format. The next feature is to capture the themes and meanings of text collections and applies advanced analytics to the text. This process does topic discovery and modelling.

1.3   State of the Art Research

Contextual Extraction automatically pulls structured data from text-based sources which is followed by topic categorization [1]. Sentiment Analysis is one among the most spoken and researched area relating to NLP. This feature identifies the general mood, or subjective opinions, stored in large amounts of text. This is specifically useful for opinion mining [1]. Text-to-Speech and Speech-to-Text Conversion feature transforms voice commands into text, and vice versa. Document Summarization is one other imminent automatically creates a synopsis, condensing large amounts of text. Machine Translation also finds application under NLP technique which automatically translates the text or speech of one language into another [2].

Google translate is one application which is currently in use, and it used across various technology giants all over the globe. The complexity involved in the technique is because of the raw and unstructured data input and there arises the need for Natural Language Processing. With exponentially growing data every day, it would get very easy if computers can understand and process that data. We can train the models in accordance with expected output in different ways. Computers can't truly understand the human language. If we feed enough data and train a model properly, it can distinguish and try categorizing various parts of speech (noun, supporter, verb, adjective etc.) based on previously fed data and experiences. If it encounters a new word, it tried making the nearest guess which can be embarrassingly wrong few times. Hence it is very difficult for a computer to extract the exact meaning from a sentence. Solving a complex problem in Machine Learning means building a pipeline. Likewise, it means breaking a complex problem into several small problems, making models for each of them and then integrating these models. A comparable thing is done in NLP. It has numerous steps which will give the desired output at the end. The very first step in the process is to start with Sentence Segmentation [3].

Breaking the piece of text in various input sentences is the initial task. An algorithm was developed to break the sentences when a dot is encountered for sentence segmentation. Similarly for speech, a pause is made to recognize end of a sentence. Word tokenization is the next step in the pipelining process. When the input received by the machine, breaking the sentence into individual words called as tokens is the primary aim. A model has already been developed to tokenize them. This includes whenever a space is encountered. A modelling technique has also been employed. Prediction is the next process in the NLP pipeline [3, 4]. Predicting Parts of Speech for each token is one other challenging part of NLP process. Predicting whether the word is a noun, verb, adjective, adverb, pronoun, etc. This will help to understand what the sentence is talking about. This can be achieved by feeding the tokens and the words around it to a pre-trained part-of-speech classification model.

A model classifies the parts of input sentences based on its previous experience. This is where statistics play an important role. Lemmatization is feeding the model with the root word. The next step in the pipeline is to identify stop words. Few words on input sentences are categorized as stop words which is intended to be filtered while doing statistical analysis. The list of stop words are already fed using NLP algorithms. This list of words may also vary from input to input. The next step is to do the Dependency Parsing which means finding out the relationship between the words in the sentence and how they are related to each other. A parse tree is created in dependency parsing with root node being main verb in the sentence. Numerous machine learning model has already been made to identify various parts of speech and train to identify the dependency between words by feeding many words.

Google for instance has already released McParseface which is used for deep learning in NLP method also. Finding noun phrases step in the pipeline groups the words that represent the same idea. Named Entity Recognition (NER) maps the words with the real-world places [4]. NER system is separately an area of research interest. The final step is Coreference Resolution which does higher level language and literature understanding. This pipelining of NLP technique applies to applications with higher complexity also. Various libraries to support these steps have already been researched and are in use in various complex systems.

2 Mathematics in NLP

2.1 Artificial Intelligence

Natural language processing is an interesting offspring of Artificial Intelligence domain. NLP is also built on fundamental principles of mathematics like Calculus, Linear Algebra, Probability, Statistics, and Optimization [5]. Set theory is a branch of mathematical logic that studies set theory, which informally are collections of objects. Before taking NLP system into account, following definitions in set theory are be considered. Membership is set theory is defined as "If B is a set and x is one of the objects of B, this is denoted as x B, and is read as "x is an element of B", as "x belongs to B", or "x is in B". If y is not a member of B, then this is written as y B, read as "y is not an element of B", or "y is not in B" [6]. Likewise, a subset is defined as "If every element of set A is also in B, then A is said to be a subset of B, written A B (pronounced A is contained in B). Equivalently, one can write B A, read as B is a super set of A and B includes A, or B contains A". A partition of a set S is a set of nonempty subsets of S such that every element x in S is in exactly one of these subsets.

Set operations are also important aspects when it comes to NLP as the process involved in NLP relates to group of words. Two sets can be "added" together. The union of A and B, denoted by A B, is the set of all things that are members of either A or B. The intersection of A and B, denoted by A B, is the set of all things that are members of both A and B. Two sets can also be "subtracted". The relative complement of B in A denoted by A B (or A B), is the set of all elements that are members of A but not members of B. A new set can be constructed by associating every element of one set with every element of another set. The Cartesian product of two sets A and B, denoted by $A \times B$ is the set of all ordered pairs (a, b) such that a is a member of A and b is a member of B. The inclusion–exclusion principle is a counting technique that can be used to count the number of elements in a union of two sets, if the size of each set and the size of their intersection are known De Morgan's laws is one other important theory in set membership.

The triple of fuzzy intersection, union and complement form a De Morgan Triplet. That is, De Morgan's laws extend to this triple. Based on the mathematically definition of set theory and set membership, it is evident that set theory forms an integral part of natural language processing. NLP system runs on algorithms written based on set theory. The group of input into tokens falls under this mathematical calculation. The primary step in NLP itself forms a mathematical principle. Regular expressions are also useful because they enable us to specify complex languages in a formal, concise way. Fuzzy set is a mathematical model of vague qualitative or quantitative data, frequently generated by means of the natural language.

The model is based on the generalization of the classical concepts of set and its characteristic function [6]. A fuzzy subset A of U is defined by a membership function A mapping U into a closed unit interval [0,1], where for xU. Fuzzy set also plays equally important role in NLP system. If we consider bird's-eye view of a forest for example, the following gives an insight on Fuzzy set logic and how it is applied in NLP systems. Is location A in the forest? If the answer is Certainly yes, then forest(A) = 1. Is location B in the forest? If the answer is Certainly not, forest(B) = 0. Is location C in the forest? If the answer is Maybe yes, maybe not. It depends on a subjective opinion about the sense of the word "forest". The fuzzy set theory and related branches are widely applied in the models of optimal control, decision-making under uncertainty, processing vague econometric or demographic data, behavioral studies, and methods of artificial intelligence that is NLP. For example, there already exists a functional model of a helicopter controlled from the ground by simple" fuzzy" commands in natural language, like "up", "slowly down" "turn moderately left", "high speed", etc., which is also an offspring of NLP application. The processing of fuzzy sets generalizes the processing of the deterministic sets. Namely, if A, B are fuzzy sets with membership functions A, B, respectively, then also the complement A, union AB and intersection AB are fuzzy sets, and their membership functions are defined for xU. Fuzzy set theory is a mathematical theory whose program is to provide us with methods and tools which may make us possible to grasp vague phenomena instrumentally. Binary algebraic operation xy is extended to fuzzy quantities by extension principle, i.e., ab, where xR.

2.2 Mathematical Logic in NLP

As the concept of sets is present at the background of many fields of mathematical and related models, it is applied, e.g., to mathematical logic in NLP where each fuzzy statement is represented by a fuzzy subset of the objects of the relevant theory, or to the computational methods with vague input data where each fuzzy quantity or fuzzy number is represented by a fuzzy subset of R. Therefore, it is appropriate for using in modelling of natural language semantics. Humans, as well as NLP systems, interpret the meaning of individual words. Several types of processing contribute to word-level understanding – the first of these being assignment of a single part-of-speech tag to each word. In this processing, words that can function as more than one part-of-speech are assigned the most probable part-of speech tag based on the context in which they occur. This another example of how mathematics is involved in almost all the current NLP systems and the foundation that it has with the system's evolution.

Natural Language processing systems are now being employed in wide range of applications. These systems are found in simple websites to complex mathematical and statistical application. This paper also discusses the wide range of applications covered by NLP systems. Biologists are expected to answer large-scale questions that address processes occurring across broad spatial and temporal scales, such as the effects of climate change on species. This motivated the development of a new type of data-driven discovery focusing on scientific insights and hypothesis generation through the novel management and analysis of preexisting data. The information collected about effects and causes of change in biodiversity were captured through systems. However, the problem aroused with the data which was in human-readable form and too much for a person to transform into a digital data pool.

Scientists came to conclusion that machines can better handle the volume but cannot determine which elements of the text have value. To mobilize the valuable content in the literature, an innovative algorithm was needed to translate the entirety of the biological literature into a machine-readable form, extract the information with value, and feed it in a standards compliant form into an open data pool. Natural language processing was chosen for this problem as solution. The scientists with help of data scientists and machine learning experts employed complex NLP algorithms on enormous data that was in hand. Data Driven Discovery was the methodology handled by NLP experts to make the human readable input to machine understandable format. NLP approaches can extract large amounts of information from free text. However, biology text presents a unique challenge when compared to news articles to machine learning algorithms due to its ambiguity, diversity, and specialized language. Successful Information Extraction strategies. Research addressing the transformation of natural language text into a digital data pool is generally labeled as "information extraction" (IE). An IE task typically involves a corpus of source text documents to be acted upon by the IE algorithm and an extraction template that describes what will be extracted [7].

A variety of NLP techniques have been used in IE applications, but most only progress to syntactic parsing. More sophisticated techniques higher in the stack (semantic parsing and discourse analysis) are rarely used in IE applications because they are highly specialized that cannot be reliably applied in general applications and are more computationally expensive. Multiple tools were employed for fuzzy matching of terms, automated annotation, named-entity recognition, and morphological character extraction that use a variety of approaches. [10] Considering the constantly changing nature of biodiversity science and the constraints of NLP algorithms, best results were achieved by drawing information from high quality modern reviews of taxonomic groups rather than repositories of original descriptions. However, such works can be rare or nonexistent for some taxonomy. Thus, issues such as proper aggregation of information extracted from multiple sources on a single subject were few problems faced during the research on this application.

Natural Language Processing in recent times have been employed in various technology sectors. This includes Banking systems, Customer services and Online Education systems. Chat bots are notable application of NLP. Chat-bots can mimic human conversation and can offer personalized services. There are two types of chatbot application. First type of chat-bot is web-based chat-bot which runs on the cloud, and it can be accessed through web interface. Second type of chat-bot is a standalone chat-bot application which can be accessed on a single computer. Natural Language Processing platforms namely Dialogflow.com (formerly Api.ai), Wit.ai, Luis.ai and Pandorabots.com are assessed based on their Natural language understanding abilities and complex feature development ability.

A research study suggest that majority of the survey respondents are aware of educational use of mobile phone. Using NLP techniques in online educational forums have now become a common application. The intelligent chat-bot is capable of handling sub-intent goals of the user's input texts. The intelligent chat-bot is also capable of engaging in small talks with the learners. The intelligent chat-bot is accessible through the web browsers and android application. Node.js is used as runtime environment to process the server side and client-side requests. BotUI javascript framework is used to develop the User Interface in such applications. Machine Translation is other application of NLP. The idea behind MT is simple to develop computer algorithms to allow automatic translation without any human intervention.

3   Implication to the Community

3.1   NLP Contribution

Google translate is one existing application widely in use. Google translate is based on Statistical Machine Translation [8]. Google translate gathers as much text as it can find that seems to be parallel between two languages, and then it crunches data to find the likelihood that something in language. Machine translation is challenging given the inherent ambiguity and flexibility of human language. Sentiment analysis also known as opinion mining or emotion AI is an interesting type of NLP application that measures the inclination of people's opinions. The task of this analysis is to identify subjective information in the text. Sentiment analysis helps to check whether customers are satisfied with goods or services. Classical polling method have been replaced by these systems involving NLP. In shrunken world, people willingly share their opinions on social networks. The search for negative texts and the identification of the main complaints significantly helps to change concepts, improve products and advertising, as well as reduce the level of dissatisfaction. In turn, explicit positive reviews increase ratings and demand. NLP has contributed a lot by dealing with terabytes of raw data found in social media, customer websites. Marketers also use NLP to search for people with a likely or explicit intention to make a purchase. Behavior on the Internet, maintaining pages on social networks and queries to search engines provide a lot of useful unstructured customer data. Selling the right ad for internet users allows companies like Google to make the most of its revenue.

Advertisers pay Google every time a visitor clicks on an ad. At its core, market intelligence uses multiple sources of information to create a broad picture of the company's existing market, customers, problems, competition, and growth potential for new products and services. Sources of raw data for that analysis include sales logs, surveys, and social media, among many others. Automatic Summarization is one other application of NLP. Question answering (QA) is concerned with building systems that automatically answer questions posed by humans in a natural language. All of them have a few NLP-applications or functions to understand speech is only half of the path and another one naturally is to give a response. A spell checker is a software tool that identifies and corrects any spelling mistakes in a text. Most text editors let users check if their text contains spelling mistakes. This application has NLP as its base. Character Recognition systems also have numerous applications like receipt character recognition, invoice character recognition, check character recognition, legal billing document character recognition, and so on which also adds to the bucket of NLP systems. The first in-depth science mapping analysis, as well as an STM analysis of the NLP-enhanced clinical trial research, is being conducted, especially for disclosing its structure and evolution. There are several research studies presenting discussions on issues regarding NLP-enhanced clinical trial research. For example, by including all the applications or methodological articles leveraging texts to support healthcare and meet customers' needs, researchers presented a systematic review of articles concerning the application of NLP to clinical and consumer-produced textual data [9, 10].

Their study indicated that although clinical NLP continued to facilitate practical applications with an increasing number of NLP approaches being adopted to process large-scale live health information, there was a need to enable the adoption of NLP in clinical applications available in our daily life. Firstly, keyword searches, rule-based approaches, and supervised NLP techniques based on machine learning were the most adopted approaches. Second, well-defined keyword search and rule-grounded systems usually achieved high accuracy.

Third, classifications based on supervised methods could achieve higher accuracy and was easier to conduct, as compared to unsupervised methods. An analysis of scientific collaboration was made by scientists to find novel approach to deal with clinical data.

Strategic diagrams were drawn to explore the research themes of high importance in the NLP-enhanced clinical trials research field, where the sphere size was proportional to articles of each research theme. From the perspective of countries/regions, several countries such as the USA, the UK, Australia, Canada, and China seemed to show balanced interest in diverse topics in the research. However, comparatively, the others showed particular interest in one or more topics. Their study demonstrated that deep NLP had potential at speeding up the curation of diseases. This study is the first to apply the STM approach to the NLP-enhanced clinical trial research, which was able to get rid of the limitations by using manual coding or word frequency analysis.

3.2   The Consequences

Text classification in medical domain could result in an easier way of handling large volumes of medical data. They can be segregated depending on the type of diseases, which can be determined by extracting the decisive key texts from the original document. Due to various nuances present in understanding language in general, a requirement of large volumes of text-based data is required for algorithms to learn patterns properly. The problem with existing systems such as MedScape, MedLinePlus, Wrappin, and MedHunt is that they involve human interaction and high time consumption in handling a large volume of data. By employing automation in this proposed field, the large involvement of manpower could be removed which in turn speeds up the process of classification of the medical documents by which the shortage of medical technicians in third world countries can be addressed.

This is another research area employing NLP. Classification of the document into generic tags enables the users of the application or product to easily access the desired information within the application without facing any kind of problems or troubles. The introduction of Text classification in healthcare could turn out to be productive if they are implemented carefully. One of the major sources for data related to healthcare is the Internet. The number of pages available on the Internet is significantly growing every year which in turn increases the amount of data related to health.

Medical information has attracted a wide range of audience since its introduction in the electronic form. Health related information is now widely available for use to the public. They are being provided by several health organizations, medical universities and government institutions in a validated form which can be readily used by the public users. The users of the medical domain can either be end-users or the professionals involved in medicine. The end-users navigate through the medical documents when they must search for the desired information related to a particular disease. On the other hand, professional experts require Text classification to ease their hectic processes of locating the documents which may be required during case studies of similar diseases. The introduction of Text classification in healthcare could turn out to be productive if they are implemented carefully. One of the major sources for data related to healthcare is the Internet.

The number of pages available on the Internet is significantly growing every year which in turn increases the amount of data related to health. Medical information has attracted a wide range of audience since its introduction in the electronic form. Health related information is now widely available for use to the public. They are being provided by several health organizations, medical universities and government institutions in a validated form which can be readily used by the public users. The users of the medical domain can either be end-users or the professionals involved in medicine. The end-users navigate through the medical documents when they must search for the desired information related to a particular disease. On the other hand, professional experts require Text classification to ease their hectic processes of locating the documents which may be required during case studies of similar diseases. Various NLP algorithms are being developed to address this problem in medical industry.

4   Conclusion

NLP become very popular and important part of our life because this technology which is increasing day by day reduces human effort in terms of work mainly in communication. As we have seen that, NLP provide a wide set of techniques and tools which can be applied in all the areas of life. NLP techniques help us improving our communication, our goal reaching and the result we receive from every day. By using NLP provides the feature of questioning-answering which play a vital role in communication. This paper basically serves or give the detail communication between the machine and user which can happen through the different levels in NLP. While using NLP features the main advantage is time consumption for taking input as well as giving the output.

References

1. Johri, P., Khatri, S.K., Al-Taani, A.T., Sabharwal, M., Suvanov, S., Kumar, A.: Natural language processing: history, evolution, application, and future work. In: Abraham, A., Castillo, O., Virmani, D. (eds.) Proceedings of 3rd International Conference on Computing Informatics and Networks. Lecture Notes in Networks and Systems, vol. 167. Springer, Singapore (2021). https://doi.org/10.1007/978-981-15-9712-1_31J
2. Podrebarac, A.A.: Introduction to Natural Language Processing. Fragmentation of the Photographic Image in the Digital Age, pp. 204–211. Routledge (2019)
3. Garbade, M.J.: A simple introduction to natural language processing. Becoming Hum. Artif. Intell. Mag. 15 (2018)
4. Roy, S.: Math, Stats and NLP for Machine Learning: As Fast As Possible, 9 February 2018. https://medium.com/meta-design-ideas/math-stats-and-nlp-for-machine-learning-as-fast-as-possible-915ef47ced5f. Accessed 4 Apr 2020
5. Galassi, A., Lippi, M., Torroni, P.: Attention in natural language processing. IEEE Trans. Neural Netw. Learn. Syst. 32(10), 4291–4308 (2020)
6. Zhao, R., Mao, K.: Fuzzy bag-of-words model for document representation. IEEE Trans. Fuzzy Syst. 26(2), 794–804 (2017)
7. Nova'k, V.: An Introduction to Fuzzy Logic Applications in Intelligent Systems. https://link.springer.com/chapter/10.1007/978-1-4615-3640-68. Accessed 6 Apr 2020
8. Kharkovyna, O.: Natural Language Processing (NLP): Top 10 Applications to Know, 19 December 2019. https://towardsdatascience.com/natural-language-processing-nlp-top-10-applications-to-know-b2c80bd428cb. Accessed 6 Apr 2020
9. Dutta, D.: Developing an Intelligent Chat-bot Tool to assist high school students for learning general knowledge subjects. Accessed 10 Apr 2020
10. Polyakova, E.V., Voskov, L.S., Abramov, P.S.: Generalized approach to sentiment analysis of short text messages in natural language processing. Accessed 11 Apr 2020

Non-fungible Tokens and Their Applications

Jeet Patel, Delicia Fernandes, Darshkumar Jasani, Kunjal Patel,
and Muhammad Abdul Basit Ur Rahim[✉]

Computer Engineering and Computer Science, California State University,
Long Beach, CA, USA
{jeetatul.patel01,deliciadomnic.fernandes01,
darshkumarpravinbhai.jasani01,
kunjalashvinbhau.patel01}@student.csulb.edu, m.basit@csulb.edu

Abstract. We are moving toward digitization, and the word used most frequently while describing it is "money." This is backed up by "Bitcoins," a type of virtual currency that can be traded via Blockchain, a distributed ledger technology. Bitcoins are essentially extended sequences of code that have been changed to take on monetary value. However, Non-fungible tokens (NFTs) were created in late 2017 as a brand-new class of distinct and undivided blockchain-based currency. While fungible tokens have enabled new use cases, such as initial coin offerings, it is unclear if NFTs will be applicable. NFTs advance the reinvention of this infrastructure by enabling digital representations of physical assets. Because this technology is distributed and immutable, it is difficult to modify, replicate, or fake transactions. This study attempts to assist and educate the research community about this new disruptive technology and direct them in the correct steps for blockchain technology adoption.

Keywords: Blockchain · Tokenization · Smart Contract · Cryptocurrency · Non-Fungible Token · Applications

1 Introduction

Non-fungible tokens (NFTs) have attracted a lot of media coverage and increased public interest. A flood of media coverage helped NFTs gain popularity, leading to more than billions in transactions in 2021. The rapid development of NFTs has raised more fundamental problems, such as what they are, what it means to "possess" a work of digital art that is freely available online, and why anyone would pay for such ownership. At the same time, the media focuses on these startling statistics. NFTs have received much publicity, but because they are a new phenomenon, there are misconceptions about them. The Non-Fungible Token (NFT) is an essential application of blockchain technology [1].

A non-fungible token is, according to Merriam-definition, Webster's "a unique digital identifier that cannot be copied, substituted, or divided, that is recorded in a blockchain, and that is used to certify authenticity and ownership (as of a specific digital asset and specific rights relating to it)" [5]. The term "non-fungible token" (NFT) refers to a non-transferable data unit is kept on a blockchain, a digital ledger. Digital items such as images, videos, and audio are included in NFT data units. NFTs are distinct from blockchain cryptocurrencies like BTC and LSK since each token can be uniquely identified. NFTs are tokens that can be used to signify ownership of certain goods [8]. Thanks to them, we can tokenize items like works of art, valuables, and even real estate. The Ethereum blockchain secures them so that no one can change the ownership record or create a new NFT by copying and pasting an existing one. They can only have one official owner at a time. The high-vibe NFT marketplace is an online venue where people may build and advertise their NFTs.

The current traditional arrangement does not resolve the problem of digital art ownership. For illustration, let's imagine someone has a digital artwork or image worth a few thousand dollars. He showcases him on social media to show off his new item. This creates a situation where anyone with internet access might screenshot this image and use it to their advantage. As a result, there are increasing ownership difficulties with digital art. Since NFTs can be traced back to their owners and are valued for their data rather than the content (such as images, audio, or videos), they solve the ownership problem.

2 Background

2.1 Blockchain and Its Types

Blockchain refers to a shared, immutable, digitally decentralized ledger that facilitates recording transactions in a business network. The way we gather and distribute information will alter as a result of the new technology known as the blockchain. In simple words, an online database that anyone with internet access can approach from any location at any time all around the globe. Unlike a conventional database, it is not controlled by any central organization or institution like banks and governments, unlike conventional databases. Therefore, it becomes practically impossible to hack or tamper with the overall network based on this technology by falsifying transactions, records, and any information [11].

Blockchain technology is a radical innovation that can challenge or even replace existing business models relying on third parties for trust [3]. Blockchain technology integrates other technologies, including consensus processes, peer-to-peer connections, smart contracts, and cryptography, to produce a brand-new database. Additionally, it records each transaction's time, date, participant information, and other legislative or contractual aspects. Cryptocurrencies like Bitcoin and Ethereum are primarily powered by blockchain technology, which makes digital trading secure by verifying and keeping transaction records in a distributed, time-stamped manner. In his article titled "Bitcoin: A Peer-to-Peer Electronic Cash System," Satoshi Nakamoto first used the term "Bitcoin," a cryptocurrency, in 2008 [12].

The Blockchain provides fast, shareable, and entirely transparent information but is only permitted to operate with the approval of authorized members. It is important to remember that Blockchain Technology has applications in tracking orders, payments, accounts, and manufacturing. The end customers can occasionally see transaction details here as well. Blockchain technology has applications in various fields and industries, including finance, trade, entertainment, government, education, and health and medical systems. The number of these applications is steadily growing. Because of this, blockchain technology is immune to fraud and can change the financial industry in a sophisticated and transparent way. Additionally, while blockchains are ideal for smaller works of art, additional support is needed when the artwork is larger than is feasible to store on the blockchain [2].

Public and Private blockchains are the two main types of blockchain technology, but there are also two other varieties: Hybrid and consortium or Federated. Following is a quick discussion of each form of blockchain listed above:

Public Blockchain: Publicly accessible Blockchains are known as public Blockchains. A Public Blockchain is a distributed ledger technology that allows anybody to join and perform transactions (whether a node or an end user) [11]. They can access, read, write, and update the Blockchain without requiring unique authentication (login with user id and password). Cryptocurrencies like Bitcoin, Ethereum, and Litecoin, among others, employ this technology.

Private Blockchain is a form of restricted or permission-based Blockchain technology that can only be used in a closed network. Typically, this Blockchain is used inside a company or organization, and only authorized users can access and conduct network transactions. To access, read, write, and update on the blockchain, members of this network must first log in with a user ID and password. These blockchains typically have lower levels of trust than public blockchains. A private blockchain can be used for supply chain administration, digital identification, and elections, among other things [11, 13].

Consortium Blockchain: Permissioned blockchain is another name for consortium blockchain. Instead of being controlled by a single organization or network node, this blockchain network is administered by several companies or nodes. It is the perfect remedy, but everyone must work together to make it work [11]. For instance, a brand-across alliance would be necessary for the supply chain, food, and medication. For example:- Ripple, the Energy Web Foundation, etc.[13].

2.2 Smart Contracts

Smart contracts are a type of computer program that specializes in the automation and execution of an agreement to ensure that all members of the blockchain can trust the outcome of the results without any time delay. They are programs that are stored on a blockchain and are executed when a specific condition that is set beforehand has been met. At the moment, there is a multitude of cryptocurrency blockchains such as Neo, Tezos, Tron, etc. out of which Etherium is the most pre-eminent. Any individual on a blockchain can create and deploy a smart contract. One of the main use cases of a smart contract on a blockchain is eliminating the middleman to facilitate easy trade between anonymous and/or identified parties. In summary, smart contracts help us implement a transparent system without the requirement of a middleman, and all blockchain members can trust the outcome. As a direct result of this, there is a reduction in cost and latency.

Table 1. Comparison table based on the Focus Area, Applications, and Limitations of NFT.

Focus Area	Applications	Limitations
Digital Art	Royalty sharing Expose Art to Larger Markets Versatile Utilization	Immutability Environmental Impact
Licenses & Certifications	Course completion certificates Degrees Licenses	Intellectual Property Rights
Artifact	Integrity Availability Privacy Secondary Market	Security and Privacy Issues (Transparency as a curse)
Fashion	Luxurious digital wearables	Environmental Impact
Sports	Digital autographs Avatars Stickers Tickets Game Highlights	Uncertain Value
Domain Names	Candy.com Sushi.com	Need to own ETH to purchase Uncertain value
Virtual World	Metaverse Decentral and	Environmental Impact Highly energy intensive
Tokenized Digital Assets	Cryptokitties Unique Digital Assets Tradable Digital Assets	Physical Art cannot be digitized Fluctuating value
Collectibles	Trading cards Memes Tweets	Uncertain Value Environmental Impact

3 Analysis

Table 1 discusses the focus area, applications, and limitations of NFTs that provide a new opportunity for digital content creators to make their art available to a broader community, eventually helping them make more money by digitizing assets, commercializing intellectual property, and certifying the ownership of real assets online. NFT domain name systems are a collection of smart contracts, a fancy word for software published on the blockchain. This means that power is returned to you, the user, instead of a single entity owning your data online. And, because it is constructed on the blockchain, everyone may access the data stored there, resulting in complete openness and transparency. Art and collectible NFTs were primarily ideas with "potential" a year ago, but no one outside the crypto world took them seriously. What categories are currently at that early speculation stage and are likely to erupt? Table 1 gives an overview of the application of NFTs and their limitations, respectively.

NFT Token Standards by OpenZeppelin: OpenZeppelin is an open-source platform for developing safe decentralized applications (dApps). The framework includes the tools needed to construct and automate Web3 applications. Furthermore, businesses of any size can use Open-Zeppelin's audit services to uncover the best practices in the market. It performs security audits on your behalf and applies security measures to protect the security of your dApps. After identifying potential faults in the code, they deliver a report comprising best practices and recommendations to remove the system's weaknesses.

Furthermore, OpenZeppelin Defender is a web application developed by OpenZeppelin. It is a security and automation platform for smart contract operations. You may use Defender to collaborate with your team, build multiple workflows, deal with contracts manually, and execute financial transactions. Furthermore, Defender provides access to a user-friendly UI and the infrastructure required for sending transactions and establishing automated scripts. Furthermore, OpenZeppelin has amassed an impressive library for developing smart contracts, which now power over 3,000 public projects, details given in Table 2.

Table 2. ERC721 Functions and methods explained

Function name	Uses
balance(address owner)	Returns total number of tokens
owner(uint256 tokened)	Returns address of token owner
safe transfer from(address from, address to, uint256 tokened, bytes data)	Safely transfers tokens of "tokened" from from-Address to to-Address
safe transfer from(address from, address to, uint256 tokenId)	Transfer tokens: "token" from "from"-Address to "to"-Address
transfer from(address from, address to, uint256 tokenId)	Transfers "token" token from owner-Address tokened-Address
approve(address to, uint25tokenId)	safe transfer from to-Address
setApprovalForAll(address operator, bool tokenId)	Approves or removes the operator
approved(uint256 tokened)	Returns the account safe transfer from approved for token token
isApprovedForAll(address own, address operator)	Returns if operator is allowed to manage all the assets of owner

ERC721: OpenZeppelin provides three NFT standard ERC721 which is an OpenZeppelin smart contract standard. It comes pre-configured with token minting (token creation) and auto-generation token ID and URI. It allows holders to burn their tokens and permits them to halt all token transfers. The contract employs access control to manage access to the minting and halting functions. The minter and pauser roles, as well as the default admin role, will be granted to the account that deploys the contract [25].

ERC-721 is a free and open standard for creating non-fungible or unique tokens on the Ethereum network. While most tokens are fungible (every token is identical to every other token), ERC-721 tokens are entirely distinct. ERC-721 specifies the minimum interface a smart contract t must provide for unique tokens to be managed, owned, and sold. It does not impose a standard for token metadata or limit the addition of extra functions. The following are the limitations of ERC-721:

- It does not support batch transfers.
- ERc721 has only support for the creation of non-fungible tokens.
- Individual transactions and smart contracts are required for each token type.
- It is impossible to revert transactions after transferring assets to the wrong address using ERC721 [26].

ERC1155: It is a novel token standard that aims to take the best from previous standards to create a fungibility-agnostic and gas-efficient token contract. It addresses the limitations of the ERC721 and tries to overcome them. This specification defines a smart contract interface that can represent an unlimited number of fungible and non-fungible token kinds (Table 3).

Table 3. ERC1155 Functions and methods explained

Function name	Uses
balanceOf(address account, uint256 id)	Returns number of tokens of "id" type, where "account" cannot be null
balanceOfBatch(address accounts, uint256 ids)	It is a batched version of "balanceOf"
setApprovalForAll(address operator, bool approved)	Allows or revokes permission to "operator" according to "approved"
isApprovedForAll(address account, address operator)	Returns true if "operator" has been approved to transfer "account" tokens
safeTransferFrom(address from, address to, uint256 id, uint256 amount, bytes data)	Transfers token type "id" with "amount" tokens from "from" to "to"
safeBatchTransferFrom(address from, address to, uint256 ids, uint256 amounts, bytes data)	Batched version of "safeTransferFrom"

Existing standards, like ERC-20, require individual contracts for each coin type. The token ID in the ERC-721 standard is a single non-fungible index, and the collection of these non-fungibles is deployed as a single contract with settings for the entire collection. On the other hand, the ERC-1155 Multi Token Standard permits each token ID to represent a new customizable token type, each with its information, supply, and other features.

The following are the advantage of ERC1155 over ERC721

- Single, smart contracts of ERC1155 tokens are capable of supporting multiple functions

- It supports batch transfers
- It supports the conversion of fungible tokens to non-fungible tokens and vice-versa
- For the security of assets safe transfer function enables verification of transaction validity and reverses the transactions [26].

4 Applications

One of the fundamental building blocks of many blockchain transfers from smart contracts is crucial in many application-focused blockchains like Ethereum. These digital contracts are transparent, decentralized, autonomous, and trustless. Once implemented, they are frequently irrevocable and unchangeable. The benefits of smart contracts include the elimination or reduction of the need for middlemen and contract enforcement in a deal or transaction.

NFTs Use Cases in Finance: Decentralized finance (Defi) dApps are a vital substitute for conventional financial services. Their acceptance is rising due to the immutable, transparent, and trustless properties of blockchain and smart contract technology. Users can take advantage of this new generation of financial services without paying intermediary costs or for centralized custody.

Blockchain and NFTs in Gaming: The billion-dollar global gaming business is expanding fast, but value creation and distribution within the sector can sometimes be unequal. Non-fungible tokens (NFTs), distinctive digital assets for in-game content, are frequently the engine behind blockchain technology in the gaming industry. Smart contracts are crucial to NFTs because they enable player ownership and proved scarcity, interoperability, and immutability.

NFTs and Blockchain in the Legal Industry: With the introduction of e-signatures for legally binding agreements, technology has been driving innovation in the legal sector. Smart contracts could soon provide parties to legal agreements with a choice and reduce the expense of engaging lawyers. Some U.S. states have started to approve the use of smart contracts in certain circumstances.

NFTs Uses in Corporate Structures: Legislation in Delaware paved the way for the growth of decentralized autonomous organizations (DAOs). DAOs operate as businesses and allow for the inclusion of ownership and compensation in smart contracts. DAOs can enable complex, automatically enforced incentive structures within a corporate framework. They can also generate savings in administrative costs like office space, hiring, and payroll.

NFTs in Emerging Technology: The capacity of blockchain technology to simplify difficult computing operations like those involved in machine learning and artificial intelligence is one of the most interesting uses of these technologies (AI). Applications for smart contracts will need to develop into increasingly complicated systems as more and more sectors of the economy adopt them. Many professionals believe that the distinctive qualities of blockchain and AI could complement one another.

Potential NFTs Advantages in Other Industries: In the future, smart contract technology has the potential to advance countless additional fields. Many scientists are keen to take advantage of its advantages to meet the expanding Internet of Things demands. The world of digital agreements is set to continue being revolutionized by smart contracts and dApps.

Blockchain in Real Estate: By adding blockchain to real estate deals, smart contract technology can remodel the documentation and transaction processes. Some experts assert that smart contracts can benefit parties by simplifying complex credit or mortgage agreements. The necessity for legal advice or other advisory services becomes less critical with using smart contracts and blockchain in real estate.

5 Conclusion and Future Work

Non-fungible tokens are one-of-a-kind digital representations of assets on a blockchain. As the world investigates how distributed, immutable ledgers might make transactions safer and speedier, NFTs play an important role. These assets have a preserved transaction history and the potential to expedite commerce and are a cornerstone in the rising digital world.

In this study, we aimed to raise awareness of NFTs and their ever-expanding use cases for assisting researchers. In future work, we plan to investigate how NFTs have no environmental impact but how the minting of NFTs can have significant environmental consequences. Furthermore, we intend to address the issue of the high gas fee that needs to be paid for its creation, which is the money paid to blockchain miners for processing transactions embedded in the blockchain. In the energy transition, there is an urgent need to decrease overall carbon emissions [4].

References

1. Chen, H., Cheng, Y., Deng, X., Huang, W., Rong, L.: AB-SNFT: securitization and repurchase scheme for non-fungible tokens based on game theoretical analysis. In: Eyal, I., Garay, J. (eds.) Financial Cryptography and Data Security. FC 2022. LNCS, vol. 13411, pp. 407–425. Springer, Cham (2022). https://doi.org/10.1007/978-3-031-18283-920
2. Basu, S., Basu, K., Austin, T.H.: Crowdfunding non-fungible tokens on the blockchain. In: Chang, SY., Bathen, L., Di Troia, F., Austin, T.H., Nelson, A.J. (eds.) Silicon Valley Cybersecurity Conference. SVCC 2021. CCIS, vol. 1536, pp. 109–125. Springer, Cham (2021). https://doi.org/10.1007/978-3-030-96057-58
3. Regner, F., Schweizer, A., Urbach, N.: NFTs in practice – non-fungible tokens as core component of a blockchain-based event ticketing application (2019). https://www.researchgate.net/publication/336057493. NFTs in Practice - Non-Fungible Tokens as Core Component of a Blockchain-based Event Ticketing Application
4. Babel, M., Gramlich, V., Körner, M.F., et al.: Enabling end-to-end digital carbon emission tracing with shielded NFTs. Energy Inform. 5(Suppl. 1), 27 (2022). https://doi.org/10.1186/s42162-022-00199-3
5. Merriam-Webster NFT definition; meaning. In: Merriam-Webster. https://www.merriam-webster.com/dictionary/NFT. Accessed 19 Nov 2022

6. Hofstetter, R., de Bellis, E., Brandes, L., et al.: Crypto-marketing: how non-fungible tokens (NFTs) challenge traditional marketing. Mark. Lett. (2022). https://doi.org/10.1007/s11002-022-09639-2

7. Kallet, R.H.: How to write the methods section of a research paper. Respir. Care **49**(10), 1229–1232 (2004). PMID: 15447808

8. Felin, T., Lakhani, K.: What problems will you solve with blockchain? MIT Sloan Manag. Rev. **60**, 32–38 (2018)

9. Lidén, E.: Potential Advantages and Disadvantages of NFT-Applied Digital Art (Dissertation) (2022). http://urn.kb.se/resolve?urn=urn:nbn:se:uu:diva-478464

10. Blockchain Technology and its Applications Across Multiple Domains: A Survey by Wajde Baiod, Janet Light, Aniket Mahanti (2021). https://scholarworks.lib.csusb.edu/cgi/viewcontent.cgi?article=1482

11. Suman, A., Patel, M.: An introduction to blockchain technology and its application in libraries. Library Philos. Pract. (e-journal) **6630** (2021). https://digitalcommons.unl.edu/cgi/viewcontent.cgi?article=12649

12. A Detailed Study of Blockchain: Changing the World by Shweta Singh, Anjali Sharma, Dr. Prateek Jain. https://www.ripublication.com/ijaer18/ijaerv13n1426.pdf

13. Blockchain Technology and its Types by P K Aithal, P S Saavedra, Sreeramana Aithal, Surajit Ghoash December 2021 (IJASE) (2021)

14. Chohan, U.W.: Non-fungible tokens: blockchains, scarcity, and value (March 24, 2021). Critical Blockchain Research Initiative (CBRI) Working Papers (2021). Available at SSRN: https://doi.org/10.2139/ssrn.3822743

15. Popescu, A.-D.: Non-Fungible Tokens (NFT) - innovation beyond the craze (2021)

16. Nadini, M., Alessandretti, L., Di Giacinto, F., et al.: Mapping the NFT revolution: market trends, trade networks, and visual features. Sci. Rep. **11**, 20902 (2021). https://doi.org/10.1038/s41598-021-00053-8

17. Rehman, W., Zainab, H., Imran, J., Bawany, N.: NFTs: Applications and Challenges (2021). https://doi.org/10.1109/ACIT53391.2021.9677260

18. Khan, S.N., Loukil, F., Ghedira-Guegan, C., et al.: Blockchain smart contracts: applications, challenges, and future trends. Peer-to-Peer Netw. Appl. **14**, 2901–2925 (2021). https://doi.org/10.1007/s12083-021-01127-0

19. Survey on blockchain-based smart contracts: technical aspects and future research. IEEE Access (2021). https://doi.org/10.1109/ACCESS.2021.3068178

20. Wang, S., Yuan, Y., Wang, X., Li, J., Qin, R., Wang, F.Y.: An overview of smart contract: architecture, applications, and future trends. In: 2018 IEEE Intelligent Vehicles Symposium (IV) (2018). https://ieeexplore.ieee.org/document/8500488

21. An Overview of Smart Contract: Architecture, Applications, and Future Trends. https://ieeexplore.ieee.org/document/8500488

22. Smart Contracts on the Blockchain – A Bibliometric Analysis and Review. https://papers.ssrn.com/sol3/papers.cfm?abstractid=3576393

23. A Review of Smart Contracts Applications in Various Industries: A Procurement Perspective. https://www.hindawi.com/journals/ace/2021/5530755/

24. An Introduction to Smart Contracts and Their Potential and Inherent Limitations. https://www.skadden.com/insights/publications/2018/05/an-introduction-to-smart-contracts-and

25. Documentation OZ ERC721. In: OpenZeppelin Docs. https://docs.openzeppelin.com/contracts/4.x/erc721. Accessed 19 Nov 2022

26. Howell, J.: ERC 1155 vs. ERC 721 – key differences. In: 101 Blockchains. https://101blockchains.com/erc-1155-vs-erc-721/. Accessed 19 Nov 2022

Blockchain and Distributed Ledger Technology (DLT) in Business

DLT-Based Central Bank Digital Currency Key Concepts

Elcelina Carvalho Silva[1,2(✉)] ⑩ and Miguel Mira da Silva[2] ⑩

[1] Universidade de Cabo Verde, Campus Palmarejo Grande, CP. 379-C, Praia, Cabo Verde
Elcelina.silva@docente.unicv.edu.cv
[2] Instituto Superior Técnico, Avenida Rovisco Pais, CP 1049-001 Lisbon, Portugal
mms@tecnico.ulisboa.pt

Abstract. CBDC could be a decentralized application (Dapp) if it is based on decentralized ledger technology (DLT), such as blockchain, but it is not yet standardized with a common language. Central banks, other supervisory agencies, international policymakers, industry, technology providers, and software developers use different terminology to describe CBDC, instigating ambiguities and mistakes in the CBDC understanding. This research aims to survey the key concepts by which CDBC operates and analyze the relationships between key concepts. The key concepts of CBDC design could be a blueprint and a research tool to promote dialog between scientific communities and practitioners, simplify the collaboration between researchers and policymakers, namely central banks and other regulators, industry, and software developers in CBDC Design and, improve the secure implementation of the CBDCs with a schematic representation of the actual knowledge of CBDC produced by researchers and practitioners.

Keywords: Decentralized Ledger Technology · Central Bank Digital Currency · Decentralized Application · Concepts

1   Introduction

Distributed ledger technology (DLT) has established itself as an umbrella term to designate multi-party systems that operate in an environment with no central operator or authority [1]. This technology is transforming payment, clearing, and settlement (PCS) processes, including how funds are transferred and how securities, commodities, and derivatives are cleared and settled [2], combining several other technologies and computing concepts to create modern cryptocurrencies: electronic cash protected through cryptographic mechanisms instead of a central repository or authority [3].

Central Bank Digital Currency (CBDC) is described as a digital form of central bank money that can be exchanged in a decentralized manner, known as the peer-to-peer network [4], similar to cash (or coin) but in digital form, meaning that transactions can occur directly between the payer and the payee without the need for a central intermediary. It is defined as central bank-issued digital money denominated in the national unit of

A. Rocha et al. (Eds.): WorldCIST 2023, LNNS 802, pp. 299–308, 2024.
https://doi.org/10.1007/978-3-031-45651-0_30

account that represents a central bank liability in a digital form, unlike cash (physical coins) and banknotes [5].

CBDC represents a technologically advanced representation of central bank money that, if it is well designed, could offer a safe, neutral, and final means of settlement for the digital economy [6]; it presents an opportunity to design central bank money that preserves the core features of central bank money: finality, liquidity, and integrity [7].

The CBDC research challenges include international cooperation and harmonization of regulation to mitigate the potential risks of cryptocurrency; standards for card-based payments and micropayments, security standards in the financial sector applied to blockchain-based financial transactions in 5G networks, and standards for accounting and tax [8]. These authors emphasize that the main issues of CBDC research are solutions for managing CBDC through distributed networks; the design features such as value-based, account-based, wholesale, retail-oriented, interest-bearing, non-interest bearing; risks of unregulated digital currencies on monetary policy; categorization and identification of factors that could influence the intention to use CDBC.

The CBDC research comprises several dimensions, such as financial, legal, technological, and social [9]. As digital money, all dimensions of its regulation connect to the technological theory and principle by which the digital currency operates. Still, the technological concept of CBDC is not yet systematized with a common language that facilitates its understanding, instigating the use of different terminologies to describe CBDC by researchers and practitioners. This fact is a challenge for academics and industry to achieve robust CBDC development with the increasing need for a common language to improve the safe and secure development of CBDCs.

At this moment, the early stage of development and several initiatives is being created by policy institutions, Non-Governmental Organizations (NGOs), and IT providers to research CBDC. The Bank of International Settlement (BIS) has been publishing reports since August 2020 reporting the central bank's research and experiments on retail, wholesale and cross-board use cases. According to the last report published on July 2022[1], 29 pilots and 72 central banks have communicated publicly about their CBDC research.

In this context, we propose to survey the key concepts of CBDC and analyze the relationship between these concepts using the literature review research method proposed by [10].

The paper is organized into five sections: Here, in Sect. 1, we introduce the research objective, methodology, and structure. Section 2 explains the research method process. The results of the research questions are presented in Sect. 3. Finally, concluding remarks are made in Sect. 4.

2   Research Method

How citizens use money and make payments is changing in response to new technologies and more demanding expectations from consumers and businesses. On the one hand, the decline in the use of cash as consumers switch to card and electronic payments

---

[1] https://www.bis.org/publ/work880.htm.

is challenging governments and central banks to ensure a fast, efficient, and flexible payment system. On the other hand, new alternative digital currencies (DTL-Based digital currency or cryptocurrency) are being created with characteristics quite different from conventional money by companies that operate on a global scale, like bitcoin, demanding a response in terms of regulation.

To survey the key concepts of CBDC Design and their relationships, we chose to identify the most common terminology and concepts used by researchers and practitioners like central banks, policymakers, industry, technology providers, and software developers to describe CBDC. From this focus, we defined the following research questions (RQ):

- **RQ1:** What are the key concepts of the CBDC domain?
- **RQ2:** How do the concepts of the CBDC relate to one another?

We select scientific papers from different data sources to retrieve a maximum of studies using several keywords in the **search string**:

```
("central bank") AND ("central bank digital currency") AND
("decentralized application").
```

To find all relevant scientific literature, we adapted the search string in all the most important electronic digital libraries in software engineering: Google Scholar, semantic scholar, ACM, IEEE, web of science, science@Direct, Scopus, and Springer to find scientific and conference papers. In addition, we adjust our search query for each database with an advanced search string.

We also selected the grey literature using the google search engine and websites of Arxiv, and policymakers' institutions that are conducting research in CBDC, such as the International Monetary Fund (IMF), European Union (EU), Bank for International Settlement (BIS), World Economic Forum (WEF), International Telecommunication Union (ITU), and Central Banks. We also researched on websites of technology providers like Hyperledger, consensus R3/Corda, and others to find reports and white papers on CBDC research practice.

We used the selection criteria proposed by [10] to identify studies that provide direct evidence for the research questions. We defined inclusion and exclusion selection criteria as follows:

- **Inclusion criteria:** The selected studies must be in English, with author and date information, accessible, not duplicated, the content should be focused on CBDC design or implementation, and the document's quality needs to be available. The paper could be included through the snowballing process (Heuristics).
- **Exclusion Criteria:** Studies not written in English, without date, author unidentified, title and abstract not related to the CBDC, document type like site news, social network post, PowerPoint file, video, poster, and newspaper, duplicated, content not related to the research questions, quality of the document not available.

The selection studies were done according to the PRISMA guideline [11] which includes the identification, screening, eligibility, and inclusion steps (See Fig. 1).

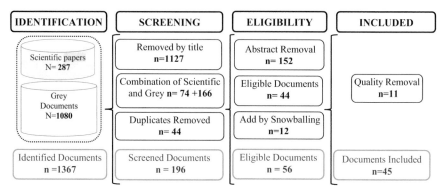

Fig. 1. Studies Selection process based on the PRISMA guideline [11]

3 Results

In this section, we will report the review results presenting the key concepts of the CBDC and the relationships between these concepts.

3.1 Constructs of the CBDC Design (RQ1)

This section presents a list of concepts used in CBDC design and their characteristics (see Table 1). We identified the key concepts of CBDC Design using the multivocal literature review. These concepts are function, proposal, participant, layer, life cycle, principle, model, access, digital asset, architecture, infrastructure, type, use case, and governance.

Table 1. Concepts of the CBDC and Their Relationships

Concepts	Characteristics	References
Function	Medium of Exchange, Store of Value, Unit of Account	[12–16]
Proposal	Monetary Policy, Financial Stability, Safe Payment System, Digital Innovation, Digital Inclusion, Digital Transformation	[17–22]
Participant	Central bank, government regulatory agencies, accounting office, audit bodies, commercial banks, clearing institutions, non-banked financial institution, Payment Service Provider, business, household, citizen	[12, 23–26]

(continued)

Table 1. (*continued*)

Concepts	Characteristics	References
Layer	Regulatory layer, Regulated Layer, User layer	[14, 23, 26, 27]
Life cycle	Issuance, management, distribution, circulation, withdraw, redeem	[14, 24, 26, 28–30]
Model	Direct, hybrid, indirect	[28, 29, 31–33]
Digital Asset	Digital form, issued by a central bank, algorithm-based, tokenized, encrypted, decentralized	[12–15, 21, 22, 34–37]
Types	Wholesale, Retail	[12, 15, 17, 38, 39]
Use Case	Wholesale Payments, Retail Payments, Cross-border Payment, Cross-Currency Payment, Machine-to-Machine Payment, transactions	[15, 17, 24, 33, 38–42]
Architecture	Application Layer, Service Layer (API Layer), Smart contract Layer, Protocol Layer, Network Layer	[14, 24, 25, 29, 42, 43]
Infrastructure	DLT-Based, Non-DLT-based	[31, 44]
Access	Account-based (valued-Based), Token Based (wallet-based), IoT Based-access	[31]
Principle	Finality, Convertible, Interest Rate, Transferable, Convenient, Available, Secure, Resilient, Scalable, Extensible, Flexible, Interoperable, Private, Compliant, Auditable, Robust	[12, 14, 16, 17, 19, 20, 25, 35, 38, 45–49]
Governance	National laws, supervisory policy, Digital Identity Integration, Technology neutral regulation, GDPR, AML/CFT, Tax regimes, regulation-by-design	[14, 16–20, 33, 37, 38, 48–50]

(*continued*)

Table 1. (*continued*)

Concepts	Characteristics	References
Governance	National laws, supervisory policy, Digital Identity Integration, Technology neutral regulation, GDPR, AML/CFT, Tax regimes, regulation-by-design	[14, 16–20, 33, 37, 38, 48–50]

In the next section, we will present the relationships between the concepts.

3.2 CBDC Concepts Relationships (RQ2)

Cryptocurrency regulation via CBDC implementation involves financial, technological, legal, and social dimensions, drivers by the need to solve problems related to crypto-economy, disruptive innovation, crypto market, data economy, financial technology (fintech), crypto transactions, crypto-wallets, money laundering (crypto laundering), cyber-crime economy, tax and accounting of digital currency, and data protection [9]. We used this reference to analyze the relationships between the key concepts found, assuming that these concepts need to be integrated into the listed CBDC dimensions of regulations.

Our focus on finding the key concepts was the technological scope. Still, CBDC, legal digital money issued by a central bank, must be aligned with the legal and financial area of digital currency and social usage by citizens and businesses. In this context, we consider that concepts relate one to another and can be linked to one or more dimensions of CBDC regulation.

In Table 2, we show the relationship between concepts and dimensions proposed by [9]. The legal dimension relates to the digital currency goal, law, supervisory policy, standards, and guidelines to develop digital currency is emphasized; in this scope, the currency function, proposal, governance, participant, principle, and type relates to the legal dimension of CBDC.

The social dimension relates to day-to-day digital currency use by citizens and businesses to purchase goods and services. The social dimension concepts are participant, principle, use case, and access.

The financial dimension of CBDC is responsible for digital currency creation, distribution, and circulation, including layers, lifecycle, model, principle, type, and use case. The technological dimension of CBDC is associated with technical procedures to create digital currency concepts, includes principle, type, use case, digital asset, architecture, infrastructure, and access.

The **principle** concept is associated with legal, social, financial, and technological dimensions because is legal via governance relationships and la interconnected to participants that need to use CBDC with security. It is also associated with financial via currency lifecycle relationships and technology via technical implementation of principles.

Table 2. Table captions should be placed above the tables

Concepts	Related Concept (s) Relationship (s)	Legal	Social	Financial	Technological
Function	Have proposal, includes Participant	X			
Proposal	Includes Participant, guides Governance	X			
Governance	Supports currency function, composes principle, is composed by layer	X			
Participant	Composes layer	X	X		
Layer	Composes architecture	X	X	X	
Life cycle	Uses layer, includes model, issue digital asset, includes principle			X	
Model	Circulates CBDC via access			X	
Principle	Composes architecture	X	X	X	X
Type	Is used by use cases	X		X	X
Use Case	Is support by Function		X	X	X
Digital Asset	Have type				X
Architecture	Includes type, includes use case				X
Infrastructure	Supports use case				X
Access	Is supported by infrastructure, uses digital asset, gives access to participant		X		X

4 Conclusion

The key concepts of CBDC are function, proposal, participant, layer, life cycle, principle, model, access, digital asset, architecture, infrastructure, type, use case, and governance. These Concepts are related to the legal, social, financial, and technological dimensions of the CBDC regulation.

The legal dimension of the CBDC regulation involves law, supervisory policy, standards, and guidelines to develop a digital currency, which must comply with national and international laws and best practices. The social dimension of the CBDC regulation relates to the participant, principle, use case, access of digital currency by citizens and businesses. The financial dimension of the CBDC regulation is responsible for digital currency creation, distribution, and circulation, which is ensured by the system implementation. The technological dimension of the CBDC regulation is associated with the technical procedure to create a digital currency, the technology used to operate the application, and to give participants access to CBDC for use.

CBDC adoption will change national and international financial market operations, and the payment system ecosystem, and involves social, legal, financial, and technological dimensions; therefore, this circumstance justifies why CBDC is being researched by several organizations such as academics, industry, central banks, international financial institutions, policymakers, technology providers, and technology-based non-profit organizations to find a better solution of regulation for each dimension. To find the response to the financial and technological regulation of CBDC as a decentralized-based application that operates in a decentralized market infrastructure, it needs to include the cooperation between economists, lawyers, and information technology professionals in a CBDC project of research and experimentation.

For future work, we plan to propose a conceptual model of CBDC design using all key concepts found in the literature.

Acknowledgment. The research work reported in this publication was supported by Fundação para Ciancia e Tecnologia (Foundation for Science and Technology) of Portugal through the individual research grant SFRH/BD/151432/2021.

References

1. Rauchs, M., et al.: Distributed ledger technology systems: a conceptual framework. SSRN Electron. J. (2018)
2. Mills, D.C., et al.: Distributed ledger technology in payments, clearing, and settlement. Financ. Econ. Discuss. Ser. **2016**(095) (2016). https://doi.org/10.17016/FEDS.2016.095. https://papers.ssrn.com/abstract=2881204. Accessed 2 Aug 2021
3. Yaga, D., Mell, P., Roby, N., Scarfone, K.: Blockchain technology overview. National Institute of Standards and Technology, no. NISTIR 8202, p. 59 (2018). https://csrc.nist.gov/publications%0Ahttps://csrc.nist.gov/CSRC/media/Publications/nistir/8202/draft/documents/nistir8202-draft.pdf
4. Bech, M., Garratt, R.: Central bank cryptocurrencies (2017). https://papers.ssrn.com/sol3/papers.cfm?abstract_id=3041906. Accessed 18 Feb 2019
5. Reserve Bank of New Zealand, The Future of Money–Central Bank Digital Currency Te Moni Anamata, Central Bank of New Zealand, pp. 1–47, December 2021. https://www.rbnz.govt.nz/notes-and-coins/future-of-money/cbdc
6. Carstens, A.: Central bank digital currencies: putting a big idea into practice (2021). https://www.bis.org/speeches/sp210331.pdf
7. Shin, H.S.: Central bank digital currencies: an opportunity for the monetary system. BIS Annu. Gen. Meet. **948**, 1–12 (2021)
8. Carvalho Silva, E., Mira da Silva, M.: Research contributions and challenges in DLT-based cryptocurrency regulation: a systematic mapping study. J. Bank. Financ. Technol. **6**, 63–82 (2022). https://doi.org/10.1007/s42786-021-00037-2
9. Carvalho Silva, E., Mira da Silva, M.: Motivations to regulate cryptocurrencies: a systematic literature review of stakeholders and drivers. Int. J. Blockchains Cryptocurrencies **2**(4), 360 (2021). https://doi.org/10.1504/ijbc.2021.120381
10. Kitchenham, B., Charters, S.: Guidelines for performing systematic literature reviews in software engineering (2007). http://citeseerx.ist.psu.edu/viewdoc/summary?doi=10.1.1.117.471. Accessed 26 Nov 2019

11. Liberati, A., et al.: The PRISMA statement for reporting systematic reviews and meta-analyses of studies that evaluate health care interventions: explanation and elaboration. J. Clin. Epidemiol. **151**(4), e1–e34 (2009). https://doi.org/10.1016/j.jclinepi.2009.06.006. https://pubmed.ncbi.nlm.nih.gov/19631507/

12. CBDC Research Center, CBDC Research Center Overview and Conceptual Model, R3, pp. 39–49, March 2019. https://doi.org/10.1007/978-3-030-01759-0_3

13. Shoaib, M., Ilyas, M., Hayat Khiyal, M.S.: Official digital currency. In: 8th International Conference on Digital Information Management, ICDIM 2013, pp. 346–352 (2013). https://doi.org/10.1109/ICDIM.2013.6693982

14. Han, X., Yuan, Y., Wang, F.Y.: A blockchain-based framework for central bank digital currency. In: Proceedings of International Conference on Service Operations and Logistics, and Informatics, SOLI 2019, pp. 263–268 (2019). https://doi.org/10.1109/SOLI48380.2019.8955032

15. Danezis, G., Meiklejohn, S.: Centrally Banked Cryptocurrencies (2015)

16. Duffie, J.D., Mathieson, K., Pilav, D.: Central bank digital currency: principles for technical implementation. SSRN Electron. J. 1–17 (2021). https://doi.org/10.2139/ssrn.3837669

17. Grothoff, C., Moser, T.: How to issue a privacy-preserving central bank digital currency. SSRN Electron. J. (2021). https://doi.org/10.2139/ssrn.3965050

18. de V. Burgos, A., et al.: Distributed ledger technical research in Central Bank of Brazil, p. 32, August 2017. https://www.bcb.gov.br/htms/public/microcredito/Distributed_ledger_technical_research_in_Central_Bank_of_Brazil.pdf

19. B. of Thailand, Central Bank Digital Currency: The Future of Payments for Corporates (2021)

20. W. G. on E.-C. R. and D. of the P. B. of China, Progress of Research & Development of E-CNY in China (2021)

21. Riksbank, E-krona pilot Phase 1, p. 21 (2021). https://www.lifo.gr/now/greece/koronaios-metadidetai-kai-me-salio-kai-me-koytali-dilonei-i-epistimoniki-epitropi-toy

22. B. for I. S. BIS Innovation Hub and N. Swiss Bank, Project Helvetia: Settling tokenised assets in central bank money (2020)

23. Jin, S.Y., Xia, Y.: CEV framework: a central bank digital currency evaluation and verification framework with a focus on consensus algorithms and operating architectures. IEEE Access **10**, 63698–63714 (2022). https://doi.org/10.1109/access.2022.3183092

24. Saripalli, S.H.: Transforming Government banking by leveraging the potential of blockchain technology. J. Bank. Financ. Technol. **5**(2), 135–142 (2021). https://doi.org/10.1007/s42786-021-00035-4

25. Bank of Thailand, Inthanon Phase I: An application of Distributed Ledger Technology for a Decentralised Real Time Gross Settlement system using Wholesale Central Bank Digital Currency (2018)

26. Han, J., et al.: Cos-CBDC: design and implementation of CBDC on Cosmos blockchain. In: 22nd Asia-Pacific Network Operations and Management Symposium APNOMS 2021, pp. 303–308 (2021). https://doi.org/10.23919/APNOMS52696.2021.9562672

27. Abdulkader, O., Bamhdi, A.M., Thayananthan, V., Elbouraey, F.: IBMSDC: intelligent blockchain based management system for protecting Digital Currencies Transactions. In: Proceedings of 3rd World Conference on Smart Trends in Systems Security and Sustainability WorldS4 2019, pp. 363–367 (2019). https://doi.org/10.1109/WorldS4.2019.8904003

28. Sasongko, D.T., Yazid, S.: Integrated DLT and non-DLT system design for central bank digital currency. In: ACM International Conference Proceeding Series, pp. 171–176 (2020). https://doi.org/10.1145/3427423.3427447

29. Bhawana, Kumar, S.: Permission blockchain network based central bank digital currency. In: 2021 IEEE 4th International Conference on Computing, Power and Communication Technologies, GUCON 2021, pp. 1–6 (2021). https://doi.org/10.1109/GUCON50781.2021.9574020

30. Wu, Y., Fan, H., Wang, X., Zou, G.: A regulated digital currency. Sci. China Inf. Sci. **62**(3), 1–12 (2019). https://doi.org/10.1007/s11432-018-9611-3
31. Auer, R., Böhme, R.: The technology of retail central bank digital currency. SSRN, pp. 85–100, March 2020
32. Zhang, J., et al.: A hybrid model for central bank digital currency based on blockchain. IEEE Access **9**, 53589–53601 (2021). https://doi.org/10.1109/ACCESS.2021.3071033
33. Bianco, S.D.: Central bank digital currency: towards a composable standards-based implementation. In: The Economics of Cryptocurrencies, pp. 77–82, November 2020. https://doi.org/10.4324/9780429200427-13
34. F. G. D. C. including D. F. C. F.-D. ITU-T, Taxonomy and definition of terms for digital fiat currency (2019)
35. Lee, Y., Son, B., Park, S., Lee, J., Jang, H.: A survey on security and privacy in blockchain-based central bank digital currencies. J. Internet Serv. Inf. Secur. **11**(3), 16–29 (2021). https://doi.org/10.22667/JISIS.2021.08.31.016
36. Yao, Q.: A systematic framework to understand central bank digital currency. Sci. China Inf. Sci. **61**(8) (2018). https://doi.org/10.1007/s11432-017-9294-5. Accessed 6 Feb 2019
37. C. for C. Markets, Digital Assets: a framework for regulation to maintain the United States' status as an innovation leader, pp. 1–62 (2021)
38. Bianco, S.D.: Central bank digital currency opportunities, challenges and design. In: The Economics of Cryptocurrencies, pp. 77–82, March 2020. https://doi.org/10.4324/9780429200427-13
39. Lannquist, A.: Central Banks and Distributed Ledger Technology: How are Central Banks Exploring Blockchain Today? World Econ. Forum, March 2019. http://www3.weforum.org/docs/WEF_Central_Bank_Activity_in_Blockchain_DLT.pdf
40. Bech, M.: Payments without borders. BIS Q. Rev., no. March, pp. 53–65 (2020)
41. Pocher, N.: Towards CBDC-based machine-to-machine payments in consumer IOT, vol. 1, no. 212 (2022)
42. Franko, A., Olah, B., Sass, Z., Hegedus, C., Varga, P.: Towards CBDC-supported smart contracts for industrial stakeholders. In: Proceedings - 2022 IEEE 5th International Conference on Industrial Cyber-Physical Systems, ICPS 2022 (2022). https://doi.org/10.1109/ICPS51978.2022.9816857
43. Digital Asset, The Digital Asset Platform Non-technical White Paper, Digit. Asset, pp. 1–29 (2016). https://hub.digitalasset.com/hubfs/Documents/DigitalAssetPlatform-Non-technicalWhite Paper.pdf
44. Goel, A.K., Bakshi, R., Agrawal, K.K.: Web 3.0 and Decentralized Applications (2022)
45. BIS, Central bank digital currencies: foundational principles and core features, vol. 1 (2020)
46. Avi, E., Gschwind, K., Monsalve, F., Urvantsev, V.: Partial anonymity in central bank digital currencies : a survey parameters of a CBDC, May 2021
47. Bank for International Settlements, Central bank digital currencies: system design and interoperability, vol. 2 (2021). www.bis.org
48. Gross, J., Sedlmeir, J., Babel, M., Bechtel, A., Schellinger, B.: Designing a central bank digital currency with support for cash-like privacy. SSRN Electron. J. (2021). https://doi.org/10.2139/ssrn.3891121
49. Jung, H., Jeong, D.: Blockchain implementation method for interoperability between CBDCs. Futur. Internet **13**(5), 133 (2021). https://doi.org/10.3390/fi13050133
50. Pocher, N., Veneris, A.: Privacy and transparency in CBDCs: a regulation-by-design AML/CFT scheme. IEEE Trans. Netw. Serv. Manag. **19**(2), 1776–1788 (2022). https://doi.org/10.1109/TNSM.2021.3136984

Artificial Intelligence for Technology Transfer

An AI-Based Approach for the Improvement of University Technology Transfer Processes in Healthcare

Annamaria Demarinis Loiotile[1,6](✉) [iD], Davide Veneto[2] [iD], Adriana Agrimi[4],
Gianfranco Semeraro[3,7] [iD], and Nicola Amoroso[5] [iD]

[1] Interateneo Department of Physics, University of Bari Aldo Moro, 70125 Bari, Italy
annamaria.demarinis@uniba.it
[2] Digital Innovation srl, 70125 Bari, Italy
[3] Department of Computer Science, University of Bari Aldo Moro, 70125 Bari, Italy
[4] Directorate for Research, Third Mission and Internationalization,
University of Bari Aldo Moro, 70125 Bari, Italy
[5] Department of Pharmacy-Pharmaceutical Sciences, University of Bari Aldo Moro,
70125 Bari, Italy
[6] Department of Electrical and Information Engineering, Polytechnic of Bari, 70125 Bari, Italy
[7] University School for Advanced Studies IUSS Pavia, Palazzo del Broletto,
Piazza della Vittoria, 15, 27100 Pavia, Italy

Abstract. Universities are today required to make an ever-increasing effort to bridge the gap between research, innovation, and marketable solutions. The goal of the paper is to contribute to enhancing the process of exploiting research results of universities through the proposition of an AI-based search process, the experimentation of the process by means of a prototype web portal named "UNIBA TT Skills Portal" that implements it and the evaluation of the perceived user experience of the web portal by using the System Usability Scale questionnaire. The paper illustrates a case study conducted on the University of Bari Aldo Moro, a large university in southern Italy, using technology transfer data in the health domain. The collection of data in a unique "place" aims at fostering technology transfer activities and the creation/acceleration of new innovative enterprises (start-ups and spinoffs) in the Healthcare sector.

Keywords: Technology Transfer · Artificial Intelligence · Healthcare · Recommendation systems · Usability

1 Introduction

European policy, as one of its strategic goals, increasingly aims to harness the research results to be able to provide new answers to the challenges and opportunities facing the EU. Increased efforts are needed to transform scientific knowledge into solutions that promote citizens' well-being and economic prosperity. Europe must maximize the value of R&I results by fostering a culture of knowledge valorization, ensuring that knowledge-based institutions know how to manage their intellectual capital and improving the connection between academia, industry, citizens and policymakers [1].

The transformation of the knowledge generated in research institutions and universities can take place through different channels, from the creation of spinoffs and innovative startups to intellectual property management, to industry-academia collaboration and citizen involvement [2, 3].

Modern policy requires a shift in focus from intellectual property management in knowledge transfer activities to knowledge exploitation and value creation. More efforts are needed in universities to bridge the gap between research, innovation, and marketable solutions. Although there is a large skilled workforce and strong collaboration between universities and businesses, there is a need to promote, especially in universities, a culture of knowledge valorization, to ensure that universities effectively manage their intellectual capital, and to improve links between universities, industry, citizens, and policy makers [4].

The ultimate goal of the paper is to contribute to enhancing the process of exploiting research results of universities through:

– the proposition of an AI-based search results process,
– the validation of the process by means of a prototype web portal that implements it;
– the evaluation of the perceived user experience of the web portal.

The web portal, containing all the information concerning the technology transfer activities carried on in a university, represents a tool that facilitates the valorization of research results in universities for two reasons: 1) it systematizes all technology transfer activities by collecting them in a single "place," and 2) it proposes a recommendation system that helps maximize the exploitation. The paper illustrates a case study conducted on the University of Bari Aldo Moro (UNIBA), a large university in southern Italy, using technology transfer data in the health domain. The web portal developed, named "UNIBA TT Skills Portal", using an Artificial Intelligence (AI) approach, would allow the recommendation of similar "results" present in the university database starting from the technology transfer activity searched. In this way, information retrieval of congruent activities becomes quick and easy. This platform also allows the preservation of information by going to bring to digitize even documentation that is currently found only in paper format, scattered on different web platforms or multiple offices.

This paper is organized as follows: Sect. 2 discusses the state of the art related to the use of AI for improving technology transfer processes in the university context. The experimental protocol of the proposed approach (Sect. 3) and the properties of the dataset (Sect. 4) are then explained. The discussion related to the usability are presented in Sect. 5 and Sect. 6 illustrates the conclusion of the work.

2   Related Works

Technology transfer refers to the dissemination of information, knowledge, matching the needs and the technology item adaptation to new users [5]. Nowadays, the importance of the technology is highly sustained by several stakeholders, from the economy to public administration. Furthermore, it is possible to state that technology is becoming to the society like the water to human body.

Khan J., Haleem A. and Husain Z. presented a model to identify barriers to the technology transfer [6]. Specifically, the work [6] complains about the green and traditional technology transfer in order to treat the technology transfer including the sustainability. The presented model built a hierarchy of barriers in order to provide insights to decision makers. In this way, the strategy formulation of technology acquisition and development could be done basing the assessments on factors included in the model.

In [5, 6] one of the main factors that could help to accelerate the technology transfer is related to the management. In fact, works as [5] confirm that the support from the management is fundamental for the organization that wants to actuate the technology transfer. Furthermore, both [5, 6] highlights how culture is a key factor that could accelerate or decelerate the technology transfer.

In fact, Nguyen T.D.N and Atsushi A. studied [7] an organizational culture from two points of view: whether it could enhance the efficient technology transfer and whether it could be facilitated by significant management practice factors. The study [7] gathered empirical results about 223 Japanese manufacturing subsidiaries in Vietnam. The presence of Japanese manufacturing in another country, namely Vietnam, allows to also analyze a cross-culture technology transfer case. Hence, the empirical results obtained verify that all five factors of management practices (quality practices, training, management commitment, sharing and understanding, and teamwork) constitute the systematic approach to facilitate the formation of corporate culture that improves the efficiency of cross-cultural technology transfer. The implementation of the best management practice opens new perspectives, new opportunities, and new challenges to develop the organizational culture.

The importance of an appropriate organizational culture is one of the important factors that can enable the technology transfer. Often, one of the biggest problems in the actuation of technology transfer is related to a lack of visibility of the results of the academic research. Indeed, often university is lacking industrial partners. Furthermore, the increasing competition in the global markets push forward the minimum level of technology knowledge required to develop good product. Schuh G., Aghassi S. and Valdez A.C. had studied [8] the concept of web-based platforms to support technology transfer.

Technology and knowledge mapping is an important process or method for capturing information from university researchers about their knowledge, skills, abilities, relationship, etc. It can also be defined as a process of surveying, evaluating, and linking the information, knowledge, skills, and abilities possessed by individuals and groups within an organization [8].

Novikova I., et al. had analyzed in [9] different platforms, including social platforms, that help to disseminating the new discoveries and research results. These platforms could be enablers to activate markets of the intellectual work that came from university.

Jishu S., Zuwei Y. and Tao H. [10] had implemented a micro innovation service platform and after the authors had analysed the empirical data obtained by the platform usage. Using a Neuronal Network Algorithm and the research of social perception, the authors highlighted that the number and the volume of technology transfer, technology consulting, technology development and technology service transactions increased along the improvement of the capability of small and medium enterprise of industrial innovation.

In conclusion of this section, it is possible to argue that a web platform, where universities could publish new discoveries and new research results (new algorithms of artificial intelligence as [11], or new discoveries about cybersecurity domain as [12, 13]), is a key enabler to the technologies transfer. Furthermore, a web platform could help, as highlighted by [10], the innovation of the micro and medium enterprise enhancing their know-how, the velocity of acquisition of the know-how and the relationship with Sustainable Development Goals (SDGs). In fact, as described by Henrik S. S. in [14], the impact of AI (and the innovation enabled by research results) on SDGs may be assessed, also considering the indirect impacts on multiple SDGs, and the impacts on micro, meso and macro levels.

3   Methods

The proposed system, named "UNIBA TT Skills Portal", was built following the structure of a typical client-server architecture. The system can be imagined as consisting of 3 modules, represented in Fig. 1.

The modules perform specific tasks. The "Client" module is the one in charge of interfacing with the platform users. The "Server" module is the heart of the whole system. In it will be the management logic of the platform along with the functionality proposed for this work. A more detailed explanation can be found below. The third module, called "Database," is the module that deals with the management of the database that contains the information of the universities' activities.

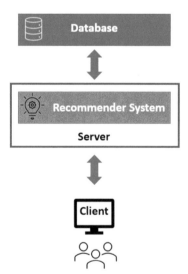

Fig. 1.   System architecture of "UNIBA TT Skills Portal"

The core of the architecture of the proposed system is the "Server" module. This module contains the artificial intelligence system suitable for the recommendation of the activities recorded by the University of Bari.

A content-based recommendation system has been implemented within the "Server" module. The choice of this type of system is due to the fact that recommendations should be made based on the similarity between the contents of different activities registered within the university database.

The selected type of recommender system is the content-based filtering. Such a system suggests items to a user, similar with respect to those in which the user has shown interest in the past. This type of recommendation system starts from the basic assumption that, for example, if an individual has searched or clicked to a specified item, in this case it could be a patent, the system recommends all items (not only patents) that are similar to that item.

Specifically, content-based recommendation systems suggest an item to a user based on the description and characteristics of the product and the individual's interest profile [14]. Some early work employing this type of recommendation used queries to build models for users. In this way, in fact, a user model was based on one or a set of queries, which could retrieve additional or new information for the user. Later, methods based precisely on information retrieval were adopted (information retrieval), such as Rocchio's algorithm and term frequency-inverse document frequency (TF-IDF). The TF-IDF is a statistical measure that weights the importance of a word in a document with respect to a collection of documents.

On a theoretical level Content-based recommendation systems, perform an analysis of an item's characteristics to identify those products that might fit with the user's interests. Items, however, show up differently depending on the system and the dataset in which they are found: in fact, they can be in the form of structured or unstructured data. In the former case, each item is characterized by the same set of attributes, variables or fields, and the values that the variables can take are known. In addition, an ID code makes it possible to distinguish between items that may have some of the same attributes.

In this specific case, because the description of the items is composed of words and sentences, it has been chosen to convert descriptions into a vectorial representation made of 128 elements. This vectorial representation allows using the cosine similarity metrics to provide recommendations.

For doing so, pre-trained Bidirectional Encoder Representations from Transformers (BERT) was used. BERT is a natural language processing deep neural network model described by Google [15], often used for converting sentences into vectorial representation. It is the first architecture in Natural Language Processing (NLP) to be pre-trained using unsupervised learning on pure unstructured text (about 2.5 billion words), in addition, it achieved state-of-the-art results on a variety of NLP tasks.

In practice, we pre-process each description and title of each item. The resulting vectors are concatenated, creating a vector of 256 numbers. This vector is stored in the database and associated with the id of the original item. As specified previously, the Cosine similarity is used to present the most similar items. The cosine similarity formulation is presented in Eq. (1):

$$\cos(\theta) = \frac{A \cdot B}{\|A\| \cdot \|B\|} = \frac{\sum_{i=1}^{n} A_i B_i}{\sqrt{\sum_{i=1}^{n} A_i^2} \sqrt{\sum_{i=1}^{n} B_i^2}} \tag{1}$$

Were A and B are two vectors containing each 256 numbers representing the compressed context (also defined as embedding) of the title and description. The application of the recommender system is visually represented in Sect. 5.

4 Dataset

The dataset used for this case study is the University of Bari database containing different activities of knowledge/technology transfer and more widely third mission carried on the healthcare domain. In particular, the data are related to the different topic of the technology transfer in a university (Table 1):

Table 1. Data related to the different topic of the technology transfer in a university

TT Topic	Explanation	N. data
Patents	Support for the submission of patent applications	10
Research Contracts	Activities to support for signing research contracts in collaboration with industry	30
Technology transfer support	Technology transfer support activities (negotiation of confidentiality or non-disclosure agreement, data sharing/material transfer agreement)	9
Scouting for new inventions	New invention scouting activities, assistance in patenting and identifying business opportunities, technology forecasting and foresight	11
External Funding Negotiations	Support activities in negotiating and managing clinical trial contracts, sourcing external funding	29

In the dataset there is a low number of items because of the whole platform is recently implemented in a prototype form. Meanwhile, digitalization is a long and slow process that needs the interaction of multiple stakeholders.

5 Usability Discussion

The proposed platform was tested by 16 users. In Table 2, information's about users are synthetized.

The user individuated to test the web platform were all employees in the administrative section of the University of Bari Aldo Moro. In order to evaluate the web-platform, some cases of study were performed by the users. In Fig. 2, an example of one of the case study is represented.

Table 2. Users information

User Information	Number
Male	7
Female	9
Mean Age	40 ± 15
Mean Education Level	18 ± 2
Total number of Users	16

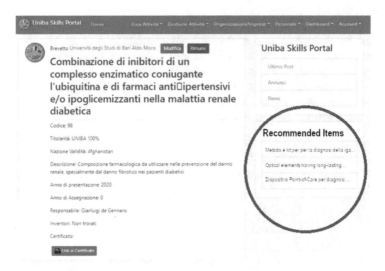

Fig. 2. Recommended items based on the currently viewed instance

The request was to find a patent ("Brevetto" in Italian language) related the "Ubiquitin-conjugating enzyme complex and antihypertensive drugs ...". In this case, the result was a patent where the University of Bari had the 100% of ownership. Furthermore, in Fig. 2 it is shown the result of the recommender system. It has correctly found patents related to the one selected and currently viewed. As predicted, the recommender system found patents related to the healthcare diagnosis area.

From a performance point of view, the recommender system is very responsive since the majority of computation, the pre-processing phase with BERT, is done when a new item is added to the system. Thus, at inference time, only the cosine similarity is computed. In addition, usability tests were conducted. Hence, 16 potential users were asked to use the system and answer certain questions. In particular, the System Usability Scale (SUS) questionnaire has been used. The SUS questionnaire provides a reliable and quick tool for measuring usability; it consists of ten statements with five possible answers: the first strongly disagrees with the statement, and the last strongly agrees with it. The range of the values of the answers is between 1 and 5. Originally created by

John Brooke in 1986, it allows for the evaluation of a wide variety of both hardware and software products and services [16].

There are several benefits of using the SUS, some of which are listed below:

– It is very easy to administer to trial participants;
– It can be used on small sample sizes with reliable results;
– It is an excellent system for ensuring whether a product is usable or not.

The statements in the questionnaire are:

1) I think I would like to use this system frequently
2) I found the site unnecessarily complex
3) I found the site very easy to use
4) I think I would need support from someone already able to use the site
5) I found the various features of the site well integrated
6) I found inconsistencies between the various features of the system
7) I think most people can learn to use the site easily
8) I found the site very difficult to use
9) I felt comfortable using the site
10) I needed to learn many processes.

The overall procedure is described by Brooke. J in [17].

The evaluation of the "SUS Score" is based on the assumption that 68 is the average score and values above means that the system is perceived usable. In Table 3 are reported the SUS Score for each participant and the mean and standard deviation of the results.

Table 3. SUS Score for each participant

User	SUS Score
1	69.85
2	69.57
3	74.08
4	71.7
5	72.9
6	68.03
7	64.47
8	63.86
9	64.64
10	72.37
11	71.37
12	68.85
13	70.53
14	76.93
15	73.01
16	71.8
MEAN	70.2475
STANDARD DEVIATION	3.51

The resulting average score is 70.25, which certifies that the website is usable according to SUS standards. Indeed, the majority of the users find the web-platform usable, and in average, the platform is considered good enough. Ofcourse, there is room for improvement. Furthermore, in the future is desirable to get SUS Score above 82, in order to have users that tend to be "Promoter" of the web-platform as Sauro J. highlighted in [18].

The majority of users also requested automatic integration with external service providers in order to avoid inserting all the information manually. In the future, API platforms, such as Google Patents, can be used in order to import all metadata directly from the web without inserting them manually into the platform.

6   Conclusion

Concluding, the collection of these activities in a unique "place" aims at fostering technology transfer activities and the creation/acceleration of new innovative enterprises (start-ups and spinoffs) in the Healthcare sector, by systematizing, integrating and providing technical support to existing Technology Transfer and Third Mission offices, through the sharing of methods, tools, resources, experiences, skills, contacts and initiatives to promote technology offerings, training of qualified professionals, the implementation of new projects of potential interest to the market (proof of concept, seed and early stage), the development and exploitation of innovations, through the most appropriate forms of protection and transfer (patents or other industrial property titles, license/option/sale agreements, etc.).

This was a first step in the activity of mapping and systematizing technology transfer data for the University of Bari and will certainly continue with the implementation of data in other areas relevant to the research activities of the University itself and with the upgrade of new artificial intelligence tools such as clustering and further classification to increase the potential of the web portal.

References

1. Commission, E., for Research, D.-G.: Innovation: towards a policy dialogue and exchange of best practices on knowledge valorisation: report about the results of the survey. Publications Office (2021). https://doi.org/10.2777/457841
2. Commission, E., for Research, D.G.: Innovation: valorisation policies: making research results work for society : industry-academia collaboration. Publications Office (2021). https://doi.org/10.2777/573275
3. Demarinis Loiotile, A., et al.: Best practices in knowledge transfer: insights from top universities. Sustainability **14**, 15427 (2022). https://doi.org/10.3390/SU142215427
4. Commission, E., for Research, D.-G.: Innovation: science, research and innovation performance of the EU 2022: building a sustainable future in uncertain times. Publications Office of the European Union (2022). https://doi.org/10.2777/78826
5. Lynn, G.: Problems and practicalities of technology transfer: a survey of the literature. https://www.indersienceonline.com/doi/abs/10.1504/IJTM.1988.025991. Accessed 21 Nov 2022. https://doi.org/10.1504/IJTM.1988.025991PDF
6. Khan, J., Haleem, A., Husain, Z.: Barriers to technology transfer: a total interpretative structural model approach. Int. J. Manuf. Technol. Manage. **31**, 511–536 (2017). https://doi.org/10.1504/IJMTM.2017.089075

7. Nguyen, N.T.D., Aoyama, A.: Achieving efficient technology transfer through a specific corporate culture facilitated by management practices. J. High Technol. Manage. Res. **25**, 108–122 (2014). https://doi.org/10.1016/J.HITECH.2014.07.001

8. Schuh, G., Aghassi, S., Valdez, A.C.: Supporting technology transfer via web-based platforms. In: 2013 Proceedings of PICMET 2013: Technology Management in the IT-Driven Services, pp. 858–866 (2013)

9. Novikova, I., Stepanova, A., Zhylinska, O., Bediukh, O.: Knowledge and technology transfer networking platforms in modern research universities. Innov. Mark. 16, 57–65 (2020). https://doi.org/10.21511/IM.16(1).2020.06

10. Shao, J., Yu, Z., Huang, T.: Innovation service platform of small and medium-sized microenterprises based on social perception and neural network algorithm. Comput. Intell. Neurosci. **2022**, 1–9 (2022). https://doi.org/10.1155/2022/8700833

11. Dentamaro, V., Giglio, P., Impedovo, D., Moretti, L., Pirlo, G.: AUCO ResNet: an end-to-end network for Covid-19 pre-screening from cough and breath. Pattern Recognit. **127**, 108656 (2022). https://doi.org/10.1016/J.PATCOG.2022.108656

12. Impedovo, D., Longo, A., Palmisano, T., Sarcinella, L., Veneto, D.: An investigation on voice mimicry attacks to a speaker recognition system. In: Demetrescu, C., Mei, A. (eds.) Proceedings of the Italian Conference on Cybersecurity (ITASEC 2022). CEUR, Rome (2022)

13. Carrera, F., Dentamaro, V., Galantucci, S., Iannacone, A., Impedovo, D., Pirlo, G.: Combining unsupervised approaches for near real-time network traffic anomaly detection. Appl. Sci. **12**, 1759 (2022). https://doi.org/10.3390/APP12031759

14. Sætra, H.S.: AI for the Sustainable Development Goals. CRC Press, New York (2022)

15. Aggarwal, C.C.: Content-based recommender systems. In: Recommender Systems, pp. 139–166 (2016). https://doi.org/10.1007/978-3-319-29659-3_4

16. Devlin, J., Chang, M.W., Lee, K., Toutanova, K.: BERT: pre-training of deep bidirectional transformers for language understanding. In: NAACL HLT 2019 - 2019 Conference of the North American Chapter of the Association for Computational Linguistics: Human Language Technologies - Proceedings of the Conference, vol. 1, pp. 4171–4186 (2018). https://doi.org/10.48550/arxiv.1810.04805

17. SUS: A "Quick and Dirty" Usability Scale. Usability Evaluation in Industry, pp. 207–212 (1996). https://doi.org/10.1201/9781498710411-35

18. Brooke, J.: SUS: a retrospective. J. Usability Stud. **8**, 29–40 (2013). https://doi.org/10.5555/2817912.2817913

Matching Knowledge Supply and Demand of Expertise: A Case Study by Patent Analysis

Vincenzo Dentamaro[1] , Paolo Giglio[1](✉) , Donato Impedovo[1] ,
and Davide Veneto[2]

[1] University of Bari, 70125 Bari, Italy
{vincenzo.dentamaro,paolo.giglio,donato.impedovo}@uniba.it
[2] Digital Innovation s.r.l., 70125 Bari, Italy
davide.veneto@dinnovation.it

Abstract. Technology transfer has the aim to analyze the mechanisms behind the value raised by knowledge in the long trip from labs to society and markets. It is a demanding task due to the potential of its refinement on one hand and the challenging mission of understanding it with detail on the other. The complication arises in terms of availability of structured data and on the complexity of the mechanisms behind the osmosis of knowledge through a heterogeneous society. This article analyzes the potentiality of a correlation paradigm between patents supply and companies' purpose of business to outline a potential hint of value through this complexity.

Keywords: Technology Transfer · Patents Analysis · Demand of expertise

1 Introduction

During the last 20 years research on Technology Transfer (TT) has put efforts on giving a comprehensive and exhaustive definition of what it is indeed. Some of the most agreed ones are in [1] by Argote et al., who defines it as "the process through which one unit is affected by the experience of another" and in [2] where Etzkovitz underlines how TT happens between the public and private sectors. Roessner in [3] defines TT as the process of diffusion of know-how between organizations. More empirical definitions of TT come from Charles et al. [4] who consider TT as the diffusion of a set of competences for a specific technology, or from Roessner in [3] who defines the same process as the transfer of ideas and knowledge from a research institution to others involved in the production of goods and services. But it seems that these definitions haven't stopped the attention on trying to circumscribe with more detail what TT is about without being too abstract. How is it possible to have a field of research whose first problem is to build a good and updated definition for it? The answer lies in the true nature of this topic, whose need to be considered is based on evidence even without an explicit definition, as it is trivial to assess that technology must result in impacting the world in any field. But the world changes, as technology and the mechanisms that produce it also evolve. The challenge to quantify this impact arises from the complexity of the system it acts on, as

A. Rocha et al. (Eds.): WorldCIST 2023, LNNS 802, pp. 321–329, 2024.
https://doi.org/10.1007/978-3-031-45651-0_32

many social and institutional entities can produce technology and many others can take advantage of it, all happening in a dynamic and evolutionary environment.

2 State of Art

Data-oriented approaches can be found in literature to cope with the aforementioned challenge, trying to propose concrete solutions that match demand and supply of technology for specific applications. In this regard, objective sources of knowledge and technology, such as patents and intellectual property in general, come to attention as quantifiable data to start from [5], which is also the case for this work. In this regard, Anuraag et al. are concerned about trying to predict yearly performance rates for 1757 technology domains by analyzing US patents [6]. Patent Analysis is also performed by Choi et al. [7] in a work where a set of criteria is established to evaluate their potential impact for technological development; Lee et al. examines patents in terms of a set of parameters (assignees, localization, originality, diversification and cycle time of technologies) to achieve similar accomplishments in terms of innovation and economic growth in [8]. In [9] Giosio et al. propose an index to rank patents and to provide charts that show the outcomes. Yang et al. propose in [10] a novel approach based on the International Patent Classification (IPC) to identify technological opportunities in the realm of patents. Chi et al. attempts instead to identify the issues involved in evaluating patent applications and infringements from existing patent databases [11], particularly it uses AI Neural Networks as a relatively innovative tool for this field of research.

Patents, as one of the most reliable objective sources of knowledge, can be used also to spot new trends of development. It is what Sobolieva et al. do in [12] by comparing the number of patent applications in different fields. A similar task is performed and presented by Bai et al. in [13]. Perez et al. also try in [14] to visualize the evolution of technology by analyzing patents owned by companies and belonging to specific technical fields. Patent activity was under attention in the work of Cairo et al. [15] to highlight trends and potential market value derived from them.

The previous approaches suggest the ability to use patent activity as thermometer of the researchers' and politics' attention on specific fields, such as on sustainability and green economy: it is what Cova et al. propose in [16] by performing an analysis on green and non-green patents granted by the European Patent Office in the period 1989–2018. Hötte et al. investigate in [17] climate change adaptation technologies in US patents to understand related patterns and eventual drivers of innovation. Mao et al. did a similar work proposed in [18] but related to Industrial Wastewater Treatment by exploiting text mining techniques on 11840 related patents. Concerning the emergent topic of Battery Electric Vehicle (BEV) transnational technologies, Yuan et al. propose in [19] to analyze patent families and priority patent applications to map the most relevant trends in the field. Patent Analysis must deal with prohibitive numbers of documents thus making a reliable analysis hard to accomplish without the aid of technologies. Artificial Intelligence has put the approach to another level, as Neural Networks are able to retrieve even higher degrees of semantic information from texts, thus allowing researchers to better map trends and clusters of innovation. Giczy et al. describe in [20] a novel dataset affordable for AI analysis, it is composed of over 13.2 million patents and pre-grant publications. Martins et al. in [21] use Machine Learning algorithms to verify the performance of classifiers in the search and retrieval of information in the domain of Patent documents. The topic is described with detail in the related survey [22] proposed by Krestel et al., while Nardin et al. aim to explore in [23] some of the issues concerning the impact of AI on innovation, i.e. as a catalyst rather than a patent analysis tool. Both of these main directions (TT definition vs. data drive analysis) seem to benefit the overall research field, as one gives the other new hints for development. For example, the works in [7, 23] about tacit knowledge and weak ties suggest the variety of psychological factors that impact knowledge acquisitions by any beneficiary. As a consequence, the organization (companies, universities etc.) in its specificity becomes a huge part of the equation. That's why intellectual property owned by private companies or trends of their activities are mostly considered in the aforementioned studies as glints of innovation. The resulting complexity deeply challenges the attempt to formalize the problem. As it is shown in this paper, this information is crucial to develop the hypotheses underlying the work here proposed. Indeed, it is important to assess what is the position of the authors with respect to these premises. In this regard, it is in the authors' opinion that determining a causal connection between technology and its impact on reality at present day is not feasible, i.e. it is epistemologically impracticable to find exact solutions that describe the transferring mechanism of knowledge and technology between entities. This results in an apparently paradoxical consequence for their approach to this research field, as it implies to focus on measuring that same transferring mechanism of technology between entities, which is considered unknowable, while abandoning any effort on trying to give an exact solution for what it really is. In other words, it is the main aim of this paper to underline the importance of shifting from a causal to a correlation paradigm when it comes to extrapolate technology transfer effects. This doesn't mean that causality is irrelevant, but it wants to suggest that causal connections of Technology Transfer are a higher level of analysis in what appears to be a complex and nonlinear system of interacting entities.

In this regard, the following work will be focused on assessing the evidence on any potential affinity between patents belonging to specific IPC classes and companies listed in the U.S. exchanges. In order to do that, the similarity shall be quantified by comparing objective information in terms of textual material and, finally, the outcome shall be used to understand if such similarities are representative of companies belonging to the same market sector and discriminative with respect to companies of different market sectors.

3   Methods

For assessing any sensitivity of companies to some patent IPC class it was necessary to compare the companies' business mission with patents' field of research. This is other than a quantitative analysis, it requires semantic analysis in order to retrieve high level information on economic activities, research fields, technologies, problems, market sector, perspectives. Nevertheless, it has to be founded on objective and official information about companies on one side and patents on the other. To do this the following were considered as the respective units of representative information to be compared for both the realms: concerning the companies, that comes as the business summary in the *Item*

1 - Business section of the *Form 10-k*, the annual report required by the U.S. Securities and Exchange Commission (SEC). Companies with more than $10 Million in assets and a class of equity securities that is held by more than 2000 owners have to file annual and other periodic reports. These companies, which are listed on stock exchanges in the U.S. and whose reports are official requirements, are the reservoir where a subset was selected for this study in terms of similar activity in the same market sector. Concerning patents, their representative content is condensed in their abstract, which therefore represents the second unit of information.

In order to perform these operations a Python source code was written for the whole analysis process. Once an Item 1 - Business and a patent abstract are retrieved, they undergo a preprocessing pipeline before the comparison: the main goal is to clean them from spurious and noisy words while organizing it in a uniform and generalizable format. The overall pipeline is therefore composed of the following stages:

Noise Canceling and Lowercasing: Python library "RE" (regular expression operations) was implemented in order to delete punctuation, quotes, links, symbols etc. It was also used for lowercasing, which is essential as the same word can be considered twice due to the presence of uppercase letters.

Tokenization and Stopwords Removal: the word tokenization process results in splitting a sample of text into a list of the single words. This allows for stopwords removal: stopwords are those words that are insignificant as they serve for syntactic coherence rather than keeping any semantic relevance. Thus, they are "stopped" or removed before any Natural Language Processing analysis, which is based on classifying and counting the single words. The Python library NLTK (Natural Language Tokenization) was implemented for these purposes.

Lemmatization: it is the process of grouping together the inflected forms of a word in order to consider them as a single item, which is indeed the word's lemma, i.e. its dictionary form. This task was performed by implementing the WordNetLemmatizer library, specifically developed for this purpose.

Vectorization: once the text is cleaned, it is structured before undergoing any analysis. Specifically, it is transformed into matrices by filling them with the text words. This was performed with the popular Sklearn library that is provided with the Tf-Idf function for this task.

For a one-to-one similarity of topics between a company activity and a patent now there are two vectors containing condensed information for them. Each word inside the vectors is assigned a different coordinate, in order to end up with vectors of numbers. The operation of "cosine similarity", a method usually adopted to quantify how similar two documents are likely to be, is then applied to measure the similarity between the vectors. The operation output is in the interval $[-1,1]$, i.e. it is a score for the single comparison between the company activity and the field of application of a specific patent. Given a single company, this operation was iterated for nearly 1 Million of patents. These patents are classified by the IPC criterion to be equally distributed in the following categories:

The number of companies considered for this work was nearly 300, all of them listed in the U.S. stock exchanges. They belong to the following 12 market sectors:

For each company it was collected a score for each patent, belonging this to one of the categories in Table 1. The overall result consists of a distribution of scores of

Table 1. IPC system with the subclasses considered for this work.

Section	Classification Symbol	Subclasses
Human Necessities	A	01,21,22,23,24,41,42,43,44,45,46,47,61,62,63
Performing Operations Transporting	B	1,2,3,4,5,6,7,8,9,21,22,23,24,25,26,27,28,29,30,31,32,33,41,42,43,44,60,61,62,63,64,65,66,67,68,81,82
Chemistry, Metallurgy	C	1,2,3,4,5,6,7,8,9,10,11,12,13,14,21,22,23,25,30,40,40B
Textiles, Paper	D	1,2,3,4,5,6,7,21
Fixed Constructions	E	//
Mechanical Engineering Lighting, Heating, Weapons	F	//
Physics	G	1,2,16H
Electricity	H	//

Table 2. Market Sectors of companies indexed in the U.S. stock exchanges.

Basic Industries	Capital Goods	Consumer Durables	Consumer Non-Durables
Consumer Services	Energy	Finance	Healthcare
Miscellaneous	Public Utilities	Technology	Transportation

similarities for the company with patents. The scores, nearly 1 Million in total, were averaged in order to give a statistical foundation to the results. Therefore, it was possible to obtain an average of the score of similarity for the company with each IPC class as the average of the scores for the patents belonging to that class. The higher the scores the higher the similarity of the company with that specific patent IPC class. In this work, the resulting averaged distribution of scores for each company will be referred to as a Technological Print for that company, i.e. the distribution of affinity of that company with certain technological fields as they are classified with the IPC system (Table 2).

A grounding hypothesis of this work to be verified is that companies operating in the same market sector exhibit similar technological prints. In order to do so, a statistical method for embedding high-dimensional data, specifically t-SNE (t-distributed stochastic neighbor embedding) was implemented. Indeed, if the hypothesis is true, t-SNE should be able to cluster them by their technological prints, without knowing the market sector they live in.

Datasets
As mentioned before, two types of source information were used, specifically the *Business* section and *Financial Statement* section of the *10-K form* of companies and the abstract of patents.

10-K forms were extracted from the SEC official database EDGAR.

Patents were retrieved from the Google Cloud dataset "Patents Public Data". The extraction was limited to patents of the last 10 years, which is considered by the authors as a reliable time window for selecting the most recent technologies whose effects are likely to have some impact on the market.

4   Results and Conclusions

Nearly 300 companies were considered for the analysis. For each of them a technological print was obtained as the average of the scores with nearly 1 Million patents, each score classified in one of the IPC patents categories. Technological prints of companies belonging to the same market sector were averaged in order to obtain a technological print of each market sector. This operation was made in order to condense the results and make them easier to be shown. In Fig. 1 it is then possible to see the features of technological prints for the "Technology" market sector and to spot what IPC patent classes are the most sensitive for it.

Nevertheless, each company's technological print can be formalized as a vector of scores, each one belonging to a specific IPC class. These vectors represent the data that

Technology

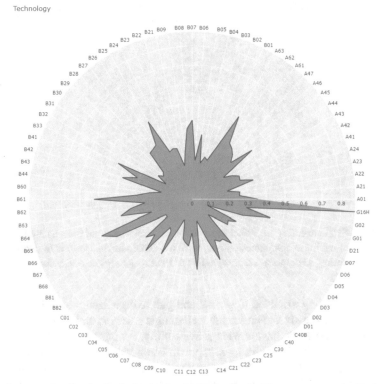

Fig. 1. An example of technological print for the Technological Sector compared with the broad spectrum of IPC patent classes. A peak corresponding to the IPC class G16H underlines the sensitivity of this patent class for the technological companies.

shall be analyzed from the t-SNE classifier in order to cluster the 300 companies. If technological prints are truly distinctive depending on the companies' market sector, then t-SNE should be able to cluster them without knowing each company's market sector. Figure 2 represents the t-SNE outcome. Different clusters are distinguishable in terms of their chromatic properties, as each color represents a specific market sector. Since, the clusters are chromatically homogeneous, this is the evidence that t-SNE was able to reconstruct the companies' market sector distribution starting from their specific technological print.

A few considerations can be made from the results above: firstly, it was possible to quantify the semantic sensitivity of different market sectors to specific IPC patent classes. This was possible by referring to official documentations of companies' activities and patents abstracts. This kind of information must be considered as binding knowledge of the firms as well as a semantic condensation of the intellectual property contents. Therefore, there must be some value in trying to match them, which is the outcome shown in the first part of this work. Another outcome of this preliminary analysis was the adoption of the *technological print* concept, which shall be considered as a marker of the knowledge that best feeds a company's needs in its activity and mission. The

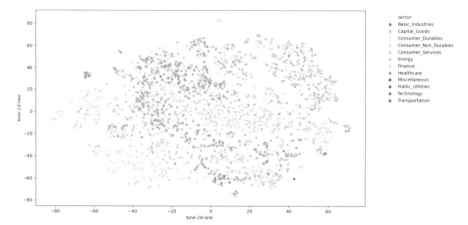

Fig. 2. The t-SNE clustering of 300 companies based exclusively on their technological print. A similarity among those companies belonging to the same market sector is highlighted by the chromatic clustering, as colors are representative of their respective market sector.

concept can be easily extended to private and public organizations, entire market sectors (as it was previously shown for the technological sector) and in general for all the potential beneficiaries of that knowledge availability. An interesting aspect of the *technological print* is the possibility to extrapolate counterintuitive information regarding that knowledge that could be of strategic significance for a specific company. Indeed, what a company needs to be competitive appears to be neither trivial nor a single binary of knowledge and skills. The core of this work analysis was anyway the stress on the correlation paradigm. As mentioned before, it was the authors' aim to underline the efficacy of a correlation analysis rather than a causal one when it comes to understanding how intellectual property affects the market.

References

1. Argote, L., Ingram, P.: Knowledge transfer: a basis for competitive advantage in firms. Organ. Behav. Hum. Decis. Process. **82**(1), 150–169 (2000). https://doi.org/10.1006/obhd. 2000.2893.S2CID7893124
2. Etzkowitz, H.: Academic-industry relations: a sociological paradigm for economic development. In: Leydesdorff & Van den Besselaar, pp. 139–151 (1994)
3. Roessner, J.D., Wise, A.: Public policy and emerging sources of technology and technical information available to industry. Policy Stud. J. **22**(2), 349–358 (1994). Charles, D., Howells, J.: Technology Transfer in Europe: Public and Private Networks. Belhaven Press (1992)
4. Charles, D., Howells, J.: Technology Transfer in Europe: Public and Private Networks. Belhaven Press (1992)
5. Kremic, T.: Technology transfer: a contextual approach. J. Technol. Transf. **28**(2), 149–158 (2003)
6. Anuraag, S., Triulzi, G., Magee, C.: Technological improvement rate predictions for all technologies: use of patent data and an extended domain description (2021). Tratto da https://www.sciencedirect.com/science/article/pii/S0048733321000950

7. Choi, Y., Park, S., Lee, S.: Identifying emerging technologies to envision a future innovation ecosystem: a machine learning approach to patent data. Scientometrics **126**(7), 5431–5476 (2021)
8. Lee, K., Lee, J.: National innovation systems, economic complexity, and economic growth: country panel analysis using the US patent data. In: Innovation, Catch-up and Sustainable Development, pp. 113–151. Springer, Cham (2021)
9. Giosio, C.: Sistemi semi-automatici per l'analisi massiva di brevetti (2013)
10. Yang, W., Cao, G., Peng, Q., Zhang, J., He, C.: Effective identification of technological opportunities for radical inventions using international patent classification: application of patent data mining. Appl. Sci. **12**(13), 6755 (2022)
11. Chi, Y.C., Wang, H.C.: Establish a patent risk prediction model for emerging technologies using deep learning and data augmentation. Adv. Eng. Inform. **52**, 101509 (2022)
12. Sobolieva, T.O., Holionko, N.G., Batenko, L.P., Reshetniak, T.I.: Global technology trends through patent data analysis. In: IOP Conference Series: Materials Science and Engineering, vol. 1037, no. 1, p. 012059 (2021). IOP Publishing
13. Bai, Y., Chou, L., Zhang, W.: Industrial innovation characteristics and spatial differentiation of smart grid technology in China based on patent mining. J. Energy Storage **43**, 103289 (2021)
14. Perez-Molina, E., Loizides, F.: Novel data structure and visualization tool for studying technology evolution based on patent information: the DTFootprint and the TechSpectrogram. World Patent Inf. **64**, 102009 (2021)
15. Cairo, P.: Analisi brevettuale del settore smaltimento rifiuti: realizzazione di un Patent landscape sulle principali aree tecnologiche.= Patent analysis of the waste disposal sector: creation of a Patent landscape on the main technological areas (Doctoral dissertation, Politecnico di Torino) (2022)
16. Cova, C.: Environmental Innovation and Open Innovation: evidence of collaboration from green patent data
17. Hötte, K., Jee, S.J.: Knowledge for a warmer world: a patent analysis of climate change adaptation technologies. Technol. Forecast. Soc. Chang. **183**, 121879 (2022)
18. Mao, G., Han, Y., Liu, X., Crittenden, J., Huang, N., Ahmad, U.M.: Technology status and trends of industrial wastewater treatment: a patent analysis. Chemosphere **288**, 132483 (2022)
19. Yuan, X., Li, X.: Mapping the technology diffusion of battery electric vehicle based on patent analysis: a perspective of global innovation systems. Energy **222**, 119897 (2021)
20. Giczy, A.V., Pairolero, N.A., Toole, A.A.: Identifying artificial intelligence (AI) invention: a novel AI patent dataset. J. Technol. Transf. **47**(2), 476–505 (2022)
21. Martins, C.A., Farias, H.C., Francisco, R.S.: Classification and extraction of information from patent data. RELCASI **13**(1), 3 (2021)
22. Krestel, R., Chikkamath, R., Hewel, C., Risch, J.: A survey on deep learning for patent analysis. World Patent Inf. **65**, 102035 (2021)
23. Hansen, M.T.: The search-transfer problem: the role of weak ties in sharing knowledge across organization subunits. Adm. Sci. Q. **44**(1), 82–111 (1999)

Anomaly Detection Using Smartphone Sensors for a Bullying Detection

Vincenzo Gattulli$^{(\boxtimes)}$, Donato Impedovo, and Lucia Sarcinella

Dipartimento di Informatica, Università degli studi di Bari Aldo Moro, 70125 Bari, Italy
vincenzo.gattulli@uniba.it

Abstract. Anomaly Detection is a fundamental process of detecting a situation different from the ordinary. The following work deals with anomalies in the human behavioral domain while filling out a questionnaire about bullying and cyberbullying. In this work, data obtained from smartphones' sensors (accelerometer, magnetometer, and gyroscope) are analyzed to apply useful Anomaly Detection techniques to detect any abnormal behaviors adopted while filling out the questionnaire implemented in an Android application. Psychology and computer science are merged to analyze and detect any latent patterns within the data set under examination to understand any polarizing content proposed during the use of the app and identify users who exhibit anomalous behaviors, possibly common to classes of users.

Keywords: Anomaly Detection · Sensors · Smartphone · Bullying · Cyberbullying

1 Introduction

Anomaly Detection is a process that tends to find anomalous *"Outliers"* patterns within a data set that do not conform to expected behavior. Anomaly detection plays a crucial role in several areas, including Fraud Detection for credit cards, the health care industry, Intrusion Detection for cyber-security, failure detection within critical systems, etc., but it is also helpful in the context of human behavior [1]. Human behavior could affect micro behaviors that might tend toward an identifying attitude of the user while performing a normal activity, such as holding the smartphone device. While filling out a questionnaire, based on the reported questions, if uncomfortable for the individual, could detect micro behaviors termed anomalies that could identify a behavioral class. Therefore, human behavior can be detected by special smartphone sensors. Smartphones are naturally equipped with different types of sensors, such as thermometers, magnetometers, barometers, microphones, cameras, ambient light sensors, GPS, proximity sensors, accelerometers, gyroscopes, and many others. In other studies, however, an attempt was made to assess factors related to daily stress and mental health by collecting sensor data such as a microphone, ambient light sensor, GPS, and accelerometer. In this case, a correlation was found between daily stress levels, mobility, and sleep periods. It is also shown that monitoring an individual's behaviors with the help of sensors can determine other mental health problems such as schizophrenia, bipolar disorder, and depression. Although many advantages are present, the use of smartphones as a sensor system has many challenges and limitations: *Incomplete or inconsistent collected data, Difficulty in measuring the reliability of the data*, the *Need to preserve resources, Ensure user privacy*. This paper is contextualized in bullying/cyberbullying, where the figure of the bully/cyberbully and that of the victim/cyber victim or the individual outside the incident manifest. In this project, were analyzed data obtained from the sensors of smartphones (*after using the appropriate app*) to apply Anomaly Detection techniques (*more generically classified as Machine Learning*) helpful in detecting any abnormal behaviors adopted during the use of the app and thus the questionnaire. In other words, this work aims to analyze and detect any latent patterns within the dataset under investigation to understand any polarizing content proposed during the use of the app and identify users who exhibit anomalous behaviors, possibly common to classes of users. Anomaly has been defined as the sudden minor change in frequency detected by smartphone sensors. They are defined as micro-behaviors that can be intensified in one of four classes (*Bullying_Victimization, Bullying_Bullying, Cyberbullying_Victimization, Cyberbullying_Cybully, External*). This document is structured as follows: The second chapter, "*2. State of the Art*," discusses the literature regarding the most widely used algorithms in the field of Anomaly Detection. The third chapter, called "*3. Material*," deals with the description of the dataset created through an Android application. The fourth chapter, called "*4. Methods*," discusses the models used for the experimental phase. The fifth chapter, called "*5. Experimental Setup*," deals with the pipeline of the experiment performed. The sixth chapter, called "*6. Results*," reports the experiment's results with attached observations. Finally, conclusions are given in chapter "*7. Conclusion*".

2 State of the Art

Before defining and classifying the most widely used algorithms in the field of Anomaly Detection, it is also necessary to differentiate between the anomalies they are found. Thus, one can mainly find three types of anomalies: *Contextual Anomalies, Collective Anomalies*, and *Point or Global Anomalies*. Concerning the type of algorithm used, one can obtain either a label (label) to indicate whether the instance is an anomaly or a value relative to the score (score) that stands for the degree of confidence about the abnormality of the anomaly. Based on the algorithms that are mentioned below, a label is often used for supervised ones. In contrast, the score is usually binary for semi-supervised and unsupervised anomaly detection algorithms $(-1, 1)$. In practice, algorithms with unsupervised approaches are of more common use, having data sets where anomalous instances are not known a priori, and one wants to find the data that differ significantly from the norm. It should also be specified that these anomalies must be associated with their context, as the same data might be anomalous for some subjects and normal for others.

Supervised Algorithms: These types of algorithms need a labeled data set. In doing so, it can distinguish normal instances from anomalous ones. Standard algorithms include Decision Trees, K-Nearest Neighbors, Random Forests, and Logistic Regression. Although it is challenging to have labeled data available, they are necessary because, very often, unsupervised approaches do not achieve the desired performance. Thus, there are semi-supervised algorithms that provide a middle ground.

Bayesian Network-Based Algorithms: These algorithms are suitable for the anomaly detection task, as they can handle high-dimensional data that are humanly difficult to interpret. Bayesian networks operate in a multi-class setting to estimate the posterior probability for instances as normal or anomalous. There are several variations for detecting anomalies: An approach based on the calculation of the anomaly score and its definition, using relative density, an Approach using the distance between the datum in question relative to a particular neighbor as the anomaly score, A hybrid approach that calculates the average distance of n objects as the neighborhood distance and averages the number of neighbors within the average distance. Other less common variants are currently being studied [2].

Clustering-Based Algorithms: This algorithm is used to create a model that clusters data to create a pattern that is later used to detect anomalies in new data. Several clustering techniques can be categorized into Partitioning Clustering, Hierarchical Clustering, and Density-based Clustering. Each of them has its advantages and disadvantages. Indeed, you may have such anomalies because they do not belong to any cluster further from the centroid or directly belong to more minor, scattered clusters.

Statistical Models: However, according to a stochastic model, the basic idea is to exploit rules and properties of inferential statistics according to which normal instances distribute in regions of higher probability. In comparison, anomalies distribute in areas of lower probability. A statistical model is then fitted to the data to determine the predicted behavior. This model is used on the new instances to identify those that do not belong to the model. Based on the applied statistics, instances with a low probability of being generated by the learned model are classified as outliers. While parametric techniques assume knowledge of the underlying distribution and estimate parameters from the data provided, nonparametric techniques generally do not take knowledge of the underlying distribution [3, 4]. Since the literature specifically in this proposed area is restricted, approaches related to tasks that appear to be like the ones under consideration in this paper have also been inspired. One of these is undoubtedly Intrusion Detection. Intrusion Detection [3] refers to the detection of potentially malicious activities (intrusions, penetrations, and other forms of abuse of computers) in a computer system. An intrusion consists of a discrepancy from normal system behavior; therefore, anomaly detection techniques are applicable in the intrusion detection domain. It refers to Intrusion Detection using an unsupervised approach which is particularly interesting but at the same time challenging because a supervised approach requires not inconsiderable maintenance of data labels in terms of how often they are updated. The key challenge for anomaly detection in this domain (as well as in the domain proposed in this paper) is the analysis of the huge volume of data. This aspect represents an affinity with the domain in which the work illustrated in this paper is contextualized. The rate of false alarms (false positives) is a further affinity between the two domains. Since the data amounts to millions of examples, detecting the outliers (a small fraction compared to the non-outlier data) is highly complex. The work proposed in [5] is contextualized in the domain of Intrusion Detection in industrial control systems where surely a supervised approach is very limiting as the methodologies behind intrusion attempts evolve with very high frequency; therefore, an unsupervised approach is surely useful in detecting even small variants of methodologies presented in the past. The algorithms used in [5] that were also decided to be adopted in the domain in which this work is contextualized through which anomaly detection can be automated are the Elliptic Envelope, Local Outlier Factor, and Isolation Forest in their original configurations. In [6], is funded the use of the Isolation Forest algorithm in the task of anomaly detection of human behaviors in data streams, certainly has several similarities with the task under consideration in this paper. According to [7], an additional Machine Learning model used in the context of Intrusion Detection is the One-Class SVM, which is particularly useful in detecting potentially anomalous patterns in data without needing to explicitly define a threshold value to consider a data item anomalous or not (in contradistinction to methods for Intrusion Detection, for example, based on Markov Chain [7]). In terms of task formalization, on the other hand, one deals with data obtained from smartphone sensor sampling in the form of a time series. In the literature, it is possible to find in [8] the treatment for this type of data for the Anomaly Detection task. Their comparison regarding data distribution was the basis for the idea of using the Elliptic Envelope. The Isolation Forest is also discussed as a model for similar tasks. As a more generic task of Anomaly Detection on sensor data acquired from smartphones, their acquisition and representation, mainly using a Support Vector Machine, is discussed in [9]. The results obtained were the basis for the idea of using the One-Class SVM as a model.

3 Material

Two main stages, the Pre-adjustment and Post-adjustment, are performed. These two phases were necessary to clean up the dataset obtained from the experiment with the users and detect the correct anomalies.

3.1 Pre-adjustment

Before carrying out the "pre-adjustment" phase, the test was conducted with users. The test consists of an Android application that incorporates a questionnaire re-done by psychologists. The graphical layout of the existing app is simple and essential. Every choice has been made to make the graphical interface simple and appealing to the end user and, to eliminate or at least drastically reduce the possibility that the user during the activities may make mistakes, may have doubts about the actions to be performed, or other problems in general. The initial test dataset consists of 164 text files (.txt format) named "sensor_," to which the user ID was concatenated. The data for the Accelerometer, magnetometer, and gyroscope sensors are within a single file; each is sampled every 20 ms. The sensor's name has been concatenated with the name of the task where it was sampled. Also specified at the beginning and end of each line is the date and time at which the sensor data was recorded. The data used are for four videos that are shown to users, where for each of them, a question attached to the video is shown. The other data, however, are related to 83 general questions. These data were recorded using a client-server approach, where the client was the Android device sending its sampled data to the server. The format of the measurements is as follows: "*data*" - "*sensor_activity*" - "*x_value, y_value, z_value.*". Within the initial dataset are files related to users who needed to complete the test correctly. These users therefore must be excluded because their data cannot be used for testing. Ninety-nine users completed the test correctly. A table duly compiled by the psychologist who processed the data is used as a ground truth for the users, where there are classifications of users in the five classes, namely *Bullying_Victimization, Bullying_Bullying, Cyberbullying_Victim, Cyberbullying_Cybully, and External*. Although most users recorded sampling consistent with expectations, the variety of smartphones used during the experiment led to unexpected sampling. Four types of problems encountered can be listed: *Duplicate sensors*: some users sampled one sensor twice as many times as others; *Asynchronous sensors*: some users, due to the client-server architecture, did not send all three sensors at the same time; thus, resulting in a file with a total number of sensor rows different from each other; *Missing sensors*: 5/99 users registered a file with one or more missing sensors and consequently were removed from the dataset; *Missing activities*: 18/99 users did not answer the questions; for these users, the questions they were able to answer before the connection dropped were considered.

3.2   Post-adjustment

Two scripts were carried out, one related to the video tasks and attached questions and one associated with the generic question tasks, as the two tasks are different. The script related to the videos and attached questions was planned to iterate first on the tasks and then on the users, as there are no users who have yet to see all the videos. A file with all rows is initially extracted for each sensor. On the other hand, the script related to the generic questions was planned to iterate first on the users and then on the activities; since it was not known a priori whether the user had answered all the questions or not. Again, a file with all rows is initially extracted for each sensor. The maximum number of rows among all files is calculated within the video-related script. Initially, the generated files contain raw data: that is, there are files containing dual and asynchronous sensors. A check was made on the number of rows to: Identify files with several rows equal to 0 for one or more sensors related to users with missing sensors, Identify files with several rows for one sensor equal to approximately twice the number of rows for the other sensors (to identify files with duplicate sensors), Identify files with a different number of rows between them (to identify files with asynchronous sensors). Once the files containing "raw" data were identified, they were normalized according to the following logic: Files with several rows equal to 0 for one or more sensors were eliminated, and thus the related users were excluded from experimentation, Files with several rows for one sensor equal to about twice the number of rows for the other sensors saw the number of rows for the doubled sensor halved by saving the second of the two instances (i.e., skipping those of odd index and saving those of even index), Files with different numbers of rows from each other underwent the extension of the last value for sensors with fewer than the maximum number of rows since this value is the one closest to reality. The final output is, therefore, that of a file with a ".csv" extension for each sensor, user, and activity, all having the same number of rows after trimming the empty rows. The three files related to the Accelerometer, magnetometer, and gyroscope sensors are merged within a single file, keeping the ".csv" extension. Within the script related to the questions, on the other hand, given the significantly lower number of rows, it was decided to optimize the process and generate files having the correct number of rows directly, thus without empty rows. The process is like the video script, except that it iterates first over the users and then over the tasks (so that the files and folders are created only for the questions the user answered), and all subsequent steps for sensor normalization were combined within a single script. The final output is the same. The last number of users for the post-adjustment dataset is 94 for video tasks and 93 for questions.

4   Methods

The models used in the experimental phase, named *"Anomaly Detection with ML models,"* are the following: *Elliptic Envelope, Isolation Forest, Local Outlier Factor, One-Class SVM,* and *Descriptive Statistics.* The first models are the standards and are among the most widely used state-of-the-art models. For each user, machine learning algorithms were run for each task, and the returned predictions were saved in a file having a ".csv" extension. The returned predictions were organized into ".csv" files where 1 = Normals and -1 = Anomalies. This was done by sensor analysis (*Accelerometer, Magnetometer, Gyroscope*): *Elliptic Envelope, Isolation Forest, Local Outlier Factor*, and *One Class SVM*. Given many users and activities, leading to a total number of predictions exceeding 8000, it is difficult to make inferences about the data or visualize them. Therefore, it was also decided to use descriptive statistics techniques. A script was created ad hoc and used to generate tables (in the form of.*xlsx* files) containing the outputs of the algorithms' predictions in the form of the percentage of anomalies out of the total. In these tables, is possible to calculate the percentage of anomalies in users, activities, and user categories. Is it possible to see for which users and which tasks are funded a significantly higher than an average number of anomalies (in this type of task, it is normal to find at least a minimum of anomalies; the total absence of anomalies would be equivalent to a motionless hand or a cell phone resting on a fixed surface). The designed script is created, for each model, a table with the results obtained, i.e., a number relating the anomalous rows to the total rows between 0 and 1 (where 0 is equivalent to the total absence of anomalous points and one is equivalent to the presence of only anomalous points). The users are on the table's rows, and the activities are on the columns. Once the tables for the individual models have been created with the results of the various predictions, a threshold value is set; a low, medium and high threshold are chosen. Only the anomalies that deviate from normal are considered with a higher threshold. The final Summary Table, on the other hand, was created manually; it is used to compare the various algorithms based on their performance and results on both users and activities; from the final table, graphs are generated that allow us to provide an overview of the behavior of the models. The models are on the rows, and the activities and classes are on the columns. Given many questions, it was decided to consider the first 20, and the last 20 based on the percentage of anomalies reported.

5   Experimentation Setup

This chapter provides a perspective on how the experimentation was structured:

- *Android Application:* Grant permission, View four videos that depict situations of cyberbullying, Fill in a questionnaire created with the collaboration of a psychologist, Open a debate on a Telegram group created ad-hoc, Finish the test and uninstall the app; *Data Creation:* The phase of reacting and cleaning the dataset was divided into two phases: Pre-adjustment and Post-adjustment, as explained in the chapter *"3 Dataset"*; *Anomaly Detection with ML models:* Standard ML models were implemented to identify anomalies by considering three different thresholds of experimental application, considering a wider to a narrower one; *Graphical and Descriptive Analysis*: Finally, it was possible to view the anomalies and give a visual interpretation of the psychological and behavioral situation of the actors of bullying and cyberbullying.

6   Results

This section shows how to visualize the difference in the percentage of anomalies based on activities and classes of users. On the X axis, is possible to find the activities or classes of users, while on the Y axis, is possible to find the percentage of anomalies, respectively, for each of the four models used. There are graphs for each threshold parameter used: low (Deviated from the average value by considering everything as an anomaly), medium, and high (Captured the most experimentally relevant anomalies). A graph was also created for the first 20 questions and the last 20 questions for the high parameter. The total number of users is 94 for videos (*Bullying_Victimization = 30, Bullying_Bully = 13, Cyberbullying_Victimization = 12, Cyberbullying_Cyberbully = 3, External = 53*) and 93 users for quizzes (*Bullying_Victimization = 30, Bullying_Bull = 13, Cyberbullying_Victimization = 12, Cyberbullying_Cyberbully = 3, External = 52*); A user can belong to more than one class.

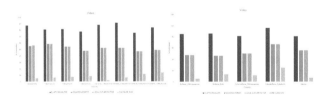

Fig. 1. Histograms for video activity (left) and video classes (right) for low threshold.

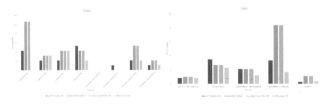

Fig. 2. Histograms for video activity (left) and video classes (right) for medium threshold

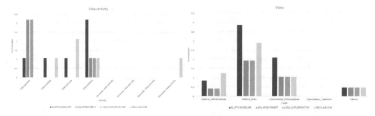

Fig. 3. Histograms for video activity (left) and video classes (right) for high threshold.

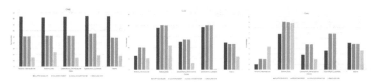

Fig. 4. Histograms for quiz classes: low threshold (left), medium threshold (center), high threshold (right).

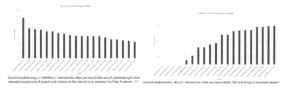

Fig. 5. Histograms for the first 20 questions with the most anomalies, averaging the scores obtained between the 4 models, high threshold (see left). Histograms for the last 20 questions with the most anomalies (therefore the first 20 with the least anomalies) by averaging the scores obtained between the 4 models, high threshold (see right).

From an algorithmic point of view, the *Elliptic Envelope* is much more discriminating toward values that deviate from the expectation of the dataset's probabilistic distribution (Figs. 1, 2, 3, and 4). The Local Outlier Factor focuses on outliers by looking locally at the data density relative to its neighborhood. The isolation forest cannot optimally identify anomalous points close to each other, tending to label them all anomalous or not (the total number of anomalies is like that of the LOF). The One-Class SVM differentiates abnormalities from normality more sharply, identifying fewer of them than the other models. From a social perspective, it is possible to trace computer data back to inferences from observing the data. The bully and cyberbully user classes encounter more anomalies on the medium. Thus, a bully and you cyberbully may be active participants in the affair, resulting in being more restless while performing this questionnaire. Even if the threshold is narrowed down, i.e., one considers a threshold other precisely that threshold that best identifies the peaks of abnormality, the bully class encounters more abnormalities. The bully is an active part of the impact; it is more involved in the affair psychologically than actively. The idea of composing a questionnaire could lead him to perform those micro-behaviors that can identify a class of belonging. The questionnaire is divided into two macro-parts: The first phase of watching four videos inherent to bullying and cyberbullying, and the second phase is composed of a 5-point Likert scale. The activities related to the videos encountered more anomalies than those related to the questions. Observing Fig. 5, the question that was part of the questionnaire, "*QuizActivityButtons74 = CYBERBULLY: Indicate how often you have DONE acts of cyberbullying 8. Send videos/photos/pictures of assaults and violence on the internet (Websites, YouTube, Facebook...)*" and "*QuizActivityButtons61 = BULLY: Indicate how often you have SUBJECTED to bullying a) I have been beaten*" were found to be the most anomalous among the other questions. The action "*I have been beaten*" elicits feelings to emphasize the fear of the situation among the victims but that among those who actively act. But the action of sending or receiving media files also arouses a different than normal and, therefore, abnormal feeling among these users.

7 Conclusion

In the paper, data obtained from smartphone sensors were analyzed to apply Anomaly Detection techniques helpful in detecting anomalous behaviors adopted during the use of the app. This work aims to analyze and detect any latent patterns within the data set under examination to understand any polarizing content proposed during app use and identify users who exhibit anomalous behaviors, possibly common to classes of users. Data from three smartphone sensors, namely Accelerometer, magnetometer, and gyroscope, were analyzed for 94 users. The anomaly was defined as the smallest abrupt change in frequency detected by the smartphone sensors. They were defined as identifiable micro-behaviors in one of the four classes (*Bullying_Victimization, Bullying_Bullying, Cyberbullying_Victimization, Cyberbullying_Cyberbully, External*). The conclusions are diverse in nature; however, comparing them with similar work is difficult because the literature specifically in this proposed area is very young and modest.

The first type of conclusion possibly relates to the models used: *The Elliptic Envelope* is much more discriminating against values that deviate from the expected probabilistic distribution proper to the dataset, and *The Local Outlier Factor* focuses on outliers by looking locally at the density of the data relative to its neighborhood: it is therefore not ideal for this application domain and specific task, *The Isolation Forest* cannot optimally identify anomalous points close to each other, tending to label them all anomalous or not (the total number of anomalies is like that of the LOF.), *The One-Class SVM* succeeds in differentiating abnormalities more sharply from normality, identifying fewer of them than the other models. The second type of conclusion possibly relates to classes of users: The bully and cyberbully classes experience more anomalies on the medium than the other user classes, The bully class encounters more anomalies when using a higher threshold, resulting in the most discriminating. However, anomalies are also found among users who are not in bully or victim classes; this could also be due to user embarrassment in answering questions. The third type of possible conclusion relates to the activities within the dataset: Activities related to videos encounter more anomalies than those related to questions, Few anomalies are found among the first questions (these questions are used to classify the category *Bullying_Victimization*), More anomalies are found among the last questions (these questions are used to classify the *Cyberbullying_Cybully* category), Some questions that with a high parameter do not find anomalies could be eliminated from the quiz: i.e., *Do you know all your friends you have in your Internet profiles? From the beginning of school to now, have you ever brought your smartphone (With Internet access using social networks such as Facebook, WhatsApp, etc...) to class to do a non-school activity?, From the start of school to know how many times have you used your smartphone to connect to the Internet while in class to do a non-school activity?, From the start of school to now, how many times have you used your smartphone to connect to the Internet while in class to do a non-school activity?, Typically, how many hours a day do you actively spend online (browsing, chatting, attending social networks, writing emails, etc.) excluding DAD activities;*

- The questions that are part of the questionnaire, *"QuizActivityButtons74 = CYBER-BULLY: Indicate how often you have DONE acts of cyberbullying 8. Send videos/photos/pictures of assaults and violence on the internet (Websites, YouTube, Facebook"* e *"QuizActivityButtons61 = BULLY: Indicate how often you have SUB-JECTED to bullying a) I have been beaten"*, were found to have a higher percentage of anomalies.

Regarding future developments, on the other hand, it might be useful to conduct a study on hyperparameters (tuning), avoiding automated Machine Learning approaches, since this domain inherently exploits unsupervised methodologies.

Acknowledgment. This work is supported by the Italian Ministry of Education, University, and Research within the PRIN2017 - BullyBuster project - A framework for bullying and cyberbullying action detection by computer vision and artificial intelligence methods and algorithms.

References

1. Thapliyal, A., Verma, O.P., Kumar, A.: Behavioral biometric based personal authentication in feature phones. Int. J. Electric. Comput. Eng. **12**, 802–815 (2022). https://doi.org/10.11591/IJECE.V12I1.PP802-815
2. Zhao, M., Chen, J., Li, Y.: A review of anomaly detection techniques based on nearest neighbor. In: Proceedings of the 2018 International Conference on Computer Modeling, Simulation and Algorithm (CMSA 2018), vol. 151 (2018). https://doi.org/10.2991/CMSA-18.2018.65
3. Prasad, N.R., Almanza-Garcia, S., Lu, T.T.: Anomaly detection. Comput. Mater. Cont. **14**, 1–22 (2009). https://doi.org/10.1145/1541880.1541882
4. Ahmed, M., Naser Mahmood, A., Hu, J.: A survey of network anomaly detection techniques. J. Netw. Comput. Appl. **60**, 19–31 (2016). https://doi.org/10.1016/J.JNCA.2015.11.016
5. Mokhtari, S., Yen, K.K., Mokhtari, S., Yen, K.K.: Measurement data intrusion detection in industrial control systems based on unsupervised learning. Appl. Comput. Intell. **1**, 61–74 (2021). https://doi.org/10.3934/ACI.2021004
6. Ding, Z., Fei, M.: An anomaly detection approach based on isolation forest algorithm for streaming data using sliding window. IFAC Proc. Vol. **46**, 12–17 (2013). https://doi.org/10.3182/20130902-3-CN-3020.00044
7. Wang, Y., Wong, J., Miner, A.: Anomaly intrusion detection using one class SVM. In: Proceedings of the Fifth Annual IEEE System, Man and Cybernetics Information Assurance Workshop, SMC, pp. 358–364 (2004). https://doi.org/10.1109/IAW.2004.1437839

8. Liu, G., Onnela, J.P.: Online anomaly detection for smartphone-based multivariate behavioral time series data. Sensors (Basel) **22** (2022). https://doi.org/10.3390/S22062110
9. Vlădăreanu, V., et al.: Detection of anomalous behavior in modern smartphones using software sensor-based data. Sensors (Basel) **20** (2020). https://doi.org/10.3390/S20102768

Machine Learning for Automotive Security in Technology Transfer

Vita Santa Barletta[1]([✉]), Danilo Caivano[1], Christian Catalano[2], Mirko De Vincentiis[1], and Anibrata Pal[1]

[1] University of Bari "Aldo-Moro", Bari, BA 70125, Italy
{vita.barletta,danilo.caivano,mirko.devincentiis,anibrata.pal}@uniba.it
[2] University of Salento, Lecce, LE 73100, Italy
christian.catalano@unisalento.it

Abstract. The new breed of vehicles comes stashed with cutting-edge technologies. The in-vehicle communication system (CAN Bus) smartly connects to all such Electronic Control Units (ECU) to enhance the safety, security, and stability of the vehicle, driver, and passengers. We need efficient and real-time Intrusion Detection Systems (IDS) to prevent threats in vehicles. Extensive research about IDS on CAN bus have been proposed, but most are offline. Further, the in-vehicle ECUs need to be more robust for implementing IDS. This paper presents an in-vehicle multi-class IDS aiming to detect cyber attacks in real-time. The goal is to transfer this technology to Automotive Industry in order to increase the safety and security of vehicles, drivers, and passengers.

Keywords: Automobile · Cybersecurity · Software Engineering · CAN Bus

1 Introduction

The new generations of vehicles are loaded with onboard computers, responsible for the smooth and glitch-free functioning of these automobiles. These vehicles have come a long way from their older counterparts, where different faults were found manually after detailed scrutiny. Nowadays, the presence of onboard computers has made life easier for both the driver and the mechanic. The computer connects with different control units across the vehicles, like the electrical control units, keyless entry, anti-lock braking system, power steering, anti-theft, traction control, telematic/navigation system, and parking assistant. One control module generally shares information with other control modules through different protocols, such as Ethernet, FlexRay, Local Interconnect Network (LIN), Media Oriented Systems Transport (MOST), and Controller Area Network (CAN). The most widely used protocol is the CAN, due to its safety and velocity of sending messages [7].

© The Author(s), under exclusive license to Springer Nature Switzerland AG 2024
A. Rocha et al. (Eds.): WorldCIST 2023, LNNS 802, pp. 341–350, 2024.
https://doi.org/10.1007/978-3-031-45651-0_34

With advanced technologies installed in the new breed of automobiles, they became subject to numerous threats from hackers. Although the CAN bus is responsible for stable, safe, and secure communication within the Electronic Control Units (ECUs), some inherent vulnerabilities can still be exploited. For example, the CAN bus is susceptible to different attack signatures like Denial of Service (DoS), eavesdropping, fuzzy, malfunction, and flooding, among others, [12]. Random forest, thus, proved to be a good candidate for multi-class classification techniques for IDS problems. Over the years, many approaches like network segmentation, authentication, Intrusion Detection System (IDS), and encryption were proposed [6], but none were efficient against the looming threats. The principal challenge is to detect attacks on the CAN bus accurately and in real-time. The computational units or ECUs existing in the vehicles are primarily microcontrollers and possess very low computational power. Therefore, we need infrastructural changes in the electronic circuitry of an automobile to attend to the goal. Real-time IDS can be achieved only with an onboard, specialized, and dedicated IDS that is powerful enough to run complex Machine Learning (ML) tasks [19].

This paper proposes an IDS approach using a cheap computational module as an onboard dedicated IDS for anomaly detection in the CAN bus. It is possible to transfer this solution to an actual vehicle to execute several experiments and validate and improve the proposed idea. An ensemble learning-based ML algorithm was loaded in the IDS component and used to perform the job. The proposed model was tested using three distinct attack signatures: Flooding, Fuzzy, and Malfunction. The attack behavior was simulated using an Arduino Uno, which sent malicious messages on the CAN bus to test the model's efficiency, carried out by a Raspberry Pie 4. The rest of the paper is organized as follows: Sect. 2 presents the related works; Sect. 3 describes the methodology used; Sect. 4 presents the dataset and the experiments, including the laboratory and experimental setup; The results are discussed in Sect. 5; and finally, Sect. 6 presents the conclusion and future works.

2   Related Works

To ensure the safety and security of drivers and passengers, extensive research has been carried out in recent years to develop efficient Intrusion Detection Systems (IDS) for vehicles. In [13], the authors defined a taxonomy of problems where machine learning is used, namely intrusion detection, malware analysis, and spam detection. Cho and Shin [9] proposed Viden, which detects the bad ECU from its output voltage in the CAN bus. In [4], the authors proposed an efficient IDS to identify attack messages in the CAN bus by integrating an unsupervised Kohonen Self-Organizing Map (SOM) network with the K-Means algorithm. Hanselmann et al. [16] proposed an unsupervised IDS to detect attacks on the CAN bus using an LSTM neural network to capture the temporal dynamics of each ID. It is possible to consider the temporal nature of the data and therefore predict future values based on previously observed values [8]. In [26], authors created a GAN-based IDS using a deep-learning generative adversarial network. Instead, Othmane et al. [23] proposed an anomaly detection capable of differentiating messages from no-attack to under-attack. In a different approach [10], the authors used Linear SVM and BDT(Bagged Decision Trees) to create a multi-class classifier to detect Masquerade and Bus-off attacks. Kennedy Okokpujie et al. [22] proposed a multi-class IDS using a Feedforward Neural Network (FNN) and a Support Vector Machine. [18] highlighted that the random forest algorithm had been used widely across different areas with considerable success. For example, in [1], it has been used to detect false basic safety messages in the Internet of Vehicles. In [14], the authors conducted a comparison of supervised machine learning algorithms and concluded that the Random Forest classifier outperforms both K-Nearest Neighbor and XG boost. In another IDS implementation [21], the authors used Random Forest with a dynamic voting technique that presented better and more stable performance. Therefore, considering the literature, Random forest proved an excellent candidate for multi-class classification techniques for IDS problems.

The propagation of ML-based IDS approaches from laboratory to market is low due to the complex algorithms and implementation difficulties. Therefore, we require better solutions to accommodate the advanced research in IDS using ML in Technology Transfer (TT) process. TT has been ideally used to help the technology flow from the laboratory to industry, developed to developing countries, or put the information to diverse usages [24]. In a coherent definition, TT can be expressed as a point-to-point phenomenon of exploitation of knowledge for implementation [28] or even might be seen as the movement of technology from one place to another [27]. For example, in a report on the reduction of carbon emissions from vehicles[1], the author stresses the exchange of carbon-free technologies in the coming decades. Furthermore, TT processes encompass diverse phases of technological innovation and its dissemination until commercialization, for example, developing a commercializable technology developed by university-based research in collaboration with an external company [25].

Considering the extensive research in the development of IDS for in-vehicle threat detection, the lack of multi-class IDS, and the application of Technology Transfer in different commercial fields, this paper proposes a methodology introducing an IDS predicting real-world in-vehicle attacks, thereby increasing security in the automotive field.

3 Methodology

Technology Transfer refers to the process of conveying results stemming from scientific and technological research to the marketplace and to wider society, along with associated skills and procedures[2]. To achieve this goal in our research

---

[1] https://www.nbr.org/publication/driving-down-emissions-the-role-of-technology-transfers-in-decarbonizing-transportation/.

[2] https://knowledge4policy.ec.europa.eu/technology-transfer/what-technology-transfer_en.

work, Fig. 1 shows the essential elements of the technology transfer process in automotive industry.

Fig. 1. Automotive Technology Transfer

In the first phase (University Research), the idea to increase automotive security using Artificial Intelligence (AI) based on a machine learning multi-class IDS was presented. Then, regarding the channels to transfer the proposed solution (Technology Transfer), the following were identified: research projects, doctoral research, consultancy, publications, and meetings. Finally, the impacted stakeholders were identified as the driver, automotive company and workshop, insurance, and all technology, electronics, and telecommunications companies involved in hardware and software integration [25].

The proposed IDS aims to design an approach to disseminate a laboratory setup for identifying CAN attacks to be available for commercial use in an automotive company. This is also in accordance with the ISO/SAE 21434 "Road vehicles - Cybersecurity engineering" regarding the secure development process [2] and hardware components in the automotive field [11]. To do this, a multi-class approach was used to identify attacks on the CAN bus using the *Survival Analysis Dataset for automobile IDS*. A Random Forest model was trained using the Survival Analysis Dataset for automobile IDS[3] [15].

Figure 2 shows how the Random Forest Classifier works. The classifier consists of N decision trees, each of which will predict a class (vote). The class with the most votes will become the model's prediction. The model can predict the following classes: Normal, Flooding, Fuzzy, or Malfunction.

The training and testing phase was performed offline because, in general, an ECU has limited computation power. The entire concatenated dataset will be used with the parameters found in the evaluation model to create a final Random Forest model since using the single training set will lose additional information. This final model will be put into the IDS ECU, whose purpose is to perform the identification phase capturing the messages from the CAN bus and verifying if the messages sent are anomalous.

An Arduino Uno was used to simulate an ECU that sent malicious messages on the CAN bus. The injected messages correspond to the Flooding, Fuzzy, and

---

[3] https://ocslab.hksecurity.net/Datasets/survival-ids.

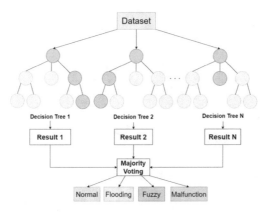

Fig. 2. Random Forest Classifier

Malfunction attacks, sending 100 messages for each attack to validate the model. To be more specific, for the Flooding attack, the CAN ID and the DATA field were set to zero; for the Fuzzy attacks, the values of CAN ID and DATA were generated randomly in hexadecimal form; and finally, for the Malfunction attack, the CAN ID corresponded to the IDs used by [15] instead, and the DATA field was generated randomly like the Fuzzy attack. The Raspberry Pi 4 analyzed the messages received from the CAN bus and determined whether the received message was an attack or not. Figure 3 shows the methodology mentioned above.

4 Experiments

4.1 CAN Protocol

The multiple control units in modern-day vehicles provide driver performance, safety, and convenience. These ECUs are connected in parallel with the CAN bus, wherein the ECUs broadcast the signals in the CAN data frame format. CAN data frame is one of the most used in-vehicular protocols in the automotive industry. It was standardized as a part of ISO 11898 (International Organization for Standardization) for road vehicles in 1993.

A typical CAN data frame consists of an arbitration field carrying the frame ID (11 bits for CAN 2.0A and 29 bits for CAN 2.0B), a one-bit Remote Transmission Request (RTR), used to distinguish between a remote frame (logic 1), and a data frame (logic 0), a six-bit control field for the control signal containing the Data Length Code (DLC) dictating the length of the Data field, and an eight-byte (64 bits) Data field containing the transmitted data [5,17].

4.2 Dataset

The Survival Analysis Dataset for automobile IDS[4] [15] was used because it contains CAN messages obtained from real vehicles: HYUNDAI YF Sonata, KIA

---

[4] https://ocslab.hksecurity.net/Datasets/survival-ids.

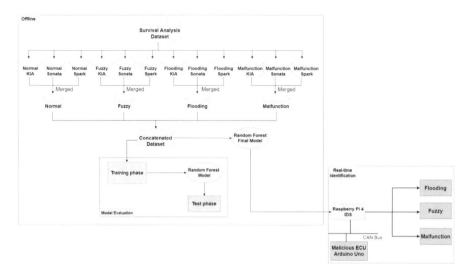

Fig. 3. Methodology phase

Soul, and CHEVROLET Spark. The authors generated three different state-of-the-art attacks that are used for intrusion detection. These are (i) **Flooding Attack**: an attacker sends a large number of messages occupying many of the resources allocated by the CAN bus. In particular, to do this, the ECU sends numerous CAN messages with a dominant ID set to 0x000; (ii) **Fuzzy Attack** an attacker generates malicious CAN messages with random ID and DATA Field. The authors randomly generated the CAN ID and the Data field from 0x000 to 0x7FF; (iii) **Malfunction Attack**: an attack will select an individual CAN ID from among the extractable CAN IDs of a vehicle. To perform this attack, the researchers chose 0×316, 0×153, and 0×18E from the HYUNDAI YF Sonata, KIA Soul, and CHEVROLET Spark vehicles as CAN ID.

The attributes contained in these datasets are **Timestamp** recorded times; **CAN ID** identifier of the CAN message in hexadecimal form; **DLC** the Data Field length from 0 to 8; **DATA** data value in hexadecimal format; **Flag**, T represents the injection message while R represents the normal message.

So starting from the three analyzed datasets, all messages were merged to obtain a single dataset for training and testing the proposed multi-class IDS Fig. 3. The features in the dataset were transformed to increase the model prediction. The values regarding the CAN ID present in the dataset were transformed into a decimal form, and then the binary representation was obtained. The DATA feature was transformed from hexadecimal to decimal form. The Timestamp and DLC were removed because they do not provide useful information, and finally, the *MinMaxScaler*[5] was applied to normalize the data in the range [0,1]. The dataset was split into training (70%) and testing (30%).

---

[5] https://scikit-learn.org/stable/modules/generated/sklearn.preprocessing.MinMaxScaler.html

To evaluate the Random Forest parameters, *HalvingGridSearchCV*[6] was used, a module of Scikit-learn that searches over specified parameter values with successive halving. After the evaluation, the algorithm found the following as best parameters: *max_depth 30*, *max_features sqrt*, and *n_estimators 9*.

4.3 Experimental Setup

To simulate an in-vehicle network, Raspberry Pi 4 Model B[7] with 4 GigaByte of RAM, Arduino MKR CAN Shield[8], and Arduino Elegoo Uno R3[9] were used. The Arduino Elegoo Uno R3 simulated an attacker who injects malicious messages on the CAN bus. Two Arduino MKR CAN Shields were used wherein one sent the messages, and the other received those and transmitted them to the IDS. The Raspberry Pi 4 Model B was used to simulate an IDS, which receives messages from the CAN bus (Receiver Arduino MKR CAN Shield) and performs the classification of messages according to the attacks defined in [15]. Figure 3 shows the real-time analysis in which the malicious ECU (Arduino Elegoo Uno R3) sent messages on the CAN bus using the MKR CAN Shield. The Raspberry Pi 4 Model B checked these messages and predicted the attack category. Figure 4 shows the laboratory setup used to emulate the in-vehicle CAN Bus IDS. The MKR CAN Shield consists of a microchip MCP2515, a second-generation stand-alone CAN Bus capable of transmitting and receiving the CAN message [20]. The Raspberry Pi 4 Model B allows a connection with a can0 interface that receives data from a physical component emulating an actual CAN protocol. The bitrate of the can0 interface was set to 500000 because it is the speed supported by the hardware.

Fig. 4. Laboratory Setup

---

[6] https://scikit-learn.org/stable/modules/generated/sklearn.model_selection.Halving GridSearchCV.html.

[7] https://www.raspberrypi.com/products/raspberry-pi-4-model-b/.

[8] https://store.arduino.cc/products/arduino-mkr-can-shield.

[9] https://www.elegoo.com/en-it/products/elegoo-uno-project-super-starter-kit.

5   Results

Prediction, Recall, and F1-Score were used as metrics to evaluate the Random Forest and the injected malicious messages. Table 1 shows the results using the traditional machine learning metrics for the Random Forest model and injected messages. Mostly, the predictions about the attack types were correct, apart from tiny aberrations that appeared in the results because of the randomness of the DATA field. For instance, the Random Forest misclassified 0.014% of Normal messages as Fuzzy, 0.01% of Malfunction messages as Fuzzy, and 0.016% and 0.052% of Fuzzy messages as Normal and Malfunction, respectively. On the other hand, the IDS always classified the flooding messages correctly.

Table 1. Results obtained with Random Forest and Arduino Elegoo Uno R3 (Injected Messages)

Technique	Class	Accuracy (%)	Precision (%)	Recall (%)	F1-Score (%)
Random Forest	Normal	99.98	99.99	99.98	99.99
	Flooding		100	100	100
	Fuzzy		99.79	99.93	99.86
	Malfunction		99.90	99.98	99.94
Injected Messages	Flooding	85.71	100	100	100
	Fuzzy		57.14	100	72.72
	Malfunction		100	25	40

In this case, the Arduino Elegoo Uno R3 only injected malicious messages to simulate the attack and validate the IDS when a vehicle is under attack. As seen in Table 1, 75% of Malfunction messages were misclassified as Fuzzy. In the case of Malfunction messages, the CAN IDs are reused from the existing ones in the dataset, and the DATA field was randomly generated. In addition, the Malfunction DATA field was randomized to avoid model overfitting. Since the Fuzzy DATA field is also generated randomly, the model probably misclassifies a Malfunction attack as Fuzzy.

6   Conclusion

This paper proposes an intrusion detection system (IDS) using Random Forest with a multi-class approach to identify attacks on the CAN bus. Two MKR CAN Shield components were used to simulate the in-vehicle network. The Arduino Elegoo Uno R3 simulated the behavior of an attacker. On the other hand, a Raspberry Pi 4 Model B simulated an IDS to identify malicious messages. Through this experiment, it was possible to see the feasibility of using an IDS inside a vehicle to detect possible attacks. This approach adopted a laboratory-based IDS system to be implemented in a real-world automotive scenario. At this moment, we are planning to transfer this solution to the technology, electronics, and telecommunication companies involved in the research project aiming to redefine the detection, response, and prevention in automotive security [3].

Acknowledgements. This study has been partially funded by the following projects: SSA (Secure Safe Apulia - Regional Security Center, Codice Progetto 6ESURE5) and KEIRETSU (Codice Progetto V9UFIL5) funded by "Regolamento regionale della Puglia per gli aiuti in esenzione n. 17 del 30/09/2014" (BURP n. 139 suppl. del 06/10/2014) TITOLO II CAPO 1 DEL REGOLAMENTO GENERALE "Avviso per la presentazione dei progetti promossi da Grandi Imprese ai sensi dell'articolo 17 del Regolamento".

References

1. Anyanwu, G.O., Nwakanma, C.I., Lee, J.M., Kim, D.S.: Novel hyper-tuned ensemble random forest algorithm for the detection of false basic safety messages in internet of vehicles. ICT Express (2022). https://www.sciencedirect.com/science/article/pii/S2405959522000923
2. Baldassarre, M.T., Barletta, V.S., Caivano, D., Piccinno, A.: A visual tool for supporting decision-making in privacy oriented software development. In: AVI 2020. Association for Computing Machinery, New York (2020). https://doi.org/10.1145/3399715.3399818
3. Baldassarre, M.T., Barletta, V.S., Caivano, D., Raguseo, D., Scalera, M.: Teaching Cyber Security: The Hack-Space Integrated Model, vol. 2315 (2019)
4. Barletta, V.S., Caivano, D., Nannavecchia, A., Scalera, M.: A kohonen som architecture for intrusion detection on in-vehicle communication networks. Appl. Sci. **10**(15), 5062 (2020)
5. Bosch GmbH, R.: Can specification 2.0. In: Proceedings of the 14th International CAN Conference (1991). http://esd.cs.ucr.edu/webres/can20.pdf
6. Bozdal, M., Samie, M., Aslam, S., Jennions, I.: Evaluation of can bus security challenges. Sensors **20**(8), 2364 (2020)
7. Caivano, D., De Vincentiis, M., Nitti, F., Pal, A.: Quantum optimization for fast can bus intrusion detection. In: Proceedings of the 1st International Workshop on Quantum Programming for Software Engineering (QP4SE 2022), pp. 15–18. Association for Computing Machinery, New York (2022). https://doi.org/10.1145/3549036.3562058
8. Catalano, C., Paiano, L., Calabrese, F., Cataldo, M., Mancarella, L., Tommasi, F.: Anomaly detection in smart agriculture systems. Comput. Ind. **143**, 103750 (2022)
9. Cho, K.T., Shin, K.G.: Viden: attacker identification on in-vehicle networks. In: Proceedings of the 2017 ACM SIGSAC Conference on Computer and Communications Security (CCS 2017), pp. 1109–1123. Association for Computing Machinery, New York (2017)
10. Choi, W., Joo, K., Jo, H.J., Park, M.C., Lee, D.H.: Voltageids: low-level communication characteristics for automotive intrusion detection system. IEEE Trans. Inf. Forens. Secur. **13**(8), 2114–2129 (2018)
11. De Vincentiis, M., Cassano, F., Pagano, A., Piccinno, A.: Qai4ase: quantum artificial intelligence for automotive software engineering. In: Proceedings of the 1st International Workshop on Quantum Programming for Software Engineering (QP4SE 2022), pp. 19–21. Association for Computing Machinery, New York (2022). https://doi.org/10.1145/3549036.3562059

12. El-Rewini, Z., Sadatsharan, K., Selvaraj, D.F., Plathottam, S.J., Ranganathan, P.: Cybersecurity challenges in vehicular communications. Vehicul. Commun. **23**, 100214 (2020)
13. Geetha, R., Thilagam, T.: A review on the effectiveness of machine learning and deep learning algorithms for cyber security. Archiv. Comput. Methods Eng. **28**(4), 2861–2879 (2021)
14. Gundu, R., Maleki, M.: Securing can bus in connected and autonomous vehicles using supervised machine learning approaches. In: 2022 IEEE International Conference on Electro Information Technology (eIT), pp. 042–046. IEEE (2022)
15. Han, M.L., Kwak, B.I., Kim, H.K.: Anomaly intrusion detection method for vehicular networks based on survival analysis. Vehicul. Commun. **14**, 52–63 (2018). https://www.sciencedirect.com/science/article/pii/S2214209618301189
16. Hanselmann, M., Strauss, T., Dormann, K., Ulmer, H.: Canet: an unsupervised intrusion detection system for high dimensional can bus data. IEEE Access **8**, 58194–58205 (2020)
17. Hartwich, F., Bosch, R.: Bit time requirements for can FD. In: Proceedings of the 14th International CAN Conference (2013). https://www.can-cia.org/fileadmin/resources/documents/proceedings/2013_hartwich_v2.pdf
18. Ho, T.K.: Random decision forests. In: Proceedings of 3rd International Conference on Document Analysis Anrecognition, vol. 1, pp. 278–282. IEEE (1995)
19. Ma, H., Cao, J., Mi, B., Huang, D., Liu, Y., Li, S.: A GRU-based lightweight system for CAN intrusion detection in real time. Secur. Commun. Netw. 2022 (2022)
20. Microchip: MCP2515 Stand-Alone CAN Controller with SPI Interface. Tech. rep., Microchip Technology Inc, Chandler (2019). https://ww1.microchip.com/downloads/en/DeviceDoc/MCP2515-Stand-Alone-CAN-Controller-with-SPI-20001801J.pdf
21. Mowla, N.I., Rosell, J., Vahidi, A.: Dynamic voting based explainable intrusion detection system for in-vehicle network. In: 2022 24th International Conference on Advanced Communication Technology (ICACT), pp. 406–411. IEEE (2022)
22. Okokpujie, K., Kennedy, G.C., Nzanzu, V.P., Molo, M.J., Adetiba, E., Badejo, J.: Anomaly-based intrusion detection for a vehicle can bus: a case for hyundai avante cn7. J. Southwest Jiaotong Univ. **56**(5) (2021)
23. Othmane, L.B., Dhulipala, L., Abdelkhalek, M., Multari, N., Govindarasu, M.: On the performance of detecting injection of fabricated messages into the can bus. IEEE Trans. Depend. Secure Comput. (2020)
24. Phillips, R.: Technology business incubators: how effective as technology transfer mechanisms? Technol. Soc. **24**, 299–316 (2002)
25. Rogers, E.M., Takegami, S., Yin, J.: Lessons learned about technology transfer. Technovation **21**(4), 253–261 (2001). https://www.sciencedirect.com/science/article/pii/S0166497200000390
26. Seo, E., Song, H.M., Kim, H.K.: Gids: gan based intrusion detection system for in-vehicle network. In: 2018 16th Annual Conference on Privacy, Security and Trust (PST), pp. 1–6. IEEE (2018)
27. Solo, R.A., Roger, E.M.: Inducing technological change for economic growth and development (1972)
28. Tidd, J., Bessant, J.R.: Managing Innovation: Integrating Technological, Market and Organizational Change, 6th edn. Wiley, Hoboken (2018)

Towards a Healthcare 4.0 Vocabulary:
A Patent-Based Approach

Annamaria Demarinis Loiotile[1,2(✉)] ⓘ, Francesco De Nicolò[1,2] ⓘ, Adriana Agrimi[3],
Giuseppe Conti[4] ⓘ, Nicola Amoroso[5] ⓘ, and Roberto Bellotti[1] ⓘ

[1] Dipartimento Interateneo di Fisica, Università degli Studi di Bari Aldo Moro, 70126 Bari, Italy
annamaria.demarinis@uniba.it
[2] Dipartimento di Ingegneria Elettrica e dell'Informazione, Politecnico di Bari, 70125 Bari, Italy
[3] Direzione Ricerca, Terza Missione e Internazionalizzazione,
Università degli Studi di Bari Aldo Moro, 70125 Bari, Italy
[4] Netval – Network per la Valorizzazione della ricerca universitaria, Lecco, Italy
[5] Dipartimento di Farmacia-Scienze del Farmaco, Università degli Studi di Bari Aldo Moro,
70125 Bari, Italy

Abstract. In modern knowledge economies, intellectual property assets are both
engines of development and drivers of social transition. One of the most widely
used approaches to identifying promising emerging technologies, record the direction of technological development and R&D activities and forecast innovation
is patent analysis. The paper analyses the patent information available in the
Knowledge-Share (KS) platform, an Italian patents database, first by using Natural Language Processing and clustering techniques, and then, by means of a
complex network analysis on the most frequent words, focuses the analysis on
those clusters inherent to health in order to investigate the technologies used and
their applications to the Healthcare 4.0 domain. The paper aims to a first attempt
of formulating a vocabulary of Healthcare 4.0 based on the patents produced by
the most important Italian universities and research centers.

Keywords: Patent analysis · Healthcare 4.0 · Natural Language Processing ·
Clustering · Complex network

1 Introduction

Under the EU valorization policy, the use of knowledge and technology, the management of intellectual property, and the involvement of citizens, academia, and industry,
through different channels, are highly promoted [1, 2]. In modern knowledge economies,
intellectual property assets are both engines of development and drivers of social transition. Industries that make intensive use of intellectual property rights (IPRs), such as
patents, trademarks, industrial designs and copyrights, generate 45% of annual GDP
(€6.6 trillion) in the EU and account for 63 million jobs (29% of all jobs) [1].

With rapid changes in technology and industry value chains, it is vital for companies
to be able to identify promising emerging technologies that can better respond to rapid

external changes and be used to launch new businesses or improve current ones. One of the most widely used approaches to identifying promising emerging technologies is patent analysis [3]. Patent documents are considered as a valuable database for understanding technology trends and design innovation strategies. They contain information about almost all relevant technological fields and record the direction of technological development and R&D activities [4]. Therefore, analysis, visualization and interpretation of patents related data are very useful in technology innovation and forecast, identification of research gaps, analysis of commercial value of technologies and competition over the markets [5].

Due to the huge volume of data, in recent years computer-aided approaches have been used to increase the speed, accuracy and objectivity of patent analysis [4]. Patent analytics is a family of techniques and tools for analyzing the technological information presented within and attached to patents. Text-mining – whose main merits are comprehensiveness, standardization and general applicability – has become more widely utilized for patent analyses because can extract technological contents effectively.

Patent analysis usually involves three steps [6]. In the first stage, patent documents are retrieved from patent databases. Next, the patent documents are transformed into structured data using text mining techniques. In the third step, working on the structured data, big data learning approaches, such as classification, regression and clustering, etc., are used for various purposes, such as patent novelty detection and patent quality identification, trend analysis and technology forecasting, R&D planning management, etc. [7].

As part of the fourth industrial revolution, one of the most revolutionized sectors is health care. Healthcare 4.0 (HC4.0) refers to the shift from traditional hospital-centered care to more virtual and distributed care with a greater focus on prevention and early intervention. Healthcare 4.0 heavily leverages the latest technologies, such as artificial intelligence (AI), deep learning, data analytics, genomics, home health care, robotics, and 3D printing of tissues and implants. By 2030 some radical changes will occur in how health care is delivered through increased access to data, additive manufacturing, artificial intelligence, and wearable and implanted devices to monitor our health [8].

The paper identifies and uses a mix of patent analysis techniques on this particular field in which technology and knowledge transfer plays a key role.

The paper analyses the information available in the Knowledge-Share (KS) platform, an Italian patents database, first by using Natural Language Processing and clustering techniques, and then, by means of a complex network analysis on the most frequent words, focuses the analysis on those clusters inherent to health in order to investigate the technologies used and their applications to the Healthcare 4.0 domain. This work addresses the following research question:

– is it possible to outline a bottom-up HC4.0 vocabulary starting from Italian patents?

The paper is organized as follows: Sect. 2 illustrates the methods that are the KS platform, the NLP-clustering approach, the complex network construction and their respective results with a final attempt of a HC4.0 vocabulary; Sect. 3 provides some conclusions.

2 Methods and Experimental Results

2.1 The Knowledgeshare Database

As recommended in the Intellectual Property Action Plan of the European Commission, published in 2020, improving easier access to and sharing of intellectual assets is "key to increase the valorisation of research results and the market uptake of innovative solutions" [9].

In order to help innovators and researchers to make the most of their results and inventions thereby generating societal impact and overcome the difficulties that Italian universities face in promoting effectively their research results and their IP assets, some tools in Italy were born in last years such as the platform Knowledge-Share (KS), developed in a joint project involving Politecnico di Torino, the Italian Patent and Trademark Office at Ministry for Economic Development and Netval (the Italian Network for the Valorization of Public Research) [10].

It showcases R&D technologies developed by Italian Universities and Research Centres, in pursuit of collaboration and commercialisation opportunities. Its goal is to convey patents in a clear and simple way, by highlighting the advantages that the technologies can bring to the sector of interest and to bring together companies, investors and innovators in contact with this research.

The platform is able to obtain three important results: to generate a real social and economic impact at national level, in accordance with the objectives of the "Third Mission"; to provide a tangible support to (not only) Italian businesses to accelerate their innovation processes; to drive economic return to be re-invested in new technology transfer activities within the public research system.

KS, recognized as best practice example in the Valorization Policies of the European Commission, aims to facilitate the interaction between university technology transfer offices, academic researchers and industry partners, by providing a portal that enables users to easily access information related to patents and technologies that represent the excellence of the scientific knowhow in Italy [11].

The KS database is composed of 1694 patents, uploaded on the platform by 89 Research Institutions (Universities, Research Centers, Scientific Institute for Research, Hospitalization and Healthcare, etc.). The patents are organized in 10 technological areas:

- Aerospace and aviation,
- Agrifood,
- Architecture and design,
- Chemistry, Physics, New materials and Workflows,
- Energy and Renewables,
- Environment and Constructions,
- Health and Biomedical,
- Informatics, Electronics and Communication System,
- Manifacturing and Packaging,
- Transports

Fig. 1. Knowledgeshare home page

The Fig. 1 shows the platform homepage and the possibility to search patents starting from a keyword, the name of the Institution, the technological area.

For each patent some info are illustrated: the inventors, the patent status, the priority number, the priority date, the license, the commercial rights and so on. The content of each patent is contained in a "marketing sheet" that provides, in simple and informative language, the technical characteristics of the patent, its possible applications and advantages. The Fig. 2 illustrates an example of marketing sheet.

Fig. 2. Example of marketing sheet in Knowledgeshare

2.2 Outlining Healthcare 4.0 Patents

We firstly investigated if the classification offered by KS was the only possible one, proposing an AI-based approach to improve classification and, thus, innovation research by potential stakeholders searching excellent technologies.

Each patent was transformed by means of NLP (Natural Language Process) techniques, then an unsupervised k-means cluster analysis was performed and, using a Silhouette score, the quality of clusters created using clustering algorithm was evaluated [12]. This analysis outlined the presence of 8 clusters (see Fig. 3) different from the original categories defined within the KS database (there are actually 9 clusters generated but one consists of no sense words). These different categories were examined and validated by the KS technology experts. In particular, it was possible to assign to each cluster a specific domain:

- Technologies 4.0 (mechanics and robotics)
- Material science
- Cancer treatment
- Optics - Image processing
- Sensor technology - ICT
- New molecules - new compounds - pharmacology
- Energy/green Technologies
- Biomedical.

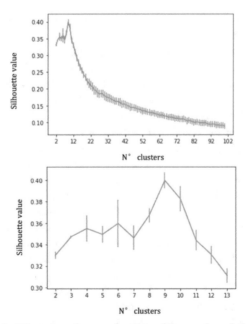

Fig. 3. The variation in Silhouette value as a function of the number of clusters (the second one shows a magnification)

It is interesting to underline that patents related to the health sector account for a large portion (about 30% of the total) and are enclosed in 4 out of 8 clusters. Accordingly, four clusters were further investigated because of their proximity to healthcare 4.0 applications: Technologies 4.0, Cancer treatment, New molecules - new compounds and Biomedical. These clusters were then considered for subsequent complex network analyses.

For each cluster, we examined the occurrence frequency of each word to narrow the research field to the most representative words: in particular, the top 5% of words was selected within each cluster and, finally, an overall amount of 106 words was determined. At this point, to highlight the relationships among these words we built a weighted complex network model whose nodes were the selected words, links were drawn between a pair of different words if they co-occurred in at least one patent and the number of co-occurrences represented the weight of an existing link.

This model was adopted to explore if particular words, and therefore concepts, could be related in significant patterns. The underlying assumption is that these patterns provide a first step towards the definition of a healthcare 4.0 ontology.

2.3 Centrality Analysis

The first research question addressed in this paper deals with the opportunity to detect which are the basic assets defining healthcare 4.0 in an unsupervised way. There is not a univocal defined ontology for healthcare 4.0, yet. Therefore, we investigate here the adoption of network metrics to address this issue.

In a complex network, by definition, the importance of node can be evaluated by means of centrality metrics, here three different options were considered: degree (d), eigenvector centrality (e) and betweenness (b). The degree measures the connections of a node, eigenvector centrality weighs this information according to the global degree distribution and betweenness evaluates the number of paths passing through each node. The reason for such choice is that, in general, three different perspectives can be used to measure centrality according to the local, global or dynamical flavour to be emphasized [13]. Moreover, as these metrics do not take into account the links' weights, we investigated the nodal strength (s). To this aim, we examined these four metrics and compared the ranking they returned in terms of Spearman correlation, see Fig. 4.

Our findings show that degree, betweenness and eigenvector centrality never show statistically significant differences, on the contrary the ranking provided by strength is significantly different independently from the number of words considered. Besides, we evaluated to which extent this result was affected by the number of considered words. We observed that, despite the varying number of words, no changes in correlation could be observed. This finding suggests that the information carried by network weights is far from being trivial and should not be neglected.

In the following Table 1 the list of the first ten words according to the importance evaluated by degree and strength is reported.

While degree accounts for the number of patents using a specific word, nodal strength accounts for the number of times that word has been used, therefore normalized strength can be easily interpreted as the percentage of words' co-occurrences. Top ten words account for 30% of existing connections in both cases. The comparison between the two

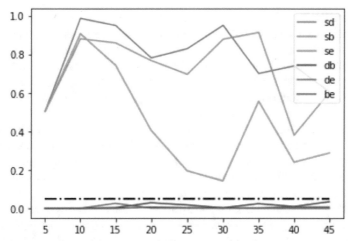

Fig. 4. Spearman correlation p-values of degree(d), betweenness (b), eigenvector centrality (e) and strength(s) varying with the number of considered words. Statistical significance 0.5 is represented with a dash-dot line.

Table 1. The top ten words by Strength and Degree

Ranking	Degree	Strength
1	Control	Gen
2	Patient	Treat
3	Method	Cellul
4	Tissue	Effect
5	System	Pharmac
6	Gen	Articul
7	Articul	System
8	Treat	Tumor
9	Disease	Effectiv
10	Effectiv	Patient

lists shows a good agreement, 60% of words are common, the others can be related in an obvious way, e.g., tumor and disease. In the following, a possible interpretation of these lists is provided.

2.4 Healthcare 4.0 Vocabulary

We attempt to give an interpretation to these ten most frequent words in order to make them part of an HC4.0 vocabulary.

Certainly, the goal of HC4.0 is to provide patients with better, value-added and cost-**effective** healthcare services [14] and improve the **effectiveness** and efficiency of the healthcare sector by trying to connect **patients**, physicians, hospitals, personal medical devices, **pharmaceutical** and medical supplier's product and service providers in order to create a smart healthcare network along the entire healthcare value chain. There is a great **control** and monitoring (and prediction) of **patient** status thanks to the enabling technologies.

As in Industry 4.0, IoT, RFID, wearable devices, robotics, and blockchain technologies in cyber-physical **system**s create a mechanism for data collection, monitoring, analysis, intervention, and feedback [15]. Linking these technologies to personalized medicine can help implement **genetic**s-based approaches to diagnosis and **treatment** and improve the effectiveness of **patient** care.

In healthcare, machine learning and artificial intelligence analyze huge amounts of information (big data) to provide accurate medical diagnosis before treatment is too late. Artificial intelligence technologies are increasingly being used to fight **disease** and save more lives by trying to diagnose **disease**s such as **tumor** early [16]. Big Data help determine the effectiveness of medical **treatments** as well as identify effective, standardized therapies for specific disorders. Big data helps improve outcomes and reduce costs through improved **disease** management strategies and the development of better diagnosis and **treatment** processes [16]. Digital health tools promise to improve individual health care delivery and increase the ability to effectively detect and **treat disease**. In addition, technologies such as cell phones, Internet applications, social networks offer new **method**s for patients to monitor their health and access information.

Through the application of communication tools to patients and medical teams, the transfer of **treatment** from the hospital to the home is intensified, without disruption in outpatient services. Through the use of technology, classical **pharma**cological therapies are replaced with the support of digital applications provided jointly to the patient and the physician, in a perspective of better and more integrated adherence to care. Nowadays, the techno-scientific evolution of medicine is essentially taking place along at least three main axes: restorative/integrative medicine, regenerative medicine and precision medicine. The three perspectives of development have several points in common, but above all they are closely linked to the possibilities offered by the information technology revolution, linked to artificial intelligence [17].

Integrative Medicine is originally "restorative" medicine, traditionally prosthetic (artificial devices designed to replace a missing body part) and more recently bionic, a branch of biomedical engineering that applies cybernetics to the reproduction of functions of living organisms.

Tissue engineering is an interdisciplinary area concerned with the development of functional 3D tissues by mixing scaffolds, **cells**, and bioactive chemicals. It is a subset of biomedical engineering in which cell biology, materials science, chemistry, molecular biology, and medicine converge. It applies to the repair, restoration, and preservation of damaged **tissue**, **articulation** or a whole organ [16].

Regenerative Medicine is based on the principles of **genetic** engineering-isolating a **gene** from the organism that possesses it and inserting it into a host even of a different species-and is developed as **gene** therapy (reconditioning of cells in vivo), use of stem

cells (ex vivo and in vivo cell regeneration), and **tissue** engineering (combination of artificial cells and materials) for anatomo-functional restoration of **tissue**s and organs [18]. The most promising regenerative medicine technique is that based on stem **cells** (SC) - embryonic, adult, and "induced" - with the ability to reproduce **tissue**s and organisms.

Precision Medicine tends to the **treatment** and prevention of **disease** based on individual variability of **gene**s, environment and lifestyle (personalization) and is based on deterministic understanding of **disease**, diagnosis of causal factors, ability to treat root causes of **disease**, using tools such as **geno**mic and post-**geno**mic biological databases, characterization methods such as "omics," cellular analysis and "mobile" technology, and bioinformatics [17].

In HC4.0, the entire healthcare **system** and its management take on an even more key role; all stakeholders in the healthcare ecosystem are actively involved in a supply chain logic. In healthcare 4.0, as automation and the use of technology increase, the participation and importance of people actually become more critical. Not only patients, physicians, and relevant and support staff are included in the system, but also nurses, physician assistants, **pharma**cists, lab technicians etc. [15]. It is important to focus on **system** and process: team, team of teams, network of care, process spread across organizational boundaries, community involvement.

3 Conclusions

Starting from the patents of the main Italian universities and research Centers, contained in the platform Knowledge Share and applying NLP, clustering techniques, we propose a classification of patents into 8 clusters, 4 of which (Technologies 4.0, Cancer treatment, New molecules - new compounds and Biomedical) were investigated for their proximity to healthcare 4.0 applications. For each cluster, we examined the occurrence frequency of each word to narrow the research field to the most representative words; the top 5% of words was selected within each cluster and, finally, an amount of 106 words was determined. At this point, a weighted complex network model was constructed to highlight the relationships among these words: the nodes were the selected words, the links were drawn between a pair of different words if they co-occurred in at least one patent and the number of co-occurrences represented the weight of an existing link. After this process, we analyzed the list of the first ten words trying to explain these words in terms of HC4.0 in order to answer to our research question namely the possibility of formulating a vocabulary of HC4.0 based on Italian patents.

The next steps of the research foresee the interpretation of a greater number of words to increase the "vocabulary" and try to understand what the emerging technologies and their applications by the main Italian research centers are.

References

1. EU Valorisation Policy. https://research-and-innovation.ec.europa.eu/system/files/2020-03/ec_rtd_valorisation_factsheet.pdf. Accessed 25 Nov 2022
2. Demarinis Loiotile, A., et al.: Best practices in knowledge transfer: insights from top universities. Sustainability **14**, 15427 (2022). https://doi.org/10.3390/SU142215427
3. Choi, Y., Park, S., Lee, S.: Identifying emerging technologies to envision a future innovation ecosystem: a machine learning approach to patent data. Scientometrics **126**(7), 5431–5476 (2021)
4. Wang, Y.H., Lin, G.Y.: Exploring AI-healthcare innovation: natural language processing-based patents analysis for technology-driven roadmapping. Kybernetes (2022)
5. Choi, J., Hwang, Y.S.: Patent keyword network analysis for improving technology development efficiency. Technol. Forecast. Soc. Chang. **83**, 170–182 (2014)
6. Abbas, A., Zhang, L., Khan, S.U.: A literature review on the state-of-the-art in patent analysis. World Patent Inf. **37**, 3–13 (2014)
7. Lei, L., Qi, J., Zheng, K.: Patent analytics based on feature vector space model: a case of IoT. IEEE Access **7**, 45705–45715 (2019)
8. Wehde, M.: Healthcare 4.0. IEEE Eng. Manag. Rev. **47**(3), 24–28 (2019)
9. Intellectual Property Action Plan. https://single-market-economy.ec.europa.eu/industry/strategy/intellectual-property/intellectual-property-action-plan-implementation_en. Accessed 25 Nov 2022
10. Knowledge-Share. https://www.knowledge-share.eu/. Accessed 25 Nov 2022
11. EU Valorisation Policy. https://op.europa.eu/en/publication-detail/-/publication/4944d5a4-c50c-11ec-b6f4-01aa75ed71a1/language-en. Accessed 25 Nov 2022
12. Jun, S., Park, S.S., Jang, D.S.: Document clustering method using dimension reduction and support vector clustering to overcome sparseness. Exp. Syst. Appl. **41**(7), 3204–3212 (2014)
13. Amoroso, N., Bellantuono, L., Monaco, A., De Nicolò, F., Somma, E., Bellotti, R.: Economic interplay forecasting business success. Complexity (2021)
14. Al-Jaroodi, J., Mohamed, N., Abukhousa, E.: Health 4.0: on the way to realizing the healthcare of the future. IEEE Access **8**, 211189–211210 (2020)
15. Li, J., Carayon, P.: Health Care 4.0: A vision for smart and connected health care. IISE Trans. Healthc. Syst. Eng. **11**(3), 171–180 (2021)
16. Haleem, A., Javaid, M., Singh, R.P., Suman, R.: Medical 4.0 technologies for healthcare: features, capabilities, and applications. Internet Things Cyber-Phys. Syst. (2022)
17. Cappelletti, P.: Medicina 4.0. Un'introduzione. La Riv. Italiana della Med. Laborat. Italian J. Lab. Med. **14**(3), 131–135 (2018)
18. Jessop, Z.M., Al-Sabah, A., Francis, W.R., Whitaker, I.S.: Transforming healthcare through regenerative medicine. BMC Med. **14**(1), 1–6 (2016)

Discrete Event Simulation for the Analysis and Re-Engineering of Production and Logistics Cycles: The Case of Master Italy Company

Francesca Intini[1], Pasquale Merla[2], Rocco Pagliara[3], Marco Partipilo[2], Giambattista Stigliano[2], and Davide Veneto[4]([✉]) [iD]

[1] CMG srl, Viale Aldo Moro, 67/L-M, 70043 Monopoli, BA, Italy
[2] Kad3 srl, Via Baione, snc, 70043 Monopoli, BA, Italy
stigliano@kad3.com
[3] Master Italy s.r.l., SP37, Km 0,5, 70014 Conversano, BA, Italy
[4] Digital Innovation s.r.l., Via E. Orabona, 4 (c/o Dipartimento di Informatica), 70125 Bari, BA, Italy
davide.veneto@dinnovation.it

Abstract. Today, demand is shifting to more customized products, which will lead to small batches, short lead times, and unique specifications. To remain competitive in this changing environment, manufacturers must improve their layout, productivity, responsiveness, flexibility, and quality. Each layout problem has its own unique characteristics. Among the tools currently available to researchers and companies to optimize production cycles, it is worth highlighting software for the processes' visualization and modeling. Simulation tools are used to connect all processes of the flexible manufacturing system into one intelligent system and analyze the production. Layouts are analyzed through computer simulations to verify the true potential of the facility and to find the most efficient long-term operating plan and the optimal structure to meet the increasing demands for customization, quick response, and overall quality, while reducing lead times. The production systems are complex systems that consists of many parts. Modeling a virtual copy of a physical system is quite a complex task and requires the availability of a large amount of information and a set of accurate models that adequately reflect the modeled reality. After a brief presentation of discrete event simulation, a methodology for analyzing and optimizing the production process and its application case is presented in this work within the Master Italy company process.

Keywords: Discrete Event Simulation · Process Optimization · Lean Production

1 Introduction

Operational problems for companies have increased due to increased global competition, scarcity of resources, higher customer expectations (in terms of higher quality, lower cost, shorter delivery times), and pressure from the government or other regulatory bodies to reduce carbon emissions and to be more energy efficient. This has led researchers

A. Rocha et al. (Eds.): WorldCIST 2023, LNNS 802, pp. 361–370, 2024.
https://doi.org/10.1007/978-3-031-45651-0_36

to identify continuous process improvement solutions to reduce waste by optimizing processes at different levels [1]. This need has become even more important in today's high variety/low volume production landscape where customer demands are extremely volatile both in terms of quantity and variety. There are numerous examples of process improvement approaches that have been applied to various manufacturing/service processes and product types, ranging from small parts/components (engines, tires, manufactured components, etc.) to whole products (aircraft, coach/bus, automotive, service processes – hospitals, banks, education, etc.) [2]. Production units are very complex systems. Nowadays, the demand is shifting towards more customized products which would result in small batch sizes, short lead times, and unique specifications. To compete in this changing environment, the manufacturer must improve its layout, productivity, responsiveness, flexibility, and quality. Each layout problem has its own unique characteristics. Flexibility is also a prerequisite for responding to shorter product life cycles, low to mid-range production volumes, evolving demand patterns, and a greater variety of product models and options. Currently, among the tools for optimizing the production cycle, available to researchers and companies, it is worth highlighting the process virtualization and modeling software [3]. The application of simulation in manufacturing is vast; it can be used to address the challenges inherent in manufacturing a wider range of products while retaining the economy-of-scale benefits of mass production. Simulation tools are used to connect all processes of a flexible manufacturing system to have an intelligent system and to analyze production [4]. The layouts are analyzed by computer simulation in order to verify the actual potential of the plants, to find the most effective plans for long-term operations (adaptively exploiting the flexibility at different levels with effective timing), and to find the best structure to satisfy the growing demands for customization, rapid response, and total quality with reduced lead times.

The present paper, in the first part, presents a brief analysis of the state of the art related to discrete event simulation (DES) and its possible applications; in the second part, it presents a methodology for the analysis and optimization of the production process and its application case within the business process of Master Italy company. Finally, a section is reported which summarizes the results obtained relating to the application of the methodology. The results show how it is possible, through DES, to thoroughly analyze and optimize a company's production process without the need to implement the various options but using a digital twin of the process. DES is a powerful tool for optimizing processes and lean production systems because it allows you to model and analyze complex systems in a controlled environment. This can help you identify bottlenecks, inefficiencies, and other factors that may be contributing to waste or reduced productivity. For the DES application, Plant Simulation was used in the paper due to its user-friendly character. Obviously, the study is also valid using other commercial and non-commercial software that apply the proposed methodology.

2   Discrete Event Simulation: State of the Art

In a period characterized by growing pressures on production, in terms of costs, times, and globalization, logistics has become a fundamental element for company success. Inefficient scheduling, local rather than global optimization, inadequate resource allocation and low productivity can lead to significant losses for companies every day. The need to perform "just-in-time" (JIT) and "just-in-sequence" (JIS) deliveries, the adoption of the Kanban methodology [5], the planning and construction of new production lines and the management of networks global production processes require the adoption of objective decision-making criteria, which help management evaluate and compare the various alternative approaches. Among the most promising tools, present in the state of the art, there is the simulation of the production process through the creation of a digital twin that describes its functioning in terms of times and interactions [6]. Discrete Event Simulation models the functioning of a system as a (discrete) sequence of events over time. Each event occurs at a particular instant in time and marks a change of state in the system. Today mainly used methodology for DES application considers that no system change occurs between consecutive events: simulation time can directly jump to the occurrence time of the next event, which is called next event time progression. In addition to next event time progression, there is also an alternative approach, the fixed increment time progression, in which time is divided into small intervals and system status is updated based on the set of events/activities that occur over time. Since it is not necessary to simulate each time interval, a next event simulation can usually run much faster than the corresponding fixed time increment simulation. Both forms of DES contrast with continuous simulation where system state continuously changes over time based on a set of differential equations defining the rates of change of state variables. The advantages of the DES are:

- It can detect and show problems that might otherwise cause costs and time-consuming corrective measures during the start-up phase.
- It offers mathematically calculated key performance indicators (KPIs) rather than the 'gut feelings' of experts.
- It allows to reduce investment costs for production lines without compromising the required output quantities.
- It allows to optimize the performance of existing production lines.
- It can incorporate machine failures and availability (MTTR- Mean Time to Repair, MTBF- Mean Time Between Failures) when calculating throughput numbers and utilization.
- It allows to verify and validate the design hypotheses (both entire production plants and individual portions) before actual implementation, minimizing the risk of making incorrect strategic decisions.

There is a variety of applications where it is useful to simulate a system that evolves over time, such as a weather system [8]. A key point, however, is that the simulated events would be continuous, which means that, for example, in a temperature against time graph, the curve would be continuous, without breaks. Suppose instead of simulating the functioning of a warehouse: purchase orders come and go, reducing inventory which is replenished from time to time [9]. A typical variable here would be the inventory itself, i.e., the number of items currently in stock for a given product. In a number against time graph, the curve would be a step function, a set of flat line segments with breaks between them. The decreases and increases in inventory events are discrete variables, not continuous. DES involves simulating such systems.

Simulation programming can often be difficult to code and debug because it is a form of parallel programming. For this reason, many researchers have tried to develop separate simulation languages, or simulation paradigms (i.e., programming styles) that allow programmers to achieve clarity in simulation code. Special simulation languages have been invented in the past, most notably SIMULA [11]. However, today the trend is to simply develop simulation libraries focusing on programming paradigms and not languages. Nowadays there is a large availability of DES software that can be easily used in different fields: from industry to supply chain, from healthcare to business management, from training to design of complex systems. Commercial DES software uses specific rules and logics for managing simulation time and events. They also provide the user with a set of libraries based on the traditional object-oriented modeling approach. Among the most promising software it must be included Plant Simulation by Siemens. It is a DES tool that allows to create, in a user-friendly way, digital models of systems, for exploring system characteristics and optimizing their performance. It is oriented towards the modeling of production plants and logistics centers but makes it possible to simulate any type of process (verification of an evacuation plan in a hospital, simulation of a public office where a counter service is provided, etc.) considering the possibility of implementing new objects and new libraries and the possibility of programming any behavior using SimTalk [12], a proprietary programming language. These digital models enable what-if scenarios to be tested without disrupting existing manufacturing systems or, when used in planning processes, long before actual manufacturing systems are installed. Comprehensive analysis tools, such as bottleneck analysis, statistics and graphs allow you to evaluate various production scenarios. The results provide the information needed to make fast and reliable decisions in the early stages of planning. Thanks to Plant Simulation [13], it is possible to model and simulate production systems and related processes. Users can also optimize material flow, resource utilization and logistics at all levels of plant planning. Using simulation, complex and dynamic business workflows are evaluated to arrive at mathematically safeguarded business decisions. The computer model allows the user to run experiments and "what-if" scenarios without having to experience the real production environment or, if applied within the planning phase, well before the existence of the real system. In general, material flow analysis is used when discrete manufacturing processes are running. These processes are characterized by non-constant material flows, which means that the part is either there or not, movement does or does not happen, the machine runs without errors or reports a failure. These processes have simple mathematical descriptions and derivations due to numerous dependencies. Before powerful computers were available, most material flow simulation problems were solved using queuing theory and operations research methods. In most cases the solutions resulting from these calculations were difficult to understand and marked by a large number of boundary conditions and restrictions which were difficult to respect in reality [13].

3 Application Case of Master Italy Company

The process of making a digital copy of the production processes and evaluating the impact of some choices to improve existing processes characterized 1) by the adoption of automatic technologies for materials storage and for the enslavement of manual assembly stations and 2) from the adoption of innovative algorithms for the planning of such enslavements, has also been performed within Master Italy, an Italian company active in the production of door and window handles. With the final aim of carrying out a discrete event analysis and simulation for the definition of the parametric process model, able of dealing with multivariability and multi- objective problems, different process analysis were initially addressed. In particular, in order to analyze the process and the information exchanged in it, the problem of knowledge formalization and representation was addressed. Mapping and reusing knowledge from the customer domain to the product domain is a major theme of product line design. Mapping and reuse of project knowledge models are the main concerns to automate and/or speed up the design phases. Both reuse and mapping involve effective knowledge communication to potential users represented by a model. In particular, the activities were aimed at the analysis and formalization of the operating processes that characterize the industrial reality of Master Italy company.

The formalization of the Master Italy processes led to the creation of a database (DB) for the assembly department made up of the following three functional blocks:

- **Resources for production**: manual stations where assembly takes place and operators who carry it out.
- **Handling system**: AMRs (Autonomous Mobile Robots) who take care of replenishing the resources for production and bringing the material back to the starting location (automatic warehouse).
- **Resources for storage**: in the specific case is an automatic warehouse (called "Silo2") in which all the products that will circulate within the portion of the plant being analyzed are stored.

The database represents a "Level 1" model, a data model that incorporates tables, data types, keys, constraints, permissions, indexes, views, and details on allocation parameters, production mixes, process variations.

Figure 1 shows the graphical schematization of the areas of interest provided by Plant Simulation software.

An analysis of the impact of the adoption of automatic technologies for the storage and transfer of material within the manual assembly department of door and window handles in Master Italy company was carried out. The department features 6 manual assembly stations with 1 operator working on each for 2 shifts, 5 days a week. At the time of the study, the storage of materials and the enslavement of the stations took place using static shelves, operators, and trolleys, according to the principles of Lean Manufacturing. The design hypothesis was to introduce automatic systems for the storage and enslavement of the line, without modifying the logic according to which these enslavements take place. The objectives of the study were 1) to verify that the automatic warehouse could at least support the requests for material necessary for the typical production of the plant and to evaluate its ability to cope with any increases in the production potential of the assembly department; 2) to provide reliable information for the sizing of the automatic material transport system. More specifically, the request was to determine the number of AMRs necessary to satisfy the requests for movements to and from the automatic warehouse.

One of the challenges to face when implementing a simulation model for scenario analysis is to be able to find the right compromise between the need to have a model able of faithfully reproducing reality and the effort necessary for the implementation.

Fig. 1. Schematization of areas of interest

It is therefore necessary to make some assumptions, hypotheses, and simplifications of reality, without however affecting the accuracy of the reproduction of the behavior of the real system and therefore the accuracy of the results obtained. In accordance with this principle, from the evaluations carried out within the production process of Master Italy company, the need to analyze the following scenarios has been highlighted:

1. **Hypothesis Rev.1**: the scenario is characterized by no SETUP time and no material returns from the manual assembly stations to the automatic warehouse. In this way it was possible to validate the model and make preliminary considerations with respect to the use of the automatic warehouse.
2. **Hypothesis Rev.2**: Following the assessments and results obtained with respect to the use of the automatic warehouse (scenario Rev.1), it was verified that the automatic warehouse was absolutely able of carrying out all the required movements. Thanks to the Plant Simulation software extreme flexibility, it was possible, with minimal implementation effort, to verify alternative scenarios in which, for example, even components necessary for the bagging phase were stored and handled within the automatic warehouse, which until then was performed in an area outside the analyzed portion of the plant. In this way it was possible to evaluate the possibility of modifying the production process, also assigning the assembly phase to the manual assembly stations. The second scenario, compared to the previous one, presents the following changes:
   a. Assembly times extended by 10 s for all stations following the introduction of the time necessary for the bagging operation.
   b. New components have been added to the BOM related to the bagging operation.
   c. Setup time was considered and new logics of advance request for material by the manual assembly stations were implemented and tested.
   d. AMRs needed to move the material was made parametric and an automatic picking system was introduced which made it possible to pick up a single tray in 20 s and the entire tray in 30 s.
   e. For the warehouse, the number of picks for single missions and the times for single and multiple missions have been changed, increasing the margin of safety against which to evaluate the performance of the warehouse itself (Fig. 2).

   For the scenario and the assumptions just set out, the following simulations were analyzed:

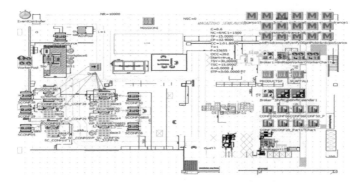

Fig. 2. Model made in Plant Simulation for "Hypothesis Rev.1" test case

- "**REV2_1AMR**" simulation: characterized by the presence of 1 AMR.
- "**REV2_2AMR**" simulation: characterized by the presence of 2 AMRs.
- "**NO PACKAGING**" simulation: characterized by the presence of 2 AMRs and the absence of the bagging operation.
- "**SATURDAY SHIFT**" simulation: characterized by the presence of 2 AMRs and the introduction of 1 shift on Saturday.The following figure shows the model, created in Plant Simulation, related to test case "**Hypothesis Rev.1**".

4   Results

As previously described, the analysis made it possible to evaluate various parameters including, for example, the commitment of the individual stations and operators and the use of automatic warehouses. The results of the simulations for each identified scenario are shown below. In particular, the following figures show the commitment of the stations and operators, while Table 1 shows the comparison of the various case studies according to the percentage of use of the warehouses.

Figure 3 shows the engagement of the stations and the commitment of their operators in differed scenarios. Green bars represent the working time, brown bars represent set-up time, blue bars represent the pauses of the operator (and consequently the stations), the grey bars represent the waiting time. Depending on the planned production, different stations and their operators work for different portions of time.

Table 1 shows how automatic warehouse works in different scenarios. In this case, the working time represents the time when the warehouse unloads materials, waiting time represents the portion of time when the warehouse is on waiting for a new unloads, and the blocked time represent the portion of time when the warehouse is "blocked" because the material is in the bay waiting to be picked up by the operator.

The following table summarizes the results of the analysis carried out in the 5 case studies identified (Table 2).

The impossibility of completing production in the period analyzed, in the case of introduction of the packaging operation and the consequent need to introduce a new shift on Saturday, is not connected to the performance of the automatic warehouse nor the transport system. This need can be explained by considering that inputs have been

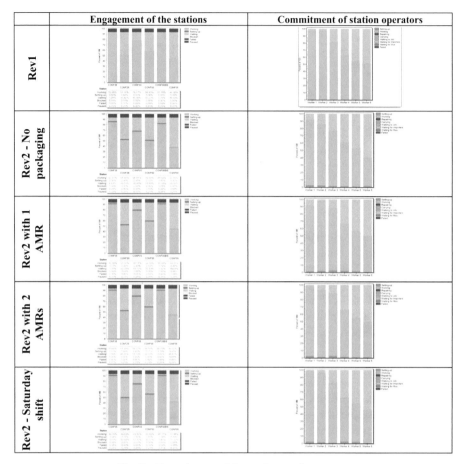

Fig. 3. Commitment of the stations and operators

Table 1. Percentage of warehouse commitment

	Rev1	Rev2 with 1 AMR	Rev2 with 2 AMRs	Rev2 - No PACKAGING	Rev2 - Saturday shift
Working	8.0%	29.5%	30.1%	31.2%	28.0%
Waiting	77.1%	50.3%	53.5%	51.9%	57.0%
Blocked	14.9%	20.1%	16.3%	16.9%	15.0%

supplied with the production results (in terms of production order and pieces produced) in a given period of time, with the stations engaged for 100% of the working time. If we consider T the time interval in which the N (about 34,000) parts produced were completed and × the time required to complete each of them, it is reasonable to hypothesize how to

Table 2. Comparison of main parameters in the different configurations studied

	Rev1	Rev2 with 1 AMR	Rev2 with 2 AMRs	Rev2 - No PACKAGING	Rev2 - Saturday shift
Parts manufactured	33655	33388	33322	34054	34054
Orders processed	345/355 (97.18%)	341/355 (96.06%)	340/355 (95.77%)	355/355 (100%)	355/355 (100%)
Warehouse unloading	1103	1098	1094	1122	1120
Total missions	1103	3164	3155	3276	3274
Transportation commitment	35.38 h	78.83 h	78.71 h	82.28 h	81.47 h

produce the same N parts, each having a new cycle time equal to $x + 10s$ (time required for packaging) a time interval $T2 > T$ is required.

5 Conclusion

Through the simulation study and with the established hypotheses, it was possible to conclude that:

- The warehouse is able to satisfy handling requests for all analyzed scenarios.
- The transport system made up of 1 AMR with a speed of 0.5 m/s is sufficient.
- With the packaging phase (both from the point of view of times and of new items to be stored in the warehouse), to complete the planned production it is necessary to consider a shift on Saturday in addition to the two shifts already foreseen from Monday to Friday.
- If packaging phase is excluded, in all analyzed scenarios, it is possible to complete the planned production in the foreseen time.

One of the key advantages of DES is that it allows for the analysis of systems and processes under a wide range of scenarios and conditions, including different input levels, resource availability, and other variables that may affect the performance of the process. This enables organizations to identify the root causes of problems and to develop targeted solutions that address these issues in a systematic and effective manner. Additionally, DES can be used to optimize the flow of materials and information within a production process, by identifying bottlenecks and other points of inefficiency and developing strategies to improve the efficiency and effectiveness of these processes. This can help organizations to reduce waste, improve efficiency, and increase the overall performance of their production processes. Some specific advantages of DES for process optimization and lean production include:

- **Better decision making**: DES allows you to test different scenarios and evaluate the potential outcomes, which can help you make more informed decisions about process improvements and changes.

- **Increased efficiency**: By identifying and addressing bottlenecks and other inefficiencies in your processes, you can improve the overall efficiency of your system.
- **Reduced risk**: Simulating process changes before implementing them in the real world can help you identify potential issues and make necessary adjustments, reducing the risk of costly mistakes or disruptions.
- **Improved productivity**: By optimizing your processes, you can increase the overall productivity of your system, potentially leading to cost savings and improved profitability.
- **Enhanced customer satisfaction**: By improving the efficiency and reliability of your processes, you can deliver a higher level of service to your customers, leading to increased satisfaction and loyalty.

Acknowledgment. The article was produced as part of the research project "Smart Processes for Mass Customization Manufacturing (SP4MCM)", Ministerial Decree 5 March 2018 - Prog n. F/190192/02/X44, CUP B48I20000040005, ASSE I Priorità di investimento 1b Azione 1.1.3, co-financed by the Ministero per lo Sviluppo Economico, and by Europe.

References

1. Oliveira, J., Sá, J.C., Fernandes, A.: Continuous improvement through "Lean Tools": an application in a mechanical company. Procedia Manuf. **13**, 1082–1089 (2017). https://doi.org/10.1016/J.PROMFG.2017.09.139
2. Dobrotă, D., Dobrotă, G., Dobrescu, T.: Improvement of waste tyre recycling technology based on a new tyre markings. J. Clean. Prod. **260** (2020). https://doi.org/10.1016/J.JCLEPRO.2020.121141
3. Javadpour, A.: Improving resources management in network virtualization by utilizing a software-based network. Wirel. Pers. Commun. **106**, 505–519 (2019). https://doi.org/10.1007/S11277-019-06176-6/FIGURES/9
4. Florescu, A., Barabas, S.A.: Modeling and simulation of a flexible manufacturing system—a basic component of Industry 4.0. Appl. Sci. **10**, 8300 (2020). https://doi.org/10.3390/APP10228300
5. Wakode, R.B., Raut, L.P., Talmale, P., Steels Pvt Ltd, B.: Overview on Kanban methodology and its implementation. IJSRD-Int. J. Sci. Res. Develop. **3**, 2321–0613 (2015)
6. Tao, F., Xiao, B., Qi, Q., Cheng, J., Ji, P.: Digital twin modeling. J. Manuf. Syst. **64**, 372–389 (2022). https://doi.org/10.1016/J.JMSY.2022.06.015
7. Sandvik, E., Gutsch, M., Asbjørnslett, B.E.: A simulation-based ship design methodology for evaluating susceptibility to weather-induced delays during marine operations. **65**, 137–152 (2018). https://doi.org/10.1080/09377255.2018.1473236
8. Abideen, A., Mohamad, F.B.: Improving the performance of a Malaysian pharmaceutical warehouse supply chain by integrating value stream mapping and discrete event simulation. J. Model. Manag. **16**, 70–102 (2021). https://doi.org/10.1108/JM2-07-2019-0159/FULL/PDF
9. Dahl, O.-J.: The roots of object orientation: the simula language. Softw. Pioneers. 78–90 (2002). https://doi.org/10.1007/978-3-642-59412-0_6
10. Bangsow, S.: Programming with SimTalk. Manufacturing Simulation with Plant Simulation and SimTalk. 85–116 (2010). https://doi.org/10.1007/978-3-642-05074-9_5
11. Bangsow, S.: Tecnomatix plant simulation. Tecnomatix Plant Simulat. (2016). https://doi.org/10.1007/978-3-319-19503-2/COVER

BOOGIE: A New Blockchain Application for Health Certificate Security

Vito Nicola Convertini$^{(\boxtimes)}$ (ID), Vincenzo Dentamaro (ID), Donato Impedovo (ID),
Ugo Lopez (ID), Michele Scalera (ID), and Andrea Viccari (ID)

Department of Informatics, University of Bari, Bari, Italy
{vitonicola.convertini,vincenzo.dentamaro,donato.impedovo,
ugo.lopez,michele.scalera,andrea.viccari}@uniba.it

Abstract. The idea of "COVID-19 passports" brings challenges in the scientific, legal, and ethical fields. Issues include counterfeiting and security of personal data, as well as durability related to the protection conferred by vaccines. In an ideal scenario, it would be highly desirable to establish international standardization for these passports, ensuring that they incorporate verifiable credentials supported by interoperable technologies. Concerns of an ethical and legal nature arise in relation to the provision of passports to those for whom immunization is either objectionable, unverified, unattainable, or unfeasible, arguing that such individuals are precluded from obtaining any vital commodities and/or services. The paper discusses the maneuvers and decrees implemented by the Italian government regarding the use of vaccination certificates and examples of fraud related to the creation of fake green passes. It also presents BOOGIE, a solution based on a research blockchain technology that ensures trust and eliminates unsuitable features for the circulation of certificates of immunity from COVID-19 and medical passports usable in the Italian context.

Keywords: Smart contract · Blockchain · Green Pass · Health certificate

1 Introduction

The dissemination of "COVID-19 passports" brings challenges in the scientific, legal, and ethical fields. Open issues are related to counterfeiting and security of personal data [1], durability related to the protection conferred by vaccines with possible option to revoke them. Also, ideally, it will be standardized internationally with verifiable credentials based on interoperable technologies; c this could occur there where vaccine hesitancy is evident (NO VAX), or to refusal among certain ethnicities or minorities, to undocumented migrants, etc. These examples signal the need for alternatives and exemptions.

This paper is divided into two parts: in the first, the maneuvers and decrees implemented by the Italian government regarding the fruition and exhibition of the vaccination certificate are presented, and some examples of fraud related to the creation of fake green passes are presented. In the second, BOOGIE (A new Blockchain applicatiOn fOr

A. Rocha et al. (Eds.): WorldCIST 2023, LNNS 802, pp. 371–380, 2024.
https://doi.org/10.1007/978-3-031-45651-0_37

health certificate security) is presented, a solution based on "blockchain" technology that ensures trust and eliminates unsuitable features for the circulation of certificates of immunity from COVID-19 and medical passports.

2 The Italian Regulatory Framework and Fraud

2.1 The Italian Regulatory

The Privacy Guarantor gave his approval to the new modalities for the revocation and use of the Green Pass in Order No. 430 dated Dec. 13, 2021. The Guarantor's indications were incorporated by the government into the final text of the Decree dated Dec. 17, 2021, which regulates several technical aspects of vaccine passport verification, including revocation and the maintenance of revocation lists [2]. The decree stipulates that the Ministry will make available, upon request, to employers, facilities and other entities required to carry out controls, specific functionalities for automated verification of compliance with the vaccination requirement, based on information processed within the National DGC Platform (PN-DGC), described regarding different work contexts and specific categories of workers. These functionalities allow verifications to be carried out in relation to workers on duty, using boolean type flags relating to compliance with the vaccine requirement (green light: vaccinated or exempt worker/red light: unvaccinated worker).

The Italian government has reproposed and revised the approach concerning the revocability of the green certificate anti COVID-19, allowing and approving it, although without implementing any mechanism capable of embodying this provision. Following the above mentioned decree (of December 17, 2021), the revocation of the Green Pass would be technically possible, following up on what was pointed out, in a formal measure, by the Data Protection Authority (June 9, 2021) in compliance with both the principle of accuracy enacted by the GDPR (679/2016) and to avert the increased risks and loss of real containment effectiveness caused by the unreliability of the conditions attested by the Green Pass itself. The Government considered it necessary to implement a revocation mechanism in the legislation appropriate to the handling of the Green Certificate, due to case histories which, with the passage of time, made it necessary to revoke it [3].

The examples formulated by the Guarantor have transformed from trivial academic hypotheses (unproven and existing in the literature) to concrete problems; among them are:

- dissemination and online sharing of hundreds of authentic but illicitly used green certificates.
- dissemination, in print, of genuine Green Passes and subsequent illicit uses.
- production and subsequent illegal sale of forged certificates, with fancy names (from Pluto to Hitler).
- dissemination of genuine but unregistered certificates acquired through the preview function.
- sale of genuine and registered certificates without the actual and successful vaccine inoculation.

- fraudulently issued and obtained green certificates (found and downloaded on several well-known apps, including Telegram).
- suspension of entire batches of COVID-19 vaccine because they were found to be defective.
- Green Passes issued to people who later tested positive for the virus.
- carrying out swabbing with positive results and subsequent quarantine measure for green pass holders.

2.2 The Fake Green Pass

On Oct. 26, 2021, a valid but, incontrovertibly, fake Green Pass began circulating online (on specialized forums and, later, on Twitter); this is because it is in the name of Adolf Hitler and shows, as his date of birth, Jan. 1, 1900. Such cunning, which at first glance may appear to be a photomontage, turns out to be in fact the result of a more complex operation, since the QR Code from which the certificate in Hitler's name was verified was for all intents and purposes functional for more than a day, until the late morning of October 27, when it was deactivated [4].

Many certificate verification apps, including the Italian development app "Verifi-caC19," released by the Ministry of Health, considered this document to be authentic: the moment the QR Code was framed, a green frame appeared on the screen, a signal of its validity (thus, someone succeeded in the attempt to generate a Green Pass that was evidently fake, but recognized by the verification apps and, for that reason, functioning).

Behind such a scam is the actualization of arduous and meticulous work; the QR codes of Green Passes, are generated from several pieces of personal information, which constitute a unique combination (examples of such data are: first and last name of the vaccinated person, the country of vaccination, the number of doses received, the date of administration, the institution that issued the Green Pass, the manufacturer of the vaccine administered, the total number of doses, the disease covered by the vaccine, the expiration of the code, and the date of generation); the peculiarity lies in the fact that all these data are not encrypted: should the code be shared on the network, the sanitized data of various individuals would be exposed [5]. The encrypted part (by crypto- graphic key) is a string of numbers, letters and symbols, functioning as a signature, attesting that the QR Code has not been forged. Such encryption employs an algorithm known as "asymmetric": the private key, held by the entity that issued the certificate, must match the one that signed the certificate itself [6]. The fact that an individual and/or group of individuals created a functioning Green Pass in the name of a historical figure who died more than seventy years ago suggests the fact that someone came into possession of these keys and used them to produce a fake certificate (Fig. 1).

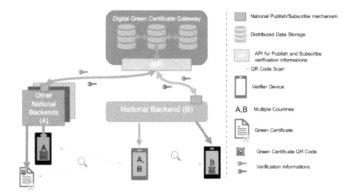

Fig. 1. Cryptographic algorithm used for the generation of valid Green Passes [14]

3 BOOGIE: A New Blockchain Application for Health Certificate Security

In the wake of the proliferation of the aforementioned virus, numerous research initiatives have been undertaken with the objective of establishing precise and prompt documentation of COVID-19 infections, with the ultimate goal of enhancing countermeasures against this illness. In this context, important help has been found within blockchain technology. A pioneering solution rooted in blockchain technology has been developed and put into operation to instill confidence and eradicate fraudulent activities associated with the distribution of unauthorized COVID-19 green certificates. Primarily, this solution utilizes Ethereum smart contracts, which can be coded to perform function calls and create events that inform involved parties about relevant medical data, updates on testing, and necessary requirements. Moreover, this design could help control COVID-19 spread using digital 'on-chain' medical passports and immunity certificates, whose immutable nature ensures trustworthiness [7] (see Fig. 2).

The work takes its basis from the research work, conducted by some scholars at Khalifa University, deputed to Science and Technology, building on pioneering work of other researchers (in block- chain field employed in other areas and/or related to the topic of COVID-19) [14], they proposed a solution based, again, on this technology, which offers:

- Tracking and tracing COVID-19 test participants is facilitated by leveraging immutable events and logs in the distributed ledger, bypassing the need for on-chain storage.
- it is manifested how self-sovereign identity (SSI) is accompanied by the above-mentioned blockchain design and an effective decentralized identity system.
- proxy re-encryption schemes are employed to integrate the blockchain-based system with the interplanetary file (IPFS) and securely store patient-related information: medical, identity, and travel-related.
- finally, security and cost analyses of the solution are performed to demonstrate its feasibility and reliability.

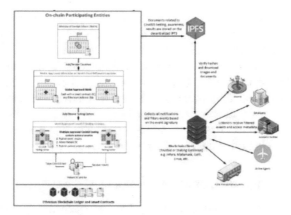

Fig. 2. Overview of the implementation of the blockchain-based solution for immunity certificate management [14]

The above figure illustrates the general schematic of the proposed solution, highlighting the on-chain participation of various entities, smart contracts, distributed storage, blockchain clients, and stakeholders. It employs four distinct types of smart contracts:

1. the MoFA smart contract;
2. the MoH smart contract;
3. the COVID-19 Testing Center smart contract;
4. the Patient smart contract.

3.1 Digital Health Passport and Immunity Certificate (Green Pass)

Digital health passports serve as a vital identification tool, aiding in the reduction of infectious disease transmission; the smart patient contract is implemented to fulfill this objective. This unalterable record contains the IPFS hash associated with vaccination and immunization records, along with an individual's medical and travel/movement history. In terms of personally identifiable information, its release is under the discretion of the information's proprietor [8].

Certificates of immunity are also provided to verify that an individual has developed relevant antibodies to mitigate COVID-19 infection, such that he or she is no longer a threat, averting the risk of infection of other individuals (a goal achieved through recovery from COVID-19 infection or through appropriate vaccination).

The immunity and, therefore, the possible exemption from physical and social restrictions, of various patients, is information that can also be part of the latter's smart contracts, using an immutable transaction by the COVID-19 Testing Center, which can announce this after an antibody test, communicating, of its own accord, the length of time the patient would remain immune (depending on the strength of the vaccine and the relevant medical pathway). To date, it is estimated that the period of immunity is still inconclusive and still a topic of discussion among scientists and researchers, whose time interval remains, essentially, as a subject of study and research.

To address this problem, through the solution that is about to be dissected, a design has been designed to include the precise instant at which a specific patient's immunity announcement was made (resulting in the issuance of the relevant certificate); it can be conducted through a notification, known, and recorded to the public. The standard procedure involves the smart contract of the COVID-19 Testing Centre triggering events aimed at notifying patients and involved parties about potential adjustments in line with the relevant medical test; these may be quarantine information or details about the clinical tests they have undergone (thus, immunity status is ensuring as an update that the Testing Centre can communicate in an event, ensuring its immutability). When revealing private information or medical test outcomes, an IPFS-only on-chain hash is utilized, and the data stored on IPFS is further encrypted for enhanced security [9].

3.2 On-Chain Participating Entities: MoHs and MoFAs

The MoH and MoFA are significant entities within the system; they embody the authoritative bodies that verify the legitimacy of the tests and the accuracy of the results, aiming for near-zero error rates. COVID-19 Testing Center necessitates an affiliation with a Ministry of Health (MoH), which subsequently requires an association with the Ministry of Foreign Affairs (MoFA). The MoH possesses the capacity to incorporate COVID-19 Testing Centers that satisfy their stipulations and concurrently. The MoH can add COVID-19 Testing Centers that meet their requirements and, at the same time, can also revoke previously added centers; this is possible through the use of immutable events and transaction logs, within the dialog box inherent in the blockchain network. Furthermore, a MoFA has the authority to add or revoke MoHs based on their stipulated requirements and regulations, playing a crucial role in curbing the spread of diseases across borders and various territories; it only borders MoHs that meet their rules and dictates. Each individual's biometric information is associated with their unique Ethereum Address (EA) on-chain, with the purpose of maintaining each individual's privacy.

3.3 On-Self-sovereign Identity (SSI)

Contrary to traditional centralized identity management (IdM) systems that rely on servers, in Self-Sovereign Identity (SSI) systems, users interact with their wallets via their Dapps, enabling continuous control over access to their sensitive data. Consequently, users of such a system are granted the autonomy to manage their own identity and credentials. In traditional identity systems, access to the organization's resources is restricted to authorized individuals only; Open Authentication (OAuth) and the OpenID Connect option serve as examples of these conventional IdM systems. An effective identity system should consist of an identity provider, a service provider, and users; the former provide authentication, registration, and identity services to users and service providers and may also be a third-party service, independent of service providers, which typically requires the identity provider to validate and authenticate the same as claimed by a user.

UPort, Sovrin, Microsoft's ION, Ontology represent some instances of blockchain-based identity management systems. These systems emphasize the development of a digital identity, operating independently from any centralized authority [9].

Blockchain-based identity management systems are characterized by their use of peer nodes to store diverse identity data, eschewing reliance on a central server, while ensuring authentication, trust, and privacy. Certain proposals for such systems maintain user anonymity, leveraging an attribute reputation model and a Self-Sovereign Identity (SSI) system. The efficacy of these blockchain-based solutions hinges on a large user community capable of requesting attestations and certificates to verify identity and individuality.

Numerous proposals exist for blockchain-based identity management, each with its own set of strengths and weaknesses. Benefits encompass zero-knowledge protocol, zero-trust model, universal detectability, selective anonymity, data transparency, and immutability. However, infrastructure costs and key management are significant considerations when selecting an SSI system.

Additionally, while password-based systems allow for easy recovery in case of loss or forgetfulness, in blockchain-based SSI systems, the loss of a private key equates to asset loss [10].

3.4 Blockchain Clients: STAKING GATEWAYS

Blockchain clients form a crucial component of the system, facilitating and maintaining communication between the blockchain network and event listeners. The storage capacity, RAM, and bandwidth are determined in advance, based on the volume of information they need to store, such as block headers or transaction data.

Blockchain clients can be categorized into types such as Infura, Linxa, Meatmask, or Geth Gateway. However, if they serve as the sole conduit for information from the blockchain network to the listeners, they risk centralizing the solution. These clients do not enhance the security of the blockchain network; their sole function is to manage data delivery to the intended listeners in a tamper-proof manner. Therefore, a more optimal solution would be the use of staking gateways.

3.5 IPFS Data Confidentiality

IPFS is utilized for the storage of off-chain documents in a decentralized manner. The extensive volume of COVID-19 related data makes on-chain storage impractical, necessitating a decentralized and secure storage solution. IPFS storage is distributed, public, and accessible to all, thus, information stored on IPFS should be encrypted, with plaintext content accessible only to authorized entities. In this system, various entities - including hospitals, testing centers, airport authorities, airline agents, employers, and academics - can access the server-stored content while maintaining confidentiality. Therefore, a mechanism that enables content sharing based on the data owner's authorization should be implemented [11].

To facilitate multi-party access to the IPFS content, a solution rooted in proxy re-encryption has been implemented. Initially, the data owner encrypts the content using a symmetric key. The encrypted data is then uploaded to IPFS, with only its hash stored in the smart contracts. Subsequently, each entity must generate a unique key for each receiver, without knowledge of the original key. As a result, participants must have a copy of the symmetrically encrypted key. Consequently, the data owner encrypts the symmetric key with their public key and disseminates it to the various entities (Fig. 3).

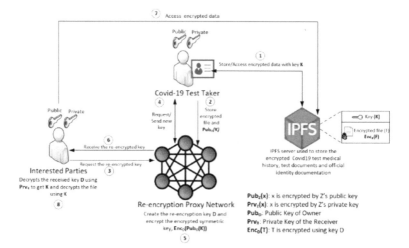

Fig. 3. Scheme, in detail, of the re-encryption process [14]

Various on-chain participants can significantly benefit from the events generated by other entities within the chain. Given the highly infectious nature of COVID-19, it's essential for all sectors where human interaction is unavoidable to maintain strong defenses against this disease.

The proposed solution enables sectors to access on-chain records and events, enabling the identification of existing and potential users of their services. Listeners, as entities within the blockchain, engage with the blockchain client to receive filtered events from the public blockchain. These transparent events are accessible to all participants in the chain, eliminating the requirement for a separate management system [12].

3.6   Implementation Details

In order to validate the concepts elucidated in the preceding section, coding was implemented using the Remix IDE [13], a platform utilized for compiling and testing smart contracts. Subsequently, four smart contracts were developed, namely, MoFA, MoH, COVID-19 Testing Center, and the patient smart contract, which is used for compiling and testing smart contracts;

The entity-relationship diagram depicted above provides an overview of the smart contracts, their functions, and their associated attributes. The MoFA smart contract can establish connections with multiple MoH smart contracts, while the latter can reference multiple smart contracts related to COVID-19 test centers. The patient smart contract, on the other hand, can be associated with one or more test center smart contracts. All the documents are stored on IPFS, with the hash values of Bytes32 types being stored as attributes in the smart contracts. The primary functions of these contracts revolve around generating events to inform the listeners about ongoing actions. This approach helps reduce on-chain costs and leverages the immutability of the available logs [14] (see Fig. 4).

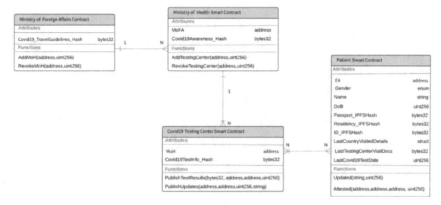

Fig. 4. Entity-relationship diagram showing interactions between smart contracts [14].

4 Conclusion

To date, there are several methodologies of deception, fraud and malfeasance regarding the green certificate/immunity management system from COVID; however, there are several automatic that allow for a scrupulous control and verification of it and a speedy and efficient revocation in the above-listed cases (fraudulent and non- fraudulent), thus preventing malicious individuals from artfully exploiting false or, at any rate, invalid certificates and/or medical passports. The proposed system helps to mitigate the spread of infectious diseases, in general, and COVID-19 Disease in particular (being the subject of eminent discussion and research in "covid years"). The reported strategy is based on the presence of four smart contracts, which rely, in turn, on negligible on-chain storage, in such a way as to exploit on-chain events and notations; building on that foundation, elements such as: self-sovereign identity (SSI), re-encryption proxies, and the biometric information associated with the unique Ethereum Address (EA) of the participating entities were incorporated. The proposed solution was developed through the utilization of carefully evaluated algorithms, ensuring the feasibility of the proposed approach. The algorithms underwent rigorous scrutiny, including cost analysis checks and vulnerability assessments conducted using SmartCheck software. It is both hopeful and imperative to contemplate how such a strategy can serve as a foundation for effective solutions in the future.

References

1. Dye, C., Mills, M.C.: COVID-19 vaccination passports. Science **371**(6535), 1184–1184 (2021). https://doi.org/10.1126/SCIENCE.ABI5245
2. Green Pass al lavoro, dopo l'OK del Garante Privacy ecco il decreto con le nuove modalità di revoca - Cyber Security 360. https://www.cybersecurity360.it/legal/privacy-dati-personali/green-pass-al-lavoro-ce-lok-del-garante-privacy-alle-nuove-modalita-di-revoca/. Accessed Jan 11 2023
3. Revoca green pass, la legge: come funziona in caso di positività, frode - Agenda Digitale. https://www.agendadigitale.eu/sicurezza/privacy/revoca-green-pass-la-legge-come-funziona-in-caso-di-positivita-frode/. Accessed Jan 11 2023

4. Il Green Pass di Adolf Hitler: il sistema è stato bucato? https://www.punto-informatico.it/green-pass-adolf-hitler/. Accessed Jan 11 2023

5. Perché il Green Pass di Adolf Hitler può essere un problema - Il Post. https://www.ilpost.it/2021/10/27/green-pass-adolf-hitler/. Accessed Jan 11 2023

6. VerificaC19, l'app che legge il Certificato Verde (Green Pass). https://www.punto-informatico.it/verificac19-app-certificato-verde/. Accessed Jan 11 2023

7. Chang, M.C., Park, D.: How can blockchain help people in the event of pandemics such as the COVID-19? J. Med. Syst. **44**(5), 1–2 (2020). https://doi.org/10.1007/S10916-020-01577-8/METRICS

8. Deng, W., et al.: Primary exposure to SARS-CoV-2 protects against reinfection in rhesus macaques. Science **369**(6505), 818–823 (2020). https://doi.org/10.1126/SCIENCE.ABC5343

9. Addetia, A., et al.: Neutralizing antibodies correlate with protection from SARS-CoV-2 in humans during a fishery vessel outbreak with a high attack rate. J. Clin. Microbiol. **58**(11), 10–128 (2020). https://doi.org/10.1128/JCM.02107-20

10. Mühle, A., Grüner, A., Gayvoronskaya, T., Meinel, C.: A survey on essential components of a self-sovereign identity. Comput. Sci. Rev. **30**, 80–86 (2018). https://doi.org/10.1016/J.COSREV.2018.10.002

11. Liu, Y., He, D., Obaidat, M.S., Kumar, N., Khan, M.K., Choo, K.K.R.: Blockchain-based identity management systems: a review. J. Netw. Comput. Appl. **166**, 102731 (2020). https://doi.org/10.1016/J.JNCA.2020.102731

12. Green, M., Ateniese, G.: Identity-based proxy re-encryption. In: Katz, J., Yung, M. (eds.) Applied Cryptography and Network Security. ACNS 2007. Lecture Notes in Computer Science, vol. 4521, pp. 288–306. Springer, Berlin (2007). https://doi.org/10.1007/978-3-540-72738-5_19

13. Remix - Ethereum IDE. https://remix.ethereum.org/#optimize=false&runs=200&evmVersion=null&version=soljson-v0.8.7+commit.e28d00a7.js. Accessed Jan 11 2023

14. Hasan, H.R., et al.: Blockchain-based solution for COVID-19 digital medical passports and immunity certificates. IEEE Access **8**, 222093–222108 (2020). https://doi.org/10.1109/ACCESS.2020.3043350

Novel Computational Paradigms, Methods and Approaches in Bioinformatics

Ontologically Enriched Rough Set Based Reasoning in Medical Databases with Linguistic Data

Krzysztof Pancerz[✉]

Academy of Zamosc, Ul. Pereca 2, 22-400 Zamosc, Poland
krzysztof.pancerz@akademiazamojska.edu.pl

Abstract. Originally, in rough set theory (RST) proposed by Z. Pawlak, approximation of sets is defined on the basis of an indiscernibility relation between objects in some universe of discourse. The problems appear if attribute values describing objects are symbolical (e.g., linguistic terms). Such a situation is natural in human cognition and description of the real world (e.g. in case of medical applications, where diseases are described in natural language terms). We can perfect rough set theory in this area by incorporating ontologies enabling us to add some new, valuable knowledge, which can be used in data analysis, rule generation, reasoning, etc. In the paper, we propose to use ontological graphs in determining approximations of sets as well as we show how ontological graphs change the look at them in case of linguistic medical data.

Keywords: reasoning · rough sets · approximation · ontologies · medical data

1 Introduction

Methodology of knowledge discovery from data is largely determined by the nature of the data. Generally, data, due to their nature, can be classified into three categories:

- quantitative (numeric) and qualitative (nominal) data,
- text data,
- image and sound data.

L. Zadeh proposed *computing with words* (cf. [15]), i.e., processing words, concepts, and natural language terms. The main idea is that words and concepts are used in place of numbers for computing and reasoning. This is justified by the fact that the ability of human is to perform many tasks without any measurements and calculations on numbers. Over time, a lot of methodologies were proposed to process data of such character. The subject of our research is a part of this trend. The main topic of research concerns decision systems, whose attribute values are words, or, in general, concepts (consisting of words) that describe objects

A. Rocha et al. (Eds.): WorldCIST 2023, LNNS 802, pp. 383–389, 2024.
https://doi.org/10.1007/978-3-031-45651-0_38

or phenomena. To effectively carry out data mining processes, we require to take into consideration the domain knowledge related to data semantics [12]. Data mining with the domain knowledge was an extensively studied research area in the past. Different forms of the domain knowledge have been used, for example:

- preference order of attribute values (see e.g. [3]),
- concept hierarchies, e.g. attribute value taxonomies (see e.g. [2,14]), attribute value ontology (see e.g. [5]),
- ontologies and semantic nets (see e.g. [1,11]).

In case of ontologies, very often we deal with domain ontologies of concepts. Then, the domain knowledge is expressed as a set of concepts together with the relationships which have been defined between them comprising the vocabulary from a given area (cf. [7]).

In [8], we proposed to incorporate the domain knowledge in a form of ontology directly into decision systems (in Pawlak's sense). In order to cover the meaning of data, the so-called simple decision systems over ontological graphs were formally defined. A topic of qualitative assessment of rough sets, defined in simple decision systems over ontological graphs, was considered in [9]. We proposed qualitative assessment of approximation from the point of view of values of a decision attribute and the domain knowledge included in an ontological graph associated with the decision attribute.

In this paper, we propose to use ontologically enriched rough sets for reasoning in medical databases with linguistic data. An illustrative example given in Sect. 3 describes the proposed idea.

2   Definitions

In this section, we recall basic definitions concerning decision systems over ontological graphs and rough sets (cf. [8,10]). Formally, ontologies can be defined on the basis of description logics, for example, in a form of axioms separated into three groups: terminological (*TBox*), assertional (*ABox*), and relational (*RBox*) (see e.g. [4]). In our approach, ontological graphs are a simplified way to represent a part of the knowledge included in ontologies.

Definition 1. *Let $\mathcal{O}$ be a given ontology. An ontological graph is a quadruple $OG = (\mathcal{C}, E, \mathcal{R}, \rho)$, where $\mathcal{C}$ is a non-empty, finite set of nodes representing concepts in the ontology $\mathcal{O}$, $E \subseteq \mathcal{C} \times \mathcal{C}$ is a finite set of edges representing relations between concepts from $\mathcal{C}$, $\mathcal{R}$ is a family of semantic descriptions (in natural language) of types of relations (represented by edges) between concepts, and $\rho : E \rightarrow \mathcal{R}$ is a function assigning a semantic description of the relation to each edge.*

We take into consideration the following family of semantic descriptions of relations between concepts (cf. [6]): synonymy $(R_\sim)$, antonymy $(R_\leftrightarrow)$, hyponymy/hyperonymy $(R_\lhd$ / $R_\rhd)$. For example, the following concepts are related to each other: $(car, automobile) \in R_\sim$, $(warm, cold) \in R_\leftrightarrow$, $(car, vehicle) \in R_\lhd$, and vice versa $(vehicle, car) \in R_\rhd$.

Definition 2. *A decision system DS is a tuple $DS = (U, C, D, V_c, V_d, f_c, f_d)$, where U is a non-empty, finite set of objects, C is a non-empty, finite set of condition attributes, D is a non-empty, finite set of decision attributes, $V_c = \bigcup_{a \in C} V_a$, where V_a is a set of values of the condition attribute a, $V_d = \bigcup_{a \in D} V_a$, where V_a is a set of values of the decision attribute a, $f_c : C \times U \to V_c$ is an information function such that $f_c(a, u) \in V_a$ for each $a \in C$ and $u \in U$, $f_d : D \times U \to V_d$ is a decision function such that $f_d(a, u) \in V_a$ for each $a \in D$ and $u \in U$.*

Approximation of sets is one of the fundamental notions of rough set theory [10]. The main idea is to approximate a given set of objects by means of other sets of objects called elementary sets, forming the so-called basic granules of knowledge. In the original Pawlak's approach, basic knowledge granules are induced by an indiscernibility relation between objects. However, the notion of rough sets can be generalized using an arbitrary binary relation between objects (cf. [13]). Let $B \subseteq A$, an indiscernibility relation IR_B is defined as $IR_B = \{(u, v) \in U \times U : \forall_{a \in B} f_c(a, u) = f_c(a, v)\}$.

Definition 3. *Let $X \subseteq U$ and $B \subseteq A$. We may characterize X with respect to B using the basic notions of rough set theory as follows. The B-lower approximation, $\underline{B}(X)$, of a set X with respect to B: $\underline{B}(X) = \{u \in U : IR_B(u) \subseteq X\}$. The B-upper approximation, $\overline{B}(X)$, of a set X with respect to B: $\overline{B}(X) = \{u \in U : IR_B(u) \cap X \neq \emptyset\}$. The B-boundary region, $BN_B(X)$, of a set X with respect to B: $BN_B(X) = \overline{B}(X) - \underline{B}(X)$.*

Definition 4. *A simple decision system SDS^{OG} over ontological graphs is a tuple $SDS^{OG} = (U, C, D, \{OG_a\}_{a \in C \cup D}, f_c, f_d)$, where U is a non-empty, finite set of objects, C is a non-empty, finite set of condition attributes, D is a non-empty, finite set of decision attributes, $\{OG_a\}_{a \in C \cup D}$ is a family of ontological graphs associated with condition and decision attributes from $C \cup D$, $f_c : C \times U \to \mathcal{C}$, $\mathcal{C} = \bigcup_{a \in C} \mathcal{C}_a$, is an information function such that $f_c(a, u) \in \mathcal{C}_a$ for each $a \in C$ and $u \in U$, $f_d : D \times U \to \mathcal{C}$, $\mathcal{C} = \bigcup_{a \in D} \mathcal{C}_a$, is a decision function such that $f_d(a, u) \in \mathcal{C}_a$ for each $a \in D$ and $u \in U$, $\mathcal{C}_a$ is a set of concepts from the graph OG_a.*

Further, we will be interested in simple decision systems in which ontological graphs are associated with decision attributes only.

3   Ontologically Enriched Rough Set Based Reasoning

In [9], we proposed qualitative assessment of approximation from the point of view of values of a decision attribute and the domain knowledge included in an ontological graph associated with the decision attribute. Assignment of the proper qualitative characteristic is based on a semantic analysis of values of a decision attribute adopted by the objects belonging to the boundary region.

In the proposed approach, we considered the worst case. Four cases can be distinguished in the process of characterizing a rough set: (*Case 1*) If there exists at least one object belonging to the boundary region of approximation, which adopts a value on a decision attribute that is a concept being an antonym of the concept defining the approximated set, then the rough set is called the significantly rough set. Otherwise: (*Case 2*) If there exists at least one object belonging to the boundary region of approximation, which adopts a value on a decision attribute that is a concept that has a far hyperonym with the concept defining the approximated set, then the rough set is called the moderately rough set. Otherwise: (*Case 3*) If there exists at least one object belonging to the boundary region of approximation, which adopts a value on a decision attribute that is a concept that has a middle-far hyperonym with the concept defining the approximated set, then the rough set is called the marginally rough set. Otherwise: (*Case 4*) If all of the objects, belonging to the boundary region of approximation, adopt values on a decision attribute that are concepts that have a close hyperonym or they are synonyms, then the rough set is called the pseudo rough set.

Example 1. Let us consider an ontological graph given in Fig. 1 that will be assigned to the decision attribute.

The meaning of semantic descriptions of relations assigned to edges is described in Sect. 2. Further, we will consider several examples of basic granules of knowledge induced by an indiscernibility relation between objects which can be stored in some medical database. For simplicity, let us take into consideration a set B of condition attributes including *Muscle pain* and *Temperature*. Moreover, let *Disease* be a decision attribute to which the ontological graph shown in Fig. 1 is assigned. By X_Δ we will denote the set of objects classified to the decision class Δ.

Let us consider the first illustrative basic granule of knowledge given in Fig. 2. This granule is covered by the B-boundary region of X_{Flu} that generates inconsistent decision rules:

- IF *Muscle pain* is *yes* AND *Temperature* is *high*, THEN *Disease* is *Flu*,
- IF *Muscle pain* is *yes* AND *Temperature* is *high*, THEN *Disease* is *Grippe*.

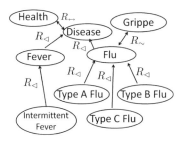

Fig. 1. A simple ontological graph

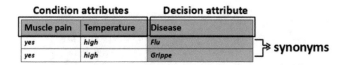

Fig. 2. The boundary region in case of synonyms

The existence of the considered basic granule in the B-boundary region proves that the set X_{Flu} is rough. However, one can see that values on a decision attribute for objects belonging to the granule are synonyms (according to the ontological graph). Therefore, X_{Flu} is the pseudo rough set. It means that decision rules generated by the B-boundary region are apparently inconsistent only.

Let us consider the second illustrative basic granule of knowledge given in Fig. 3. This granule is covered by the B-boundary region of $X_{TypeAFlu}$ that generates inconsistent decision rules:

- IF *Muscle pain* is *yes* AND *Temperature* is *high*, THEN *Disease* is *Type A Flu*,
- IF *Muscle pain* is *yes* AND *Temperature* is *high*, THEN *Disease* is *Type B Flu*.

Fig. 3. The boundary region in case of a close hyperonym

Analogously to the previous case, the set $X_{TypeAFlu}$ is rough. However, one can see that values on a decision attribute for objects belonging to the granule have a close hyperonym *Flu* (according to the ontological graph). Therefore, $X_{TypeAFlu}$ is the pseudo rough set. It means that decision rules generated by the B-boundary region are apparently inconsistent only.

Let us consider the third illustrative basic granule of knowledge given in Fig. 4. This granule is covered by the B-boundary region of $X_{TypeAFlu}$ that generates inconsistent decision rules:

- IF *Muscle pain* is *yes* AND *Temperature* is *high*, THEN *Disease* is *Type A Flu*,
- IF *Muscle pain* is *yes* AND *Temperature* is *high*, THEN *Disease* is *Intermittent Fever*.

Analogously to the previous case, the set $X_{TypeAFlu}$ is rough. One can see that values on a decision attribute for objects belonging to the granule have a far

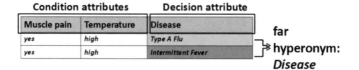

Fig. 4. The boundary region in case of a far hyperonym

hyperonym *Disease* (according to the ontological graph). Therefore, $X_{TypeAFlu}$ is the moderately rough set. It means that decision rules generated by the B-boundary region are, in fact, inconsistent.

Let us consider the fourth illustrative basic granule of knowledge given in Fig. 5. This granule is covered by the B-boundary region of X_{Flu} that generates inconsistent decision rules:

- IF *Muscle pain* is *yes* AND *Temperature* is *normal*, THEN *Disease* is *Flu*,
- IF *Muscle pain* is *yes* AND *Temperature* is *normal*, THEN *Disease* is *Health*.

Condition attributes		Decision attribute
Muscle pain	**Temperature**	**Disease**
yes	normal	Flu
yes	normal	Health

> antonyms

Fig. 5. The boundary region in case of antonyms

Analogously to the previous case, X_{Flu} is rough. However, one can see that values on a decision attribute for objects belonging to the granule are antonyms (according to the ontological graph). Therefore, X_{Flu} is the significantly rough set. It means that decision rules generated by the B-boundary region are strongly inconsistent.

One can see that qualitative assessment of rough sets from the point of view of decision attribute values and the domain knowledge included in an ontological graph associated with the decision attribute depends on the levels of abstractions (in case of semantic relations) on which decision attribute values (concepts) are considered.

4   Conclusions

We have shown that dealing with semantic relations enriches our look at rough set based reasoning in decision systems. The presented approach takes into account nuances in the interpretation of meanings of attribute values. We can create data mining tools sensitive to data semantics, which can be used in different areas, e.g., medicine, biology, economy, sociology, etc. The main task in further work is to consider more sophisticated semantic relationships between attribute values (concepts) and determine their influence on approximations of sets.

References

1. Bloehdorn, S., Hotho, A.: Ontologies for machine learning. In: Staab, S., Studer, R. (eds.) Handbook on Ontologies, pp. 637–661. Springer, Berlin, Heidelberg (2009)
2. Cagliero, L., Garza, P.: Improving classification models with taxonomy information. Data Knowl. Eng. **86**, 85–101 (2013)
3. Greco, S., Matarazzo, B., Slowinski, R.: The use of rough sets and fuzzy sets in MCDM. In: Gal, T., Stewart, T.J., Hanne, T. (eds.) Multicriteria Decision Making: Advances in MCDM Models, Algorithms, Theory, and Applications, pp. 397–455. Springer, US, Boston, MA (1999)
4. Krötzsch, M., Simancik, F., Horrocks, I.: Description logics. IEEE Intell. Syst. **29**(1), 12–19 (2014)
5. Lukaszewski, T., Józefowska, J., Lawrynowicz, A.: Attribute value ontology - using semantics in data mining. In: Maciaszek, L.A., Cuzzocrea, A., Cordeiro, J. (eds.) Proceedings of the 14th International Conference on Enterprise Information Systems, pp. 329–334. Wroclaw, Poland (2012)
6. Murphy, M.L.: Meaning Relations in Dictionaries: Hyponymy, Meronymy, Synonymy, Antonymy, and Contrast. In: The Oxford Handbook of Lexicography. Oxford University Press (2015)
7. Neches, R., Fikes, R., Finin, T., Gruber, T., Patil, R., Senator, T., Swartout, W.: Enabling technology for knowledge sharing. AI Mag. **12**(3), 36–56 (1991)
8. Pancerz, K.: Toward information systems over ontological graphs. In: Yao, J., Yang, Y., Słowiński, R., Greco, S., Li, H., Mitra, S., Polkowski, L. (eds.) Rough Sets and Current Trends in Computing. Lecture Notes in Artificial Intelligence, vol. 7413, pp. 243–248. Springer-Verlag, Berlin Heidelberg (2012)
9. Pancerz, K.: Toward qualitative assessment of rough sets in terms of decision attribute values in simple decision systems over ontological graphs. In: Peters, J.F., Skowron, A., Slezak, D., Nguyen, H.S., Bazan, J.G. (eds.) Transactions on Rough Sets XIX, pp. 83–94. Springer, Berlin Heidelberg (2015)
10. Pawlak, Z.: Rough Sets. Theoretical Aspects of Reasoning about Data. Kluwer Academic Publishers, Dordrecht (1991)
11. Ristoski, P., Paulheim, H.: Semantic web in data mining and knowledge discovery: A comprehensive survey. Web Semantics: Sci. Serv. Agents World Wide Web **36**, 1–22 (2016)
12. Witten, I.H., Frank, E.: Data Mining: Practical Machine Learning Tools and Techniques. Morgan Kaufmann (2005)
13. Yao, Y., Lin, T.: Generalization of rough sets using modal logics. Intell. Autom. Soft Comput. **2**(2), 103–120 (1996)
14. Ye, M., Wu, X., Hu, X., Hu, D.: Knowledge reduction for decision tables with attribute value taxonomies. Knowl.-Based Syst. **56**, 68–78 (2014)
15. Zadeh, L.: Fuzzy logic = computing with words. IEEE Trans. Fuzzy Syst. **4**(2), 103–111 (1996)

Modelling the Drift of Social Media Posts

Henryka Czyż[1], Andrew Schumann[2](✉), and Arkadiusz Gaweł[2]

[1] Department of Cognitive Science and Mathematical Modelling,
University of Information Technology and Management in Rzeszow,
Sucharskieho 2, 35-225 Rzeszow, Poland
`hczyz@wsiz.edu.pl`
[2] University of Information Technology and Management in Rzeszów,
Rzeszów, Poland
`andrew.schumann@gmail.com, agawel@wsiz.edu.pl`

Abstract. We have developed a theory that allows us to accurately characterize information waves in social networks such as Twitter. This theory is an extrapolation of a mathematical theory that studies particle drifts in the acoustic field. We have noticed that each post on Twitter can be considered as a particle, and its radius is the number of followers of the author of the post. Then the number of likes and retweets of this entry can be considered as a drift of this particle. We have found that our hypothesis is correct, and in this interpretation, Twitter posts behave like particles in the acoustic field. In particular, if the author of the entry has no more than 15,000 followers, then this behavior is characterized as a radiative drift in the acoustic field, and if more than 15,000, then as a viscous drift. Based on this behavior of particles (posts on Twitter), we can mathematically reconstruct an appropriate information wave.

Keywords: Twitter · radiation drift · viscous drift · particles · information wave

1 Introduction

In describing the mechanism of dissemination of information in social networks, the metaphor of an information wave is sometimes used – a source of perturbation of the wave appears in some place, for instance, the first tweet that causes public interest appears, and then from this source the "wave" spreads with some amplitude, e.g., there are multiple retweets of the first tweet or similar tweets are made, see [10–12]. Then the wave goes down until a new source of disturbance. However, there are problems in using the information wave metaphor. It is not possible to physically determine an acoustic field in which this wave propagates or a physical autowave process such as epidemics or firestorms. Nevertheless, we can see that many autowave phenomena such as firestorms [13] or heat waves [1] can be analyzed through a bisimilar process of their mentions on social media. This allows us to understand the information propagation through social media as a kind of autowave, too [14].

A. Rocha et al. (Eds.): WorldCIST 2023, LNNS 802, pp. 390–399, 2024.
https://doi.org/10.1007/978-3-031-45651-0_39

In social networks, we are dealing with individual entries, which, nevertheless, have some kind of dynamics over time – they gain likes, are shared, and are quoted. With many such posts on Facebook or Twitter, we can model their collective dynamics. It is this modelling that is of interest, since it allows us to evaluate the dynamics of public attention and interest. It turns out that we are not meant to deal with a wave in the strict sense of the word, since there is no acoustic field or another physical substrate, but nevertheless we can evaluate this group dynamics through some wave equations. Namely, we can represent individual posts as particles, for which the dynamics can be considered as a drift analogous to the particle drift in an acoustic field. In other words, we can deal only with particles (posts of social networks) and their drift, but this drift itself can already be considered through the reconstruction of the drift force in a virtual (fictitious) field. So, it turns out that based on the drift of particles (posts), we can restore and simulate the wave with its characteristics: amplitude, frequency, wavelength and propagation speed.

It should be noted that there is a theory (please see [4–8]) first proposed in [2,3] (about some applications, please see [9,15]) that, according to the characteristics of the acoustic field and the force of the wave in it, is able to determine different types of particle drifts in this field. The problem of drifts is related to the properties of the acoustic field, which consist in accelerating the movement of particles dispersed in liquid media. The acoustic field is defined as the area of a liquid medium (e.g., gas) in which mechanical waves propagate, also known as acoustic waves. In the presented model of particle drift in the acoustic field, the macro phenomena of particle transport and the formation of areas of increased concentration are analyzed.

A small particle in a fluid medium carries out, in general, complex movements in the acoustic field, caused by forces of various natures. In the drift analysis, it is assumed that the movements of particles are independent of each other, i.e., the analysis is reduced to the study of the movements of separate particles in a viscous medium under the influence of external forces.

In the acoustic field [2–8], the movement of a particle suspended in a fluid medium is a combination of a rapid oscillating motion and a slow translational movement in relation to the medium, called a drift. The main factors that cause particle drift in the acoustic field are:

1. pressure of sound radiation per particle,
2. periodic changes in the viscosity of the vibrating medium,
3. asymmetry of motion of a vibrating medium (only in standing waves).

In addition to the above-mentioned factors causing particle drift in the acoustic field, there may be other, disordered ones, causing random, irregular displacements, which have been omitted in the considerations. A typology of particle drifts was proposed in a series of papers [2–9,15]. Thus, depending on the medium, acoustic waves and particle characteristics, their drift type is determined.

In this paper, we have an inverse problem – we can describe the drift by the dynamics of individual posts in social networks (analogues of particles in an acoustic field) and, by the type of their drift, determine the characteristics of the medium and analogues of the wave. So, based on the particle drift, we are restoring the wave that caused this drift. It allows us to define information waves on social media just on seeing individual posts without tools such as Google Analytics which trace back the number of mentions of some keywords.

As in the acoustic field, we will distinguish the following two types of particle drift in social networks:

1. *Viscous drift* related to periodic changes in viscosity as a cause of the motion of particles (posts on Twitters in our experiments, made by the users, having more than 15,000 followers), see Fig. 2.
2. *Radiation drift* related to the radiation pressure of the wave as a cause of the progressive-oscillating motion of particles (posts on Twitters in our experiments, made by the users, having not more than 15,000 followers), see Fig. 3.

Thus, we do not consider an asymmetry of motion of a vibrating medium on social media.

In Sect. 2, we introduce some basic notions of the drift analysis in the acoustic field. In Sect. 3, we consider some effects of autowaves on social media by some examples. In Sect. 3, we define a drift of particles in these autowaves.

2   Drift of Particles in the Acoustic Field

The drifts of particles have at least three patterns, but in our research we focus only on the following two of them: radiate drift and viscous drift. Their force and acceleration is defined on the basis of some characteristics of the wave and the medium.

Radiative drift, related to the radiation pressure of a wave, was originally defined as a cause of the translational motion (drift) of fine particles in the acoustic field. Today it is known that it is responsible only for the progressive oscillatory motion of relatively large particles. The radiative drift force in the traveling wave field is given by:

$$F_{DR} = \frac{11}{9}\pi k^4 r_p^6 \mu_g^2 \bar{E},\tag{1}$$

where F_{DR} is a radiative drift force, k is a wave number, r_p is a radius of the dispersed phase particle, μ_g is a flow rate, $\bar{E}$ is an average value of the wave energy per unit volume.

And in the standing wave field:

$$F_{DR} = \frac{8}{3}\pi k r_p^3 \mu_g^2 \bar{E}\sin 2kx\tag{2}$$

Viscous drift force is related to periodic changes in viscosity. For small particles, the current can be determined by the formula:

$$F_{DL} = 3\pi\left(\kappa - 3\right)\frac{\eta_0}{\rho_g c}r_p^2\mu_g^2\bar{E},\tag{3}$$

where F_{DL} is a viscous drift force, c is an acoustic wave speed, ρ_g is a medium density, η_0 is a viscosity coefficient in a medium at a fixed temperature, $\kappa = \frac{c_p}{c_v}$ is an adiabatic factor (c_p is a specific heat of gas at constant pressure and c_v is a specific heat of gas at constant volume), and in the standing wave field by the formula:

$$F_{DL} = 3\pi \left(\kappa - 3 \right) \frac{\eta_0}{\rho_g c} r_p \mu_g^2 \bar{E} \sin 2kx. \tag{4}$$

In order to properly compare different types of drifts and characterize their dependence on the parameters: medium, acoustic field and particle, it is not enough to give the values of forces. It is advisable to introduce the value A_D of particle acceleration due to the drift force in the wave field which is the ratio of the maximum value of the drift force to the mass of the particle of the dispersed phase:

$$A_D = \frac{(F_D)_{MAX}}{m_p} \left[\frac{m}{s^2} \right]. \tag{5}$$

Calculating A_D allows us to compare the effectiveness of different drift types. It has an acceleration dimension. This allows for the assessment of the contribution of gravitational acceleration to the motion of particles, which is important in those issues where drift in the acoustic field and particle fall in the gravitational field are considered at the same time.

In a standing wave field, *radiative drift* is defined thus:

$$A_{DR} = 2k\rho_p^{-1}\mu_g^2\bar{E}, \tag{6}$$

where ρ_p is a particle density, and *viscous drift* is understood as follows:

$$A_{DL} = \frac{9}{4} \left(\kappa - 3 \right) \eta r^{-2} \left(\rho_p \rho_g c \right)^{-1} \mu_g^2 \bar{E}. \tag{7}$$

As we see, the values of A_{DR} and A_{DL} depend on the radius of the particle and the nature of the drift. The same dependence can be extrapolated up to the behaviour of posts on Twitter.

3   Autowaves on Social Media, Some Examples

To exemplify our model of the drift analysis on social media, we have examined some effects of information waves on Twitter – the phenomena of sharp increase in the mention of the same keywords within the framework of the same topics. The data for our examples were downloaded via the *rtweet*[1] library using authorization keys and the R language. Meanwhile, the *search_tweets*[2] function was used.

---

[1] https://www.rdocumentation.org/packages/rtweet/versions/1.0.2, 08.11.22.
[2] https://www.rdocumentation.org/packages/rtweet/versions/1.0.2/topics/search_tweets, 09.11.22.

The following two measures were applied by us: the number of tweets (count (status_id)) and the total number of tweet likes (favourites_count). Three different events that were widely commented on social media were selected: one sporting event, one political event, and one social event. They were taken from different periods.

The first event commented on in the media was the sensational loss of the Polish athlete, Iga Świątek, after a series victory in the third round of Wibledon on July 2nd, 2022[3]. The data were collected on the number of tweets in Polish and English. The dynamics in mentioning the phrase "Iga Świątek lost" in English is as follows:

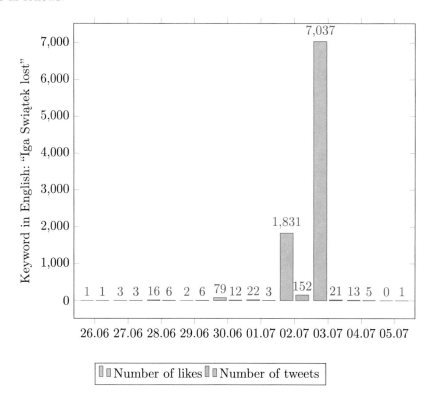

---

[3] https://www.polsatnews.pl/wiadomosc/2022-07-02/wimbledon-iga-swiatek-lost-to-alize-cornet-end-of-an-impressive-winning-series/, 08.11.22.

For the data of the phrase "Iga Świątek przegrała [Iga Świątek lost]" in Polish, please see the following chart:

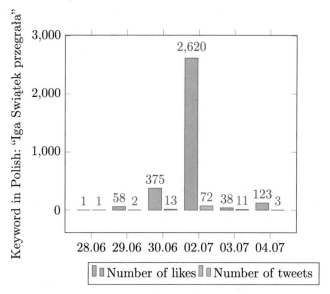

The second set was a political event, namely the resignation of Stanisław Tyszka, the Polish politician, from the Kukiz'15 political club on November 7th, 2022[4]. The results of the search for the phrase "Stanisław Tyszka" is collected there:

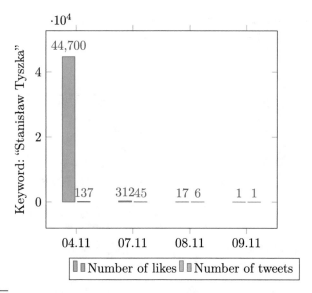

---

[4] https://www.rmf24.pl/fakty/polska/news-stanislaw-tyszka-odchodzi-z-kukiz-15-trudna-ale-pozytywna-de,nId,6395750#crp_state=1,09.11.22.

The third, social event was the disappearance of a 5-year-old child from the city of Oświęcim in Poland. An appropriate announcement "Child alert"[5] was made by the Polish police on November 5th, 2022, see its chart on Tweet:

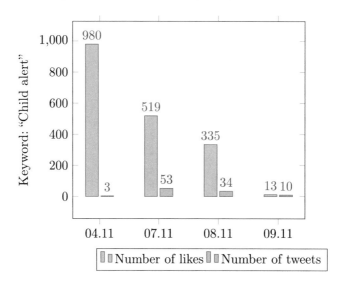

All the events were collected and combined into one diagram to show the logarithmized daily values of the number of tweets from t_0 – no tweets about it before the event – to t_{10} – blanking (no tweets after the event). All events show similar characteristics: (i) a very rapid increase in value (phase I), then (ii) reaching the maximum value (phase II), and (iii) a decrease to 0 (phase III), see Fig. 1. A small exception to the rule is the phrase "Iga Swiątek lost" where in two moments there was a characteristic change.

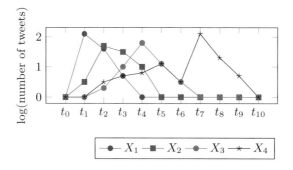

Fig. 1. In this chart, $X_1 :=$ "Child alert", $X_2 :=$ "Stanisław Tyszka", $X_3 :=$ "Iga Swiątek przegrała", $X_4 :=$ "Iga Swiątek lost".

---

[5] https://wiadomosci.wp.pl/po-jej-znikneciu-ruszyl-child-alert-news-of-5yrs-6831058274978496a, 09.11.22.

According to the chart of Fig. 1, the analyzed events on Twitter were popular only for 10 days. This allows us to calculate the acceleration of likes for this period: $\frac{\text{(number of likes of the post } p)}{10^2}$ for each tweet p, assuming that the speed of p is one like per day. Hence, in our paper, particle radii and acceleration values or wave frequency have values that differ by several orders of magnitude. The choice of a logarithmic scale in this case is very important because it enables the comparison of data characterized by values from a wide range of numbers.

4   Drift of Posts on Social Media

To demonstrate the efficiency of the drift model for analyzing posts on social media, it is enough to be grounded on very simple assumptions (evidently, this technique can be more advanced):

1. a size or radius r_p^T of a drifting particle (that is, of a tweet) p written by the author T is equal to the number of the followers of T;
2. a mass m_p^T of a drifting particle (tweet) p written by the author T is equal to the sum of the total number of likes for p and the total number of shares of p, divided by the total number of followers of T;
3. an acceleration of likes A_p^T for the particle p written by the author T is calculated for 10 d by the following expression: $\frac{\text{(number of likes of the post } p)}{10^2}$ – so, we will show that it depends on the size or radius of the drifting particle and the mass of the drifting particle.

For each analyzed event from Fig. 1, let us divide all the authors of the posts that were included in the statistics into two groups: many Twitter followers (more than 15,000) and few Twitter followers (no more than 15,000). Then, for each author, we calculate the number of likes for their post that is included in the statistics. This will also give us an acceleration in the likes of their tweet. So, for each of the four analyzed events, we have two groups of authors (many and few subscribers). For smoothing the obtained data, we will take all the data of these authors as the arithmetic mean. As a result, we get for the first group of authors Fig. 2, and for the second group – Fig. 3. The tweets of the authors who have many followers have the pattern of the viscous drift, while the tweets of the authors who have not many followers have the pattern of the radiative drift. We know that the first drift is characteristic of particles with a lower mass, and the second drift is characteristic of particles with a higher mass. This also explains the difference in drifts for Twitter. Indeed, according to our definition of post mass m_p^T, see above, authors with many subscribers may have less individual post mass than authors with fewer subscribers who are more active in discussing the post.

Knowing the acceleration of posts, we can define an appropriate drift force for them by applying expression (5): $(F_D^T)_{MAX} = m_p^T \cdot A_p^T$. And in this way we can reconstruct an appropriate information wave.

Summing up, the metaphor of information wave, given in [10–12], is wrong. We cannot directly deal with an information wave, seeing a growth of mentioning some events on social media, but we can only face a drift of some particles (posts or tweets) and analyzing them allows us to reconstruct an appropriate

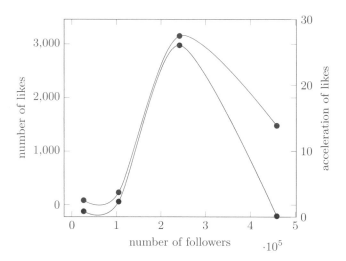

Fig. 2. The blue curve shows the relation of the number of likes to the number of followers, and the red curve shows the relation of the acceleration of likes to the number of followers. The pattern of the red curve in this chart is identical with the *viscous drift* of microparticles in the acoustic field.

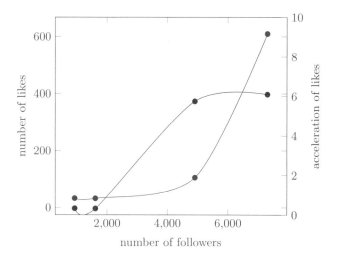

Fig. 3. The blue curve shows the relation of the number of likes to the number of followers, and the red curve shows the relation of the acceleration of likes to the number of followers. The pattern of the red curve in this chart is identical with the *radiative drift* of microparticles in the acoustic field.

information wave. The most important observation of our study is the demonstration that the patterns of drifting microparticles are invariant for different scales – these patterns are repeated even for tweets in a "virtual" dimension.

References

1. Cecinati, F., Matthews, T., Natarajan, S., McCullen, N., Coley, D.: Mining social media to identify heat waves. Int. J. Environ. Res. Public Health **16**(5), 762 (2019)
2. Czyż, H.: The aerosol particle drift in a standing wave field. Archs Acoustics **12**, 199–214 (1987)
3. Czyż, H.: On the concentration of aerosol particles by means of drift forces in a standing wave field. Acoustic **70**, 23–28 (1990)
4. Czyż, H.: Theory of the proper acoustic coagulation. In: Proceedings of the 17th International Congress on Acoustics, Rome, vol.1, pp. 35–36 (2001)
5. Czyż, H., Gudra, T., Opieliński, K.: Investigation and visualization of ultrasonic agglomeration of gas bubbles in liquid. Acta Acustica-Acustica **88**, 682–686 (2002)
6. Czyż, H.: Dispersed phase acoustics in liquid. Selected problems, (in Polish). Rzeszow University of Technology, Rzeszow (2004)
7. Czyż, H., Markowski, T.: Applications of dispersed phase acoustics, Archives of acoustics, vol. 31, No 4, pp. 59–64, Polish Academy of Sciences, Warszawa (2006)
8. Czyż, H., Gardzińska, A., Markowski, T.: Analysis of possibilities decreasing toxicity of the virus SARS-CoV-2 by acoustic methods. IOSR J. Appl. Phys. (IOSR-JAP) **13**(4), Ser. II, 36-40 (2021)
9. Dain, Y., Fichman, M., Gutfinger, C., Pnueli, D., Vainshtein, P.: Dynamics of suspended particles in a two-dimensional high-frequency sonic field. J. Aerosol Sci. **26**(4), 575–594 (1995)
10. Gradoselskaya, G., Shcheglova, T., Karpov, I.: Information waves on social networks: problematization, definition, distribution mechanisms. In: IEEE 2018 Eleventh International Conference Management of Large-Scale System Development (MLSD), pp. 1–4 (2018)
11. Gradoselskaya, G., Volgin, A.: Decomposition of a media event through the definition of information waves. In: 2019 Twelfth International Conference Management of Large-Scale System Development (MLSD), pp. 1–3 (2019)
12. Gradoselskaya, G., Shcheglova, T.: Theoretical foundation of information waves investigation in social networks. In: 2019 Twelfth International Conference Management of Large-Scale System Development (MLSD), pp. 1–3 (2019). https://doi.org/10.1109/MLSD.2019.8911027.
13. Pfeffer, J., Zorbach, T., Carley, K.M.: Understanding online firestorms: negative word-of-mouth dynamics in social media networks. J. Mark. Commun. **20**(1–2), 117–128 (2014)
14. Rani, N., Das, P., Bhardwaj, A.K.: Rumor, misinformation among web: a contemporary review of rumor detection techniques during different web waves. Concurrency Computat. Pract. Exper. **34**(1), e6479 (2022)
15. Vainshtein, P., Fichman, M., Shuster, K., Gutfinger, C.: The effect of centreline particle concentration in a wave tube. J. Fluid Mech. **306**, 31–42 (1996)

Turing Machines as Conscious Computing Machines

Jerzy Król$^{(\boxtimes)}$ and Andrew Schumann

Department of Cognitive Science and Mathematical Modelling,
University of Information Technology and Management,
ul. Sucharskiego 2, 35-225 Rzeszów, Poland
`jkrol@wsiz.edu.pl`

Abstract. We uncover certain universal features of Turing machines
(TM) as operating in a perpetually changing environment which can
have sudden and highly random influence on TMs themselves. TM adapts
and assimilates the changed environment and performs its computational
functioning in new conditions. We show that this transcends essentially
Turing computability relative to a ground model of ZFC. We distil the
formal counterparts responsible for the adaptation and assimilation of
TM and propose they may underlie the conscious behaviour of general systems including living creatures. However, in this last case more
work leading to the layer's structure of TMs is needed. We also make an
attempt toward social TMs by finding the way how TMs can group and
cooperate.

Keywords: Turing machines in models of ZFC · forcing · formal
aspects of consciousness

1 Introduction

A Turing machine is a formal concept explaining what any computability process looks like. The usual way of seeing computability as a formal process is
rooted in arithmetical constructions and rather lacks of the broader outer, i.e.
environmental, perspective. We want to fill the gap and extend formally defined
TM over external formal environment. The environment reacts on such TM in
a random way and this modifies TM. The implementation of the modification
will not destroy TM but rather enlarges its computational abilities. We base our
analysis on the well-known relation of the axiomatic Zermelo-Fraenkel set theory with the axiom of choice (ZFC) with its models (e.g. [1]). The dynamics of
models (forcing extensions) is the factor representing the reaction of TM on the
environmental stimuli. From the point of view of pure ZFC (lacking the perspective of models) the dynamics is out of the reach. It is also widely known fact that
despite the simplicity of TM, it is implemented in a way in any software of any
classical computer. Similarly the extension of TM over the formal environment
(FE) (represented by ZFC tools) we work out here, being direct and simplifying,
still it leads to universal features.

A. Rocha et al. (Eds.): WorldCIST 2023, LNNS 802, pp. 400–410, 2024.
https://doi.org/10.1007/978-3-031-45651-0_40

One important aspect of the approach is presented by the mechanism of assimilation of the FE reaction on TM such that this TM is changing from the state of completely random affecting the outer stimuli into the modified TM with the stimuli becoming its building blocks. The entire process resembles a way how a living organism is gaining its skills under the influence of the perpetually changing environment. The formal perspective proposed here seems limiting at first sight, but it might bear universal mathematical features so that they may underlie the conscious behaviour of creatures in the world. This is quite analogously to the universal TM which serves as universal computational machines in the class of all classical computers (TMs) (e.g. [2]) even though this reduction to the universal TM is not evident in each case.

2   Key Terminologies

Before we present the main construction, in this section we grasp together basic facts regarding TM as well as TM with oracles (o-TM), Turing uncomputable classes and algorithmic randomness, and set theoretic constructions like forcing. They will be needed in the following sections.

2.1   Classical Turing Machines with Oracles

Let Q be a finite set of possible (internal) states of TM, $Q = \{q_0, \ldots, q_n\}, n \geq 1$; let t be an infinite two-sided tape containing cells. Each cell has written in the symbol 1 or the symbol B, blank, $S = \{1, B\}$. Let h be a reading head which in each single step can read the content of a cell on the tape t and move to the right, if $X = R$, or to the left, when $X = L$. TM can also rewrite (change) the existed symbol $s \in S$ in the cell to a new symbol $s' \in S = \{1, B\}$.

Definition 1 ([2]). *TM comprises of the tape t, head h, and TM is in one of its internal states from S. The functioning of TM is described by the collection of steps governed by the collection of symbols (q, s, q', s', X), where $q, q' \in Q, s, s' \in S, X \in \{L, R\}$, such that TM in the state s reads the symbol s from the tape and changes its state to s' and writes down in the scanned cell a new symbol s' and moves its head to the right, if $X = R$, or to the left, if $X = L$, by a single cell in each step.*

Thus, the operation of any TM is governed by the *partial* function (not-everywhere defined, since TM may not produce any writing over a cell content and move of the head):

$$Q \times S \to Q \times S \times \{R, L\}.$$

The Turing program is a finite set of the values assigned to (q, s, q', s', X), i.e. it is a finite set of quintuples. The states S contains the leftmost state q_0 which is '1' in the cell (initial state) and the halting state (which, however, may not be attained). The detailed discussion of how TM computes, based on the above definition, can be found, e.g., in [2].

J. Król and A. Schumann

Definition 2. *TM with oracle $A \subset \mathbb{N}$, o-TM or TM^A, it is a TM with an additional infinite read-only tape A, on which there is written the characteristic function of A ($\{0,1\}$-binary sequence). The reading of the tape A is the part of the functioning of TM.*

Thus, the o-TM allows for performing computations on the data, which might not depend on any action of any TM. Given the o-TM and putting the set uncomputable but computably enumerable (c.e.) into the oracle, one gets entire hierarchy of uncomputable Turing classes. The example are $\emptyset$ sets expressing the halting problem of TM, and their n-th uncomputable jumps $\emptyset^{(n)}$ placing $\emptyset^{n-1}$ into the oracle. We refer the reader to the excellent exposition by Robert Soare [2] or in [3].

2.2 Arithmetic and Turing Classes

The concept of algorithmic randomness is based on the hierarchy $\Sigma_n^0, n = 0, 1, \ldots$ of complexity of arithmetic formulas (e.g. [2–4]). Higher arithmetic classes of formulas correspond to objects, which can be determined by a TM, however, with the increasing computational complexity. So to define a purely random binary infinite sequence $\sigma \in 2^\omega$, one requires that σ omits all or some classes. This is how the original Martin Löf (ML) test for randomness arose.

1. ML test: A sequence $\{A_n, n \in \mathbb{N}\}$ of uniformly computably enumerable (c.e.) (i.e. c.e. together with the set of its indices [3, p.11]) of Σ_1^0 classes (Σ_1^0 subsets of sequences from 2^ω) such that $\forall_{n \in \mathbb{N}} (\mu(A_n) < 2^{-n})$.
2. $A \subset 2^\omega$ is ML-null, when there exists a ML test $\{A_n, n \in \mathbb{N}\}$, such that $A \subseteq \bigcap_{n \in \mathbb{N}} A_n$.
3. $\sigma \in 2^\omega$ is ML-random, if $\{\sigma\}$ is not ML-null (for each ML test).
4. A ML test $\{A_n, n \in \mathbb{N}\}$ is *universal*, when $\bigcap_{n \in \mathbb{N}} B_n \subset \bigcap_{n \in \mathbb{N}} A_n$ for all ML-tests $\{B_n, n \in \mathbb{N}\}$.

Lemma 1. *There exists a universal ML test.*

ML-random sequence $\sigma \in 2^\omega$ is known to be 1-random. The direct modification to Σ_n^0 classes gives rise to the hierarchy of n-random sets, for all $n \geq 1$.

i. ML_n test: A sequence $\{A_k, \ k \in \mathbb{N}\}$ of uniformly c.e. of Σ_n^0 classes (Σ_n^0 subsets of sequences from 2^ω), such that $\forall_{k \in \mathbb{N}} (\mu(A_k) < 2^{-k})$.
ii. $A \subset 2^\omega$ is ML_n-null, when there exists a ML_n test $\{A_k, \ k \in \mathbb{N}\}$, such that $A \subseteq \bigcap_{k \in \mathbb{N}} A_k$.
iii. $\sigma \in 2^\omega$ is n-random if $\{\sigma\}$ is not ML_n-null (for each ML_n test).

Since the set of all subsets of natural numbers represents real numbers, the usual way how the sets of all reals are represented in models of set theory is 2^ω (or ω^ω), which is a Polish space [1,4]. That is why speaking about reals in models of ZFC is speaking about infinite binary sequences.

2.3   Forcing in Set Theory

Given B a complete Boolean algebra in a model M of ZFC, we have:

Lemma 2. *There exists a generic extension $M[r] \supsetneq M$ iff B is atomless in M.*

Definition 3. *The measure algebra (random algebra) is the Boolean algebra B which is the algebra of Borel subsets of $\mathbb{R}$ modulo the ideal of subsets of Lebesgue measure zero, $B = Bor(\mathbb{R})/\mathcal{N}$.*

Lemma 3. *The measure algebra B is the atomless complete Boolean algebra.*

It follows that there exist nontrivial random real numbers $r \in M[r] \neq M$ whenever B is the measure algebra in M.

3   Results

As we have already noticed, the oracle TM, $\mathrm{TM}^A, A \subset \mathbb{N}$, leads to the entire spectrum of Turing uncomputable classes. Starting with A as certain c.e. set, which is not Turing computable, TM^As compute the characteristic classes of other sets, belonging to the same Turing class as A itself. Then taking higher Turing classes as oracles, we repeat the computability by TMs with this oracle and so on. The idea behind finding counterparts of conscious behaviour of operating TMs is based on the following basic observations

A. Consciousness reflects self-orientation and self-understanding of a system as being in the random outer environment.
B. The environment acts on the system by random stimuli.
C. The stimuli can change the system and the changes are assimilated by it.
D. The system understands the changes and then the assimilated stimuli are no longer random or alien.
E. The effect of the stimuli on the system can be more focused and then it means a stress or it can be less focused and then it means a satisfaction. Stress and satisfaction are understood as two basic emotions. They can change the system and the changes are assimilated by it, too. They are no longer random, as well.

We try to find canonical and formal counterparts of the above points in the realm of calculational processes within TMs. First, we need to understand what can be taken as outer environment for any TM. The point is that TM is an arithmetical concept, but when for higher Turing classes TM extends Peano arithmetic (PA) in a sense, o-TMs produce also independent of PA axioms functions. That is why we propose to consider any o-TM as naturally embedded in the axiomatic ZFC theory. However, to reflect randomness of the outer stimuli coming from ZFC we propose to base our consideration on the Martin Löf notion of randomness, or more precisely, on a weaker its form, i.e. Solovay generic randomness of infinite binary $\{0, 1\}$ sequences, e.g. [3]. This Solovay randomness is a 'miniaturisation' to arithmetic of the broader concept of randomness genericity in ZFC [3]. This extended to ZFC notion of randomness has been introduced also by Solovay and is known as a forcing in set theory, which we have briefly discussed in the previous section. This last we call a ZFC-randomness and it is the proper notion for our external to the TM environment. Thus concluding, we are choosing ZFC axiomatic set theory as a formal environment for o-TMs. But this is merely the first approximation since PA independent statements can be also ZFC independent: ZFC does not prove or disprove them but they and their negations are rather consistent with ZFC. This last statement means that there is a model of ZFC where p is true and the other model where $\neg p$ is true, and both models have all provable in ZFC propositions as their true statements. That is why the method of forcing is especially well-suited for such situations. But if so, we should extend the ZFC axiomatic environment over models of ZFC, where their difference is the valid ingredient of the approach. This is precisely what we are doing when searching for the proper formal external to the TM environment.

Definition 4. *o-TM interacting with the external ZFC environment, $(o\text{-}TM)_M$, is the ordinary o-TM defined in a standard transitive model M of ZFC.*

Remark 1. Since ZFC interprets PA, so the constructions of TM are expressible in models of ZFC. The model M above is not specified at this place; one can take as M some countable transitive model (CTM) or V – the entire universe of sets, or some internal model, or others. We will discuss briefly the distinctions between the choices in what follows.

Remark 2. Given a ZFC model M, it is generally possible to add new real numbers to it. In the case of a CTM M one can add even a continuum of many different reals by nontrivial forcings from the outside of M. This generally follows from the relation between reals in V (let it be $\mathbb{R}$) and the reals in M: $R_M \subset \mathbb{R}$ and $|R_M| = \omega$ in V.

Remark 3 below explains the definition which follows.

Remark 3. TMs in different standard transitive models of ZFC with the standard natural numbers object are equivalent in the sense that PA + ZFC are equivalent in the models. The nonequivalent inputs, which extend the models and Turing computability or ZFC, can appear in oracles.

Definition 5. *The external ZFC environment interacts with $(o\text{-}TM)_M$ by adding reals into the oracles or by non-generic oracles.*

Let N be the universe of sets (e.g. CTM) for $(o\text{-}TM)_N$ and M for $(o\text{-}TM)_M$. If N is a ground model for M, i.e. $N[s] = M$ for generic s, we say N is the shrinked version of M due to the stimuli $r_s \in M \subset M[r]$. Similarly, $M[p]$ is the extended version of M due to the stimuli p generic for M. Then

Definition 6. *r_s is a stress stimuli for $(o\text{-}TM)_M$ and p is the satisfaction stimuli for $(o\text{-}TM)_M$. The resulting states of $(o\text{-}TM)_M$, i.e. N and $M[r]$ are called stress and satisfaction states respectively. The M resulting in M (without nontrivial changes) means a neutral state, see [13].*

Remark 4. Note that given two different $(o\text{-}TM)_{M_1} = TM_1$ and $(o\text{-}TM)_{M_2} = TM_2$, their 'social relations' can be also given in terms of the interactions of the external environment, since the part of this is each TM with respect to the other. In particular TMs could react on the emotions each to the other by oracles.

The reaction of $(o\text{-}TM)_M$ can be neutral (no reaction) or active, i.e. the oracle A_M becomes extended by reals in the extended by forcing model $M[r]$, i.e. $A_{M[r]}$. Thus, $A_{M[r]}$ contains generic reals.

Now we can confront the oracle TM, interacting with the external domain with the conditions A. – E. from the beginning of this section.

Theorem 1. *Let $M \to M[r]$ be the random forcing, adding the real r to M. There is a canonical formal way in which $(o\text{-}TM)_M$ fulfils conditions A. – E.*

Proof. Regarding A. that $(o\text{-}TM)_M$ reflects 'self-orientation and self-understanding of itself as being in the random outer environment'. This is in terms of Turing machines augmented by the external interaction with the environment as in Definition 5. 'Understanding' by TM is due to the ZFC, realised in the model M, where there are in use by TM internal to M real numbers, R_M. So the space of states of this TM includes also R_M. A real $r \in M[r]$ is not in M, but it will be inserted into the oracle. It is random for M by the forcing and since it is not predictable by M itself (i.e. which random real it will be). At this stage, M thinks there are *all* real numbers in M (according to understanding given by ZFC). After r is included into the oracle, TM assimilates it and changes its state to $M[r]$ so thus TM now considers r as valid real number since $R_M \subsetneq R_{M[r]} \subset \mathbb{R}$. Regarding self-orientation, this is also connected with assimilating external environmental reals as parameterising the external space. We will explain it in the Example below.

Regarding B., this is precisely stated in Definition 5.

Regarding C., the assimilation property has been already explained above as adding random r to M. The change of the state follows as $M \to M[r]$ and the state of TM after assimilation is $M[r]$.

Regarding D. that TM 'understands the changes and then the assimilated stimuli are no longer random or alien', it has been already indicated at A. above, where understanding has been given by the process (following the change of the state of TM) from ZFC_M to $ZFC_{M[r]}$. When M is in the state $M[r]$, r is no longer random (still there can be new random reals $r' \in \mathbb{R}$ to $M[r]$, $r' \notin M[r]$ and $r' \in M[r][r']$).

Regarding E., the stimuli of 'moderate focus' (a satisfaction) effects the random forcing extension $M \to M[r]$ while this of 'high focus' (a stress) results in taking a ground model N for M, i.e. $N[r] = M$, and the stimuli is not absorbed by N. The satisfaction is connected with the expansion and extension of the model while stress with the shrinking of it (see the discussion about the multiverse in the end of this section).

One could wonder whether the external stimuli which can be of arbitrary high degree of randomness can be assimilated by our TM. Let $\sigma \in 2^\omega$ be an arbitrary (of arbitrary high degree of algorithmic ML randomness) subset of $\mathbb{N}$ in V.

Proposition 1. *For any σ as above, there exists $(o\text{-}TM)_M$ with the M-random $r \in M[r]$ in the oracle which can reflect the degree of randomness of σ.*

Proof. This is based on fundamental facts from algorithmic randomness. First, randomness in arithmetic is the instance of ZFC Solovay forcing when 'miniaturised' to PA [3]. It means that we make the forcing procedure in PA theory without bothering of ZFC properties of the sequences $\sigma \in 2^\omega$. Any ML 1-random binary sequence σ_1 omits the Σ_1^0 subsets of 2^ω of arbitrary small Lebesgue measure (ML test). From the other side, given the random real r with respect to ZFC model M, r omits all measure zero subsets of $(2^\omega)_M$ which means that such r is also arithmetically 1-random with respect to M. Thus, knowing r be generic random in M, it is 1-random with respect to the pair $(M, M[r])$ ($r \notin M$). Given higher $n > 1$ ML random σ_n, it omits every subset of 2^ω of arbitrary small measure and thus a M-random r omits every measure zero subset of 2^ω coded in M. This last certainly omits every n-arithmetic subset of reals with zero measure in M; thus, such a sequence is n ML random with respect to the pair $(M, M[r])$. So, Solovay generic r can indeed reflect in the pair $(M, M[r])$ the arbitrary high degree of randomness. ∎

Remark 5. Assuming that M be the so-called ω model of ZFC, i.e. a transitive standard one with the standard natural numbers object $\mathbb{N}$, one obtains the minimal ZFC driven discrepancies between TM in M and in V. The discrepancies can be valuable by themselves, however, we do not delve in it here.

Remark 6. The above proposition works as far as there exists the generic filter of the Boolean algebra B in M. Otherwise $r \in M$ and it can not omit all measure zero subsets of 2^ω. It is known that this is always the case (generic r exists) for countable transitive models M. However, in the universe of sets V there does not exist any generic ultrafilter, hence random r as well. The usual solution is to build the Boolean model V^B in V, with the canonical embeddings $V \subset V^B \subset V$, and prove in V^B that with the value 1 there exists random real r (hence, a generic ultrafilter). This r again omits *all* measure-0 subsets of $\mathbb{R}$ in V^B with the value 1. Thus, we can assume that a random real r exists in M and V and it is n-random for $n \in \mathbb{N}$. In the multiverse approach – which will be discussed below – there is the family of models closed on the extensions and taking ground models so thus generic random rs always exist for the models.

The importance of the Proposition 1 is that $(o\text{-}TM)_M$ can assimilate arbitrary random incomes which appear in the oracle by the response to the external stimuli from V. After the assimilation the final state of TM is externally modified such that it is TM in $M[r]$ and this r is not any longer random in $M[r]$. The process how TM undergoes the changes and perceives them (refers to) from the new state is very important and requires a deeper clarification. Let us augment the Definitions 2 and 4 as

> *The state space Q for $(o\text{-}TM)_M$ contains the symbols for the forcing extensions of M, i.e. $|1|$ for $M[r]$ and $|0|$ for M. Whenever any change of*

M does not occur (trivial forcing, no external stimuli), the state of TM remains unaltered, i.e. M, and the state of TM is $|0|$. Thus, if the non-trivial forcing adding r into the oracle took place the state is recognised as $|1|$. The $|0|$ state is assigned also to the shrinking model M to N, since no generic r is added.

*Example 1 (*The modification of TM*).* Let M be a countable transitive standard model of ZFC with the standard natural numbers object. All ZFC provable statements holds true in M, but also Peano arithmetic is derivable from ZFC, so that in M there holds true the ZFC arithmetic statements. Let the external to the TM 3-dimensional spatial domain be parameterised by reals $\mathbb{R} \in V$, so the spatial external domains U are (open or not) subsets of $\mathbb{R}^3$. $\mathbb{R}$ contains both, reals R_M from a general ground model M and reals which are not in M. Among them there are generic reals with respect to all possible random forcings over M (and for other forcings of course) and new reals which are not generic and are not in M. The spatial domain, where internal TM acts (from the point of view of TM), is parameterised by R_M^3. From the outside (from the V point of view), R_M is countable, though from the M point of view, M contains *all* reals. However, the possible generic reals (with respect to various random forcing extensions) is continuum many from the external point of view, so the probability to find a generic real in $\mathbb{R}$ is much higher than for nongeneric. Let the external stimuli be generated in V and represented by some generic to the M real r. The interaction of TM in M with this stimuli leads to the overwriting on the oracle tape the binary representation of r. This r is not in M, however, the TM state is fixed to $|1|$ and TM after the entire process is internal to the forcing extension $M[r]$. At this final stage, r is no longer random in $M[r]$. The assimilation of the external random stimuli is completed. By the same process the spatial orientation can now be gained by identifying the stimuli r with the point corresponding to the external parameterisation.

This is a quite nontrivial task to decide for TM whether its actual state is $|0|$ or $|1|$. The reason is that ZFC and PA are theories in the first-order languages and as so their provability power does not allow for 'seeing' the set models of the theories (otherwise they would prove their consistency). Moreover for a CTM $M[r]$ this is always the ground model for a subsequent random forcing leading to $M[r][s]$ and so on. Still, we can assume that the interaction with the external environment gives the information about the state, e.g. a random r for M being assimilated by the oracle of TM loses its randomness and indicates the state of TM is now $|1|$. Another possibility is to refer to general results concerning a definability of the ground model M in the extension $M[r]$ (e.g. [5]). We do not elaborate on this important issue here but rather it will be addressed elsewhere. Let us resume this as: $(o\text{-TM})_M$ is in the $|1|$ state means that the oracle has been just added (in the last step) as a random real r coming from the external stimuli and there exists a model of ZFC N such that $N[r] = M$. $(o\text{-TM})_M$ is in the $|0|$ state meaning that in the last step there is no random r extending the oracle.

Given introduced TM as carrying some basic features of conscious-like behaviour, we would like to see this phenomenon more broadly. Especially, are there certain formal counterparts, already at this very basic level, which would indicate group- or 'social-like' activities of several such defined TMs? Again, guiding principles come from studying models of ZFC in this context.

As follows from the discussion above, the internal $(o\text{-TM})_M$ to M carries among its states the information about actual random forcing extensions $M \to M[r]$. However, for CTMs 'to be extended by a random forcing' is generic, i.e. it is always possible to make yet another such extension starting from $M[r]$ and this seems to be a fundamental feature of TMs. This phenomenon is deeply rooted in the foundations of set theory. One approach to set theory is based on a distinguished universe of sets, like V, which is the class containing all sets, the other approach is a set theory without the specific choice of the basic universe of sets. The first will be marked as U and the second as MV – multiverse, in what follows. Given a CTM model M, its multiverse is the family of models containing M closed with respect to taking all forcing extensions and all ground models of its members. The concept has appeared as very fruitful (e.g. [6]) and it has been shown leads to different truths values for set theory statements which can be proved in V and in all models in MV (e.g. the continuum hypothesis is true in the generalized to inner models MV [7]). The point is that the MV approach is based on the scattered truth concept depending on models of ZFC. The U approach is based on the centralised truth with respect to the distinguished universum of sets. The former one is more close to the decentralised nets point of view (scattered notion of truth) and this is attractive also for TMs interacting with the external environment and interacting with other TMs in the net.

In the context of $(o\text{-TM})_M$ let M be a CTM of ZFC and this TM changes the models along with the external random stimuli and the extension adapts this TM to the new random condition. The state space for such TM contains the positioning of TM in the *random* MV, rMV, for M. In fact this positioning is merely local, i.e. the actual pairs $(N, N[r])$ enter the game for the current state of TM. We do not require that TM is capable for any identification of models N in the entire structure rMV. Still rMV represents the space of possible paths for $(o\text{-TM})_M$ when it interacts with the environment. From the perspective of such TM it does not know about possible external V, rather rMV creates the entire 'universe' of sets. Quite similar as MV replaces a single universe of sets. Taking rMV and V simultaneously and allowing for the interactions is the place where conscious phenomena can enter the stage in the model presented here.

Now given several $(o\text{-TM})_{M_i}, i = 1, 2 \ldots$ and taking two of them, it can happen that their momentary models coincide, $M_i = M_k$ or not. More generally there can exist (or can not) a model M_{ij} containing both M_i, M_j as submodels. In general zig-zag moves (taking extensions or grounds) within the structure rMV leads to building the entire net of connections between states of M_i and M_j. Another factor in creating nets is the non amalgamation property, i.e. two forcing extensions of a CTM model are not submodels of a common model of the same hight (e.g. [8]).

The power of oracle TM computability can be also directly seen in the case of forcing [9]. It has been proved that for the oracle, Gr_0, which would be the elementary graph of M (the set of true ZF statements in M) plus the forcing partial order $\mathbb{P}$ in M, TM^{Gr_0} computes the forcing extension $M[r]$. Which indeed means that building the connections between TMs by forcing, inherently relies on the oracle Turing computability. More precise understanding of the impact of the above formal elements on true functioning of TM or nets of TMs requires much further studies also conceptual in the foundations of science.

4 Discussion

We have introduced the Turing machine interacting with the external environment, and shown formal counterparts which could be related to, if not underlie, the conscious behaviour of the real systems in our world. 'Exterior to TM' means not only as situated in different spatial regions, but also separated by different mathematics. We have shown that when TM lives in the set-theoretic world, based on the multiverse paradigm, and it is confronted with the external environment organised by the single set theoretic universe V, then the contact region of both may be the carrier of certain phenomena, allowing for developing conscious relations to the world. We think, though did not present a full justification for it here, that the structure is universal (rooted in foundations of mathematics) also for systems in real world showing conscious reactions.

The situation where one pays a bigger attention to forcing relations than to sets themselves, resembles to some degree the replacing objects by arrows – morphisms in category theory (cf. [10,12]). This kind of thinking with the priority of 'forcings over sets' became fruitful also in the context of certain fundamental problems in physics (e.g. [4,11]). This certainly requires more thorough studies and effort and partial results will be a topic for our forthcoming publication.

Also, the case of living conscious organisms could be approached from the proposed here perspective, even though it looks very simplified at first sight. One option is to introduce the structure of interacting layers. An extension of the formalism over emotions or various social phenomena is the matter of further work. Anyway one can note that the 'random' nature of some social phenomena can be embedded in the system presented here where randomness is coded formally. This is in the sharp opposition to certain previous misconceptions, like considering consciousness as a formal computational model of self-reference and claiming that formal methods would not allow the embedding.

References

1. Jech, T.: Set Theory. Springer Verlag, Berlin (2003, Third millennium edition)
2. Soare, R.I.: Turing Computability: Theory and Applications. Springer, Berlin Heidelberg (2016)
3. Downey, R.G., Hirschfeldt, D.R.: Algorithmic Randomness and Complexity. Springer, Berlin (2010)

4. Król, J., Bielas, K., Asselmeyer-Maluga, T.: Random world and quantum mechanics. Found. Sci. **28**, 575–625 (2022)

5. Laver, R.: Certain very large cardinals are not created in small forcing extensions. Ann. Pure Appl. Logic **149**, 1–6 (2007)

6. Hamkins, J.: The set-theoretic multiverse. Rev. Symbolic Logic **5**(3), 416–449 (2012)

7. Woodin, W.: The continuum hypothesis, the generic-multiverse of sets, and the Ω conjecture. In: Kennedy, J., Kossak, R. (eds.) Set Theory, Arithmetic, and Foundations of Mathematics: Theorems, Philosophies (Lecture Notes in Logic, pp. 13-42). Cambridge University Press, Cambridge (2011)

8. Habič, M.E., Hamkins, J.D., Klausner, L.D., et al.: Set-theoretic blockchains. Arch. Math. Logic **58**, 965–997 (2019)

9. Hamkins, J.D., Miller, R., Williams, K.J.: Forcing as a computational process. Mathematics ArXiv (2020). Under review. [math.LO] arXiv:2007.00418

10. Awodey, S., Heller, M.: The humunculus brain and categorical logic. Phil. Probl. Sci. **69**, 253–280 (2020)

11. Król, J., Asselmeyer-Maluga, T.: Quantum mechanics, formalization and the cosmological constant problem. Found. Sci. **25**, 879–904 (2020)

12. Król, J., Schumann, A., Bielas, K.: Brain and its universal logical model of multi-agent biological systems. Log. Univers. **16**, 671–687 (2022)

13. Schumann, A.: Behaviourism in Studying Swarms: Logical Models of Sensing and Motoring. Springer, Cham (2019)

A Formal Approach to Model Natural Phe-nomena

Maria Teresa Signes-Pont$^{(\boxtimes)}$ (iD), Joan Boters-Pitarch, José Juan Cortés-Plana, and Higinio Mora-Mora

University of Alicante, Ctra SanVicente del Raspeig s/n, 03690 Alicante, Spain
{teresa,hmora}@dtic.ua.es, {joan.boters,jj.cortes}@ua.es

Abstract. This paper presents a formal approach to the modeling of natural phenomena regardless of the mathematical, statistical or computational methods used. This framework makes it possible to face the initial difficulty of lack of knowledge in the face of little-studied or complex behavior phenomena. For this, a modular approach is carried out that guides from the external observation until reaching the level of the variables that have to appear in the future mathematical-computational treatment. This approach has been successfully tested in our previous research and provides a useful tool due to its modularity to face complex phenomena modelling.

Keywords: Natural Phenomena Modelling · Plant Pest Propagation · Epidemiology

1 Introduction

Nature offers a great variety of items that have different evolutions in time and space. The growth of malignant tumors, the expansion of infectious diseases, the spread of plant pests or the environmental changes may be some examples of the complex behavior of natural phenomena. More, unnatural events, such as the growth of the cities, the fluctuations in the world economy or the propagation of viruses in mobile computer devices can take advantage of being considered under the scope of the existing natural phenomena as far as they can be modelled by similar computational models. To deal with prevention and/or cure policies for the events, we need reliable and not too expensive tools that can account for the problems they address in a viable manner. Computational modelling joint to the most cutting-edge technologies are today the most appropriate way to produce optimal results in these cases. Recently, important research has been carried out in a broad range of works that can confirm this statement. In [1] and [2], the tumor modelling problem is approached. In [1], the authors develop a simple mathematical model to simulate the growth of tumor volume and its response to a single fraction of high dose irradiation. Reference [2] deals with the problem of uncertainty over how to best model tumor growth. The authors conclude there is a need for careful consideration of model assumptions when developing mathematical models for use in cancer treatment planning. References [3–5] present models of the spread of infectious diseases. Reference [3] presents an improvement to the standard compartmental SIR model of the Covid-19. The proposal consists on substituting the standard model by a module resulting from Padé approximation in the Laplace domain. This improvement allows developing software tools for practicing epidemiologists, and related educational resources. Reference [4] studies the SI, SIR and SIS models of disease expansion through the time-aware influence maximization problem, since there is a close similarity between the infectious disease spread and information spread in social networks. Reference [5] provides insights to the case of a significant population having asymptomatic, untested infection, because in this case model predictions are often not compatible with data reported only for the cases confirmed by laboratory tests. The authors derive from the first principle an epidemiological model with delay between the newly infected (N) and recovered (R) populations. The model also incorporates effects of social behavior, since for a highly contagious and deadly disease, herd immunity is not a feasible goal without human intervention or vaccines. In what refers to the spread of plant pests, reference [6] presents a discrete compartmental susceptible-asymptomatic-infected-dead (SAID) model to address the expansion of plant pests. The authors examine the case of *Xylella fastidiosa* in almond trees in the province of Alicante (Spain) to define the best eradication/contention protocol depending on the environmental parameters such as climatic factors, distance between trees, isolation of the plots, etc. This proposal shows how the grid architecture, along with an update rule and a neighborhood pattern, is a valuable tool to model the pest expansion. This theoretical model has also been successfully tested in the case of the dissemination of information through mobile social networks [7] and is also currently under study in the case of expansion of COVID-19. Reference [8] offers a brief overview of the current state of development in coupling pest and disease models to crop models, and discusses technical and scientific challenges. Based on the results of many years of research, our work aims to provide a formal approach of the modelling of natural event regardless of the mathematical or computational methods employed. That is to say, our goal is to provide a method of the modelling process, starting on the observation of the causes that may impact on the evolution of the parts of a phenomenon until defining the structure and parameters of the phenomenon. This approach is useful to face the initial difficulty of lack of knowledge in the face of little-studied or complex behavior phenomena and makes easier the posterior modelling process by highlighting the main parameters that intervene as well as their relationship. Following the Introduction, Sect. 2 presents methodological considerations and details the two main steps of the process. Section 3 presents the spread of a plant pest as a case study. In this section we apply our methodology of horizontal and vertical classification that leads to the identification of the parameters of the process and their relationship. Section 4 summarizes, presents conclusive remarks and proposes some future improvements.

2   Methodological Considerations: A Two-Dimensional Approach

The initial considerations are based on both an observation of the event under study and an analysis of the previous work carried out by other authors. This leads to designing the structure of the model. The parts of the structure are arranged and linked together using operators. These parts are represented by parameters.

2.1 First Stage: Characterization

Characterization is the initial process we carry out in order to store the result of the observations. It deals with the evolution of the phenomenon under the impact of several causes $c_{k,i}$, see Fig. 1. Since the $c_{k,i}$ intervene in different parts (or steps) of the unknown structure of the phenomenon, they can be used to define the p_k parts of the phenomenon. So, any part of the phenomenon is the result of the action of a particular set of causes, as shown in (1). This is called horizontal classification.

$$\forall k, p_k = \cup_{i=0}^{i=n} c_{k,i} \tag{1}$$

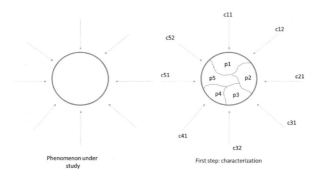

Fig. 1. Evolution of the phenomenon under the impact of several causes $c_{k,i}$,

2.2 Second Stage: Identification and Organization

The second stage deals with the structure of the phenomenon. This stage is achieved through the identification and classification of the parameters $v_{k,i}$ that define the p_k parts, see (2). In Fig. 2, we have represented the p_k parts and their overlaps caused by the dependencies of some of the parameters of different parts. The overlaps can be expanded vertically breaking down the horizontal classification, this is called vertical classification. Figure 3 shows a topographic representation of the p_k parts and we can observe some parts sharing parameters with other parts or not, see (3). Following the definition, some parameters of p2 (blue) depend on parameters p1 (green) and for this reason, p1 overlaps p2. Some parameters of p3 (pink) also depend on parameters of p2 (blue) so, p2 overlaps p3. In contrast, p5 (grey) has no overlaps so its parameters do not depend on other parameters elsewhere.

$$\forall k, p_k = p_k(v_{k,1}, v_{k,2} \ldots, v_{k,m}) \tag{2}$$

Figure 4 summarizes the proposed methodology. The scheme shows the two stages sequence of the process, from external observations (causes) to internal formalization (variables). The phenomenon is produced by causes that delimited different parts in it. The parts are defined by parameters that are depicted by variables that may or may not be independent of each other and therefore they define dependency between the variables. This two-dimensional approach (horizontal + vertical classification) provides a simple, modular and scalable framework which is useful to approach unknown phenomena that have no easy formalization.

$$\overset{k=n}{\underset{k=0}{\cap}} p_k = \emptyset \rightarrow \text{ no dependency exits between the parts } p_k$$

$$\overset{k=n}{\underset{k=0}{\cap}} p_k = \emptyset \rightarrow \text{ at least one part } p_k \text{ shares parameters with another one}$$

(3)

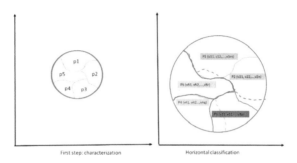

Fig. 2. Horizontal classification of the different parts.

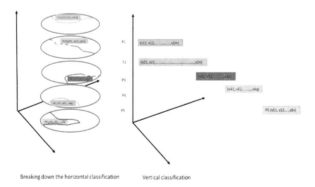

Fig. 3. Vertical classification of parameters.

3   A Case Study: The Expansion of a Plant Pest

As an application of the modelling methodology, we present the case of the spread of a plant pest, such as *Xylella fastidiosa* (Xf) that is transmitted by sap-sucking insects [6, 9]. This bacterium fatally attacks about 500 different trees and causes serious economic losses. Our initial observations as well as our knowledge acquired from the analysis of specialized publications [10, 12], attendance at specialized conferences [13, 19] and

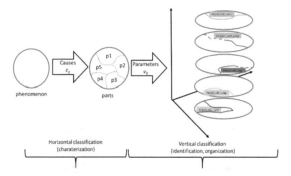

Fig. 4. Scheme of the methodology.

conversations with farmers lead to a list of environmental causes that may have an impact on the spread of the pest, apart from the tree and the bacterium. These causes are, a-the distance between trees, b-the isolation of the plots, c-the varietal susceptibility of the trees, d-the presence of vectors (insects), e-the tree care level and the f-meteorological factors (wind, temperature, humidity). Since the studied pest has no cure, it is straightforward to consider the main moments (parts) of the spread are the delays between the different states of a tree, from when it is healthy to when it dies. We set these delays as follows: ΔSE is the delay between when a tree is healthy (Susceptible) and becomes infected without having the ability to infect (Exposed), ΔEI is the delay between when a tree is Exposed and becomes Infected and finally, ΔID is the last stage of the life of a tree, from Infected and until it dies (Dead). In order to define the overlaps of the parameters of the parts, we have proposed a classification by levels (low, medium, high) of the causes, based on a qualitative estimate of the impact they may have on the spread, depending on the moment they occur. Table 1 summarizes the impact of the different causes on the delays of the spread. We justify them through the following considerations. Temperature has a high impact on proliferation of Xf population. It is rare between -1.1 and 1.7 °C. It is occasional between 1.7 and 4.2 °C and severe for temperatures higher than 4.2 °C. Nevertheless, for high temperatures proliferation is rare again as for very low temperatures. Mediterranean Basin is then suitable for the propagation of Xf [8]. The care the trees receive from the farmers makes them relatively resistant to infection. The varietal susceptibility is a crucial issue to avoid trees infection and to decrease death rate if infection occurs. Sloping terrain can be an inconvenience for the propagation of Xf, except in the case that the wind blows in favor and helps the insect to reach the tree. The distance between the trees does not represent a crucial variable since this distance does not have an extended range of variation. As an example, almond trees are planted 6 m from each other. The presence of insects (vectors) is crucial to the expansion of the pest. When considering the delays, we observe ΔSE has a strong dependence on varietal susceptibility of the tree, on presence of insects, on isolation of plots and on meteorological factors. For ΔEI the strong dependences are only on varietal susceptibility, presence of insects and meteorological factors, because they are the only ones that manage to slow down or prevent infection. Finally, for ΔID the dependences are always low or medium, since an infected tree will end up dying with an almost total probability

regardless of any consideration. Only the varietal susceptibility together with the care
level can have a greater ability to delay death.

Table 1. A qualitative estimation of the impact of the causes on the delays of the spread.

Causes	Δ_{SE}	Δ_{EI}	Δ_{ID}
1- Distance between the trees	medium	low	low
2- Isolation of the plots	**high**	medium	low
3- Varietal susceptibility of the trees	**high**	**high**	**high**
4- Presence of vectors (insects)	**high**	**high**	low
5- Tree care level	medium	medium	**high**
6- Meteorological factors (wind, temperature, humidity)	**high**	**high**	medium

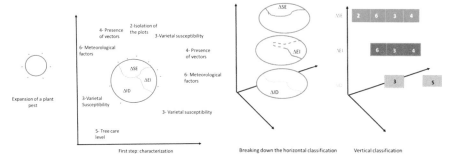

Fig. 5. Horizontal classification of the causes   **Fig. 6.** Vertical classification of the
of the spread   parameters of the spread

Figure 5 shows the horizontal classification of the causes of the spread, taking only
into account those that have a high impact on the different parts of the spread. The
overlapping of some parameters of the causes leads to the vertical classification shown
in Fig. 6. From the two-dimensional approach, we can highlight: ΔSE and ΔEI share
causes 3, 4 and 6, ΔID shares causes 3 with ΔSE and ΔEI and Causes 2 and 5 are not
shared. This means there are different relationship between the parameters that define
these causes. When the causes are shared, the parameters are related and when the causes
are not shared, the parameters are unrelated. As follows we apply the methodology to
concretely study both the impact of meteorological factors and the impact of vectors on
the spread of Xf.

3.1   Impact of the Meteorological Factors on the Spread of Xf

As follows we consider cause n° 6 (meteorological factors) shared by ΔSE and ΔEI.
The wind is defined by its direction $\theta \in [0, 2\pi]$ and its intensity $\rho \in [0, 1]$. Wind action
spreads the infection from an infected tree to neighboring trees (by enhancing vectors

flight). We place the trees in the cells of a square grid (NxN cells), and r_{ij}^k is the infection probability of a cell (i, j) at instant k ($0 \leq r_{ij}^k \leq 1$). So, at instant k we have a matrix R^k that defines the probabilities of all the cells to get infected at instant k + 1. A cell can model a single tree or a set of trees, depending on our initial assumptions. An update criterion is also needed to determine the state of the cell. The values of the cells at instant k are stored in a matrix M^k, so m_{ij}^k is the value of the cell (i, j) at instant k (value 0 = susceptible, value 1 = infected, value 2 = dead). Regarding wind power ρ and its ability to enhance the spread of infection, a partition Π of the interval [0, 1[around an infected cell is defined. The first space Π_1 indicates that the wind has no effect, thus the probability π_0 of spread is the same for all surrounding cells of the infected cell. When ρ increases, the infection reaches the space $\Pi_2,..., \Pi_n$ around the cell and the mathematical representation covers two, three, ..., n rows and columns respectively in the same direction θ of the wind. This assumption builds what we call an enlargement process. It will be carried out through a recursive formula of the evolution matrix. Note that if the angle $\theta=\xi$ is not in the first quadrant it will be sufficient to apply symmetries. The value of $f(\xi) = \tan(\xi)$ or $\cot(\xi)$ depends on the value of ξ ($0 \leq \xi < \pi/4$ or $\pi/4 \leq \xi < \pi/2$, respectively). Finally, we will aggregate all the results for each infected cell in the previous instant, as shown in Fig. 7. As assumed, the parameters that depict the wind are (ρ, θ). The value of C (loss factor for distant cells in partition Π) is determined by the values of temperature T and humidity H. When T increases, C also increases and when H increases, C decreases. This means that the elevation of the temperature together with low levels of humidity can curb the spread of the pest. So, partition $\Pi=\Pi$ (C, π_0) and loss factor C = C (T, H) are the main parameters that depict the impact of the wind. Summary in Table 2 shows the parameters that depict the meteorological factors and their level of dependency. From Table 2, it appears that the time that elapses between when the tree is healthy until it is exposed, Δ_{SE}, depends on the primitive parameters θ, T, H, π_0. Parameters T and H determine the loss factor C, and π_0 and C determine the partition Π. Finally, Π determines the extent of the spread ρ. A similar analysis may be done for Δ_{EI}.

Table 2. Parameter classification in the case of the meteorological factors

Cause n° 6: Meteorological factors (wind, temperature, humidity)	Δ_{SE}	Δ_{EI}	Δ_{ID}	level of dependency
spread = s (θ, ρ)	θ, ρ	θ, ρ		0
spread = s(θ, $\rho(\Pi)$)	θ, Π	θ, Π		1
spread = s(θ, $\rho(C, \Pi_0)$)	θ, C, π_0	θ, C		2
spread = s(θ, $\rho(C (T, H, \pi_0))$)	θ,, T, H, π_0	θ, T, H	none	3

3.2 Impact of the Presence of Vectors on the Spread of Xf

We now consider cause n° 4 (presence of vectors) shared by ΔSE and ΔEI, and study its impact on the spread of Xf. The presence of vectors is measured and controlled indirectly

$$A_1 = \begin{pmatrix} \sin\xi & f(\xi) \\ 0 & \cos\xi \end{pmatrix}$$

$$\vdots$$

$$A_n = \begin{pmatrix} a_{00} & a_{01} & \cdots & a_{0n} \\ a_{10} & a_{11} & \cdots & a_{1n} \\ \vdots & & \ddots & \vdots \\ a_{n0} & a_{n1} & \cdots & a_{nn} \end{pmatrix}$$

$$A_{n+1} = \begin{pmatrix} \frac{a_{00}}{C} & \cdots & \frac{a_{0n-2}}{C} & \frac{a_{0n-2}+a_{0n-1}}{2C} & \frac{a_{0n-1}}{C} \\ & & & & \frac{a_{0n-1}+a_{1n-1}}{2C} \\ & A_n & & & \frac{a_{1n-1}}{C} \\ & & & & \vdots \\ & & & & \frac{a_{n-1n-1}}{C} \end{pmatrix}, \quad C \in \mathbb{R}$$

Fig. 7. Matrix of the enlarged recursive process

through the evolution of the plant mass μ, since it is the main parameter that allows the nymphs of the insect to develop [15]. We use the available data from the ESA (European Space Agency) Copernicus project that provides images to be used to distinguish between different types of understory plants τ, (it is quite possible to discriminate plant mass from the vegetation of the soil of the orchards and the woody areas), as well as data on numerous plant indicators, such as leaf area index, α, leaf chlorophyll content, κ, and plant water content, ω, all of which are essential to accurately monitor plant mass, μ. In Table 3 we classify these parameters following their dependencies. The spread depends on the presence of vectors that is indirectly measured by the plant mass μ. The plant mass depends on the leaf area, α, which increases when the conditions are favorable, that is to say, the water content ω is appropriate and the sunlight λ is enough. We assume the type of plant τ is suitable for soil when the intensity of the wind ρ, temperature T and humidity H are suitable. From Table 3 it appears ρ, T, H, λ, are the primitive parameters of Δ_{SE}, and ρ, T, H the primitive parameters of Δ_{EI}. Parameter chaining can be established in the same way as in the previous case. From Tables 2 and 3 we can derive the vertical classification of the parameters of causes 6 and 4, see Fig. 8. The same pattern could be followed to analyze other causes.

Table 3. Parameter classification in the case of the presence of vectors

Cause n° 4: presence of vectors (measured through plant mass μ)	Δ_{SE}	Δ_{EI}	Δ_{ID}	level
spread $= s(\mu)$	μ	μ		0
spread $= s(\mu(\alpha))$	α	α		1
spread $= s\,\mu\,(\alpha\,(\tau, \omega, \lambda))$	τ, ω, λ	τ, ω		2
spread $= s(\mu\,(\alpha(\tau(\rho, T, H), \lambda)))$	ρ, T, H, λ	ρ, T, H	none	3

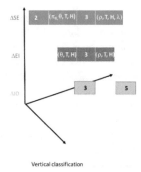

Fig. 8. Vertical classification of the parameters of the causes 6 and 4.

4 Discussion

This paper aims to provide a simple method that makes it possible to face the initial lack of knowledge in the face of little-studied or complex behavioral phenomena that must be modelled. For this, a modular approach is proposed guiding from the external observation until reaching the level of the variables that have to appear in the future mathematical-computational model. This is a useful tool in the sense it is the first sketch before the model. Quantitative experimentation and knowledge will then be required to achieve the mathematical, statistical or computational model that will solve the need of prevention and/or cure policies. Our method is based on two steps: first, characterization identifies the causes that may be responsible of the evolution of a phenomenon and second, a classification of the causes is carried out according to the place and/or time at which they occur revealing the underlying parameters that are shared or not by the causes. This approach has been successfully tested in our previous research and provides a useful tool to face complex phenomena modelling because of its modularity and scalability.

References

1. Watanabe, Y., et al.: A mathematical model of tumor growth and its response to single irradiation. Theor. Biol. Med. Model. **13**, 6 (2016). https://doi.org/10.1186/s12976-016-0032-7
2. Murphy, H., Jaafari, H., Dobrovolny, H.M.: Differences in predictions of ODE models of tumor growth: a cautionary example. BMC Cancer **16**, 163 (2016). https://doi.org/10.1186/s12885-016-2164-x
3. Nikolaou, M.: Revisiting the standard for modeling the spread of infectious diseases. Sci. Rep. **12**, 7077 (2022). https://doi.org/10.1038/s41598-022-10185-0
4. Yao, S., Fan, N., Hu, J.: Modeling the spread of infectious diseases through influence maximization. Optim. Lett. **16**, 1563–1586 (2022). https://doi.org/10.1007/s11590-022-01853-1
5. Huang, N.E., et al.: A model for the spread of infectious diseases compatible with case data. Proc R Soc A **477**, 20210551 (2021). https://doi.org/10.1098/rspa.2021.0551
6. Signes-Pont, M.T., et al.: An epidemic model to address the spread of plant pests. Case Xylella Fastidiosa Almond Trees, Kybernetes **50**(10), 2943–2955 (2021). https://doi.org/10.1108/K-05-2020-0320

7. Signes-Pont, M.T., et al.: Modelling the malware propagation in mobile computer devices. Comput. Secur. **79**, 80–93 (2018). https://doi.org/10.1016/j.cose.2018.08.004
8. Donatellia, I.M., et al.: Modelling the impacts of pests and diseases on agricultural systems. Agric. Syst. **155**, 213–224 (2017). https://doi.org/10.1016/j.agsy.2017.01.019
9. Signes-Pont, M.T., Ramírez-Martínez, D.E., García-Chamizo, J.M., Mora-Mora, H.: A multigrid approximation to the expansion of Xylella Fastidiosa in almond trees. In: WSEAS Transactions on Computers, AMATHI'19 Londres, vol. 18 (2019). E-ISSN: 2224–2872
10. Fierro, A., Liccardo, A, Porcelli, F.: A lattice model to manage the vector and the infection of the Xylella fastidiosa on olive trees. Sci. Rep. **9**, 8723 (2019). https://doi.org/10.1038/s41598-019-44997-4, www.nature.com/scientificreports/
11. Liccardo, A., et al.: A biological control model to manage the vector and the infection of Xylella fastidiosa on olive trees. PLoS One **15**(4), e0232363 (2020). https://doi.org/10.1371/journal.pone.0232363
12. Mastin, A.J., et al.: Epidemiologically-based strategies for the detection of emerging plant pathogens. Sci. Rep. **12**, 10972 (2022). https://doi.org/10.1038/s41598-022-13553-y. www.nature.com/scientificreports/
13. Signes-Pont, M.T.: Modelo computacional de la expansión de plagas. El caso de Xylella fastidiosa. BIOVEXO Xylella Forum: Status quo, expectativas e investigación 20 abril (2022)
14. Signes-Pont, M.T., et al.: Impacto económico de la Xylella fastidiosa en Comunidad Valenciana (España) "Societal Transformations and Sustainable Development with respect to Environment in the post Covid-19 Digital Era" First Annual Transform4Europe PhD conference, 8–9 December 2021
15. Cortés-Plana, J.J., et al.: Surveillance model of the evolution of the plant mass affected by Xylella Fastidiosa in Alicante (Spain), In: MMEHB 2022, Valencia, 13–15 July 2022
16. Boters-Pitarch, J., et al.: A new stochastic approach to the spread of environmental events enhanced by the wind. In: 22th International Conference Computational and Mathematical Methods in Science and Engineering, 3–7 July 2022
17. Cortés-Plana, J.J., et al.: IA- based surveillance of the plant mass to minimize the impact of Xylella fastidiosa Simposio Doctoral ELLIS, Alicante 19–23 September 2022
18. Signes-Pont, M.T., et al.: Computational modeling of the propagation of pests in plants based on behavioral considerations. In: ASETMEET 2022, Copenhague 23–25 June 2022
19. Cortés-Plana, J.J., et al.: Environmental model to manage the eradication of almond trees of Alicante (Spain) in the case of Xylella fastidiosa. 3 rd European conference on Xylella fastidiosa and XF-ACTORS final meeting, 26–30 april 2021 (poster, online) POSTER SESSION: Epidemiology and modeling of Xylella fastidiosa diseases

A Short Memory Can Induce an Optimal Lévy Walk

Tomoko Sakiyama$^{(\boxtimes)}$ (iD) and Masao Okawara

Soka University, 1-236, Tangi-Cho, Hachioji-Shi, Tokyo, Japan
sakiyama@soka.ac.jp

Abstract. In this paper, we investigate a possibility of the emergence of a Lévy walk using a single walker having a short memory. In our model, the walker recalls the past movement direction and compares it with the current direction. Based on the consistency or the inconsistency between these two information, the walker modifies the current directional rule. The walker in our model tends to move in a certain direction like a self-avoiding walk. For the walker, the coordination of the current rule implies that it can reconsider the directional rule. As a result, we found that the walker produced a power-law tailed movement called a Lévy walk which μ was close to an optimum (2.00). More importantly, the walker could not produce that characteristic movement if it could access not to a short-term memory but to a long-term memory. This finding suggests that the walker does not need any long-term memories to achieve an optimal walk like a Lévy walk.

Keywords: Lévy walk · Memory · Self-avoiding walk

1 Introduction

Theories of the movement strategy can be applied to the understanding of animal navigation [1–3]. Specifically, researchers have focused on whether or not animals demonstrated Lévy walks. The step lengths (L) of Lévy walks have a power-law distribution:

$$P(L) \sim L^{-\mu}$$

Here, an exponent μ must satisfy $1.0 < \mu < 3.0$. The motion can be ballistic if $\mu = 1.0$ while the motion can be a Brownian walk if μ is close to 3.0. Thus, Lévy walks with $\mu = 2.0$ can be an optimal Lévy walk [3]. This is because Lévy walkers with $\mu = 2.0$ can produce both ballistic motions and Brownian motions unpredictably, resulting in the achievement of the wide-area searching and the local-area searching. Thanks to this property, Lévy walkers can produce adaptive movements when walkers do not know the location of targets. An agent-based model can illustrate a cognitive process of the walker and can examine how the walker exhibits an optimal walk [4–7].

Animals have a memory capacity that impacts on the movement ability. However, little is known about a relationship between an optimal walk and the memory-use. In the previous work, one of the authors has shown that an optimal Lévy walk could be emerged

© The Author(s), under exclusive license to Springer Nature Switzerland AG 2024
A. Rocha et al. (Eds.): WorldCIST 2023, LNNS 802, pp. 421–428, 2024.
https://doi.org/10.1007/978-3-031-45651-0_42

if a walker revised its directional rule using the consecutive directional information [4]. In that paper, the walker modified its rule to produce a self-avoiding walk [8–10]. Thus, the inconsistency between consecutive two information allowed the walker to revise its rule. In this paper therefore, we investigate whether similar results can be obtained if the walker compares the current direction not with the directional information one step before but with the directional information a bit before.

Here, we develop an agent-based model where a single walker produces its movements on a two-dimensional lattice. During the walk, the walker learns the previous directional information and recalls it to check it can move in a certain direction stably by comparing with the current direction. If the inconsistency between two information occurs, the walker alters its directional rule to fix the problem. Here, we show that the walker can produce a Lévy walk with an optimal μ value. Importantly, the walker cannot produce such a characteristic movement if it can access to the past information, suggesting that the walker is not need any large memory capacity to exhibit a power-law tailed movement.

2 Methods

2.1 Simulation Environment

We set a two-dimensional lattice filed and set a single walker on the origin. As the initial rule, the walker obeys a Brownian walk. That is, the walker randomly selects one coordinate from four nearest coordinates. The simulation time was set to 100,000-time steps. We used C-language for the coding.

2.2 Direction-Comparison Model

In this section, we explain the proposed model named as the Direction-Comparison (DC) model. In this model, the walker compares the past information with the current one and modifies its directional rule. At time t, the walker at (x_t, y_t) calculates following information using the current directional rule.

$$x_{t+1} - x_t, y_{t+1} - y_t$$

Also, it calculates following information at that time.

$$x_{t+1-count} - x_{t-count},$$

$$y_{t+1-count} - y_{t-count}$$

Then, if one of following equations is satisfies, the walker coordinates its directional rules: the probabilities for moving in each direction $rule_{+x}$, $rule_{-x}$, $rule_{+y}$, $rule_{-y}$ as follows.

If $x_{t+1} - x_t \neq x_{t+1-count} - x_{t-count}$ AND $x_{t+1} - x_t \neq 0$,
if $x_{t+1} - x_t = -1$,

$rule_{-x} = r,$

$$rule_{+x} = (1 - r)/3,$$

$$rule_{-y} = (1 - r)/3,$$

$$rule_{+y} = (1 - r)/3$$

else if $x_{t+1} - x_t = 1$,

$$rule_{-x} = (1 - r)/3,$$

$$rule_{+x} = r,$$

$$rule_{-y} = (1 - r)/3,$$

$$rule_{+y} = (1 - r)/3$$

else if $y_{t+1} - y_t \neq y_{t+1-count} - y_{t-count}$ AND $y_{t+1} - y_t \neq 0$,
if $y_{t+1} - y_t = -1$,
$rule_{-x} = (1 - r)/3,$

$$rule_{+x} = (1 - r)/3,$$

$$rule_{-y} = r,$$

$$rule_{+y} = (1 - r)/3$$

else if $y_{t+1} - y_t = 1$,
$rule_{-x} = (1 - r)/3,$

$$rule_{+x} = (1 - r)/3,$$

$$rule_{-y} = (1 - r)/3,$$

$$rule_{+y} = r$$

Here, the walker can access to the information *count* time steps before. The parameter r is a random number satisfying following condition [4].

$$r \in [0.25, 1.00] = \{r | 0.25 \leq r \leq 1.00\}$$

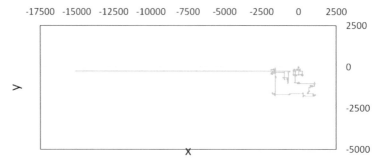

Fig. 1. An example of trajectories of the walker. Here, *count* = 5.

3 Results

Here, we set *count* = 5 as a default value. As shown in Fig. 1, the DC model seems to produce both ballistic motions and zig-zag motions.

To check the detailed movement properties, we calculated the diffusion and investigated the relationship between the mean squared displacement and time. In random walk theories, there is a relation between the mean squared displacement and time as follows:

$$< R^2 > \sim t^\alpha$$

Here, the exponent α determines the diffusive property of the walker. If $\alpha > 1.0$, then, the movement follows super-diffusive movements. On the other hand, the movement follows normal diffusive movements when $\alpha = 1.0$ [2].

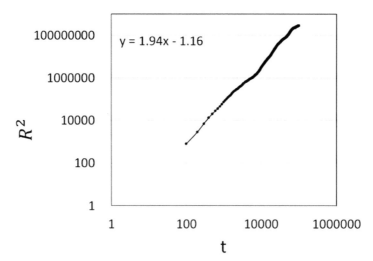

Fig. 2. The relationship between the mean squared displacement and time. Here, *count* = 5. Data were averaged over 100 trials.

According to Fig. 2, the proposed model presents super-diffusive movements (α = 1.94). Here, the squared displacements were obtained every 100-time steps from 100 trials. To investigate whether the walker replaced its directional rule regularly or not, we also calculated the time interval between two consecutive rule changes.

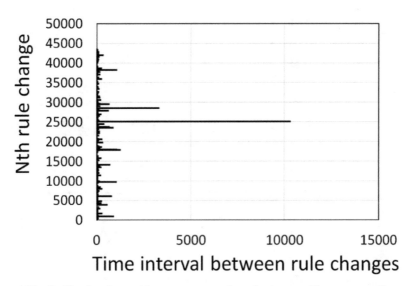

Fig. 3. The time interval between consecutive rule changes. Here, *count* = 5.

Figure 3 demonstrates an example of this calculation. As you can see, the time interval between two rule changes seems to occur unregularly. This result is linked to the fact that the walker produces both ballistic walks and zigzag walks unpredictably. For further evaluations, we obtained step lengths along each axis. Thus, the step length can be calculated as the travel distance until the walker makes a U-turn along each axis. Figure 4 presents examples of cumulative distributions of step lengths. As shown in this figure, the walker succeeded in producing Lévy walks in both directions (x-axis: AIC weights for a power-law = 1.00, μ = 1.95, y-axis: AIC weights for a power-law = 1.00, μ = 1.96). Interestingly, we could confirm that the walker was always able to produce optimal Lévy walks ($\mu \approx 2.00$).

Lastly, we checked the replacement of the parameter *count*. Here, we replaced that value from 5 to 10 or 20. According to Fig. 5, we found that the walker was likely to produce Brownian walks more often as the parameter *count* increased. These findings suggest that the walker can produce optimal Lévy walks when it accesses to the recent information. However, this tendency is disappeared as it accesses to the past information.

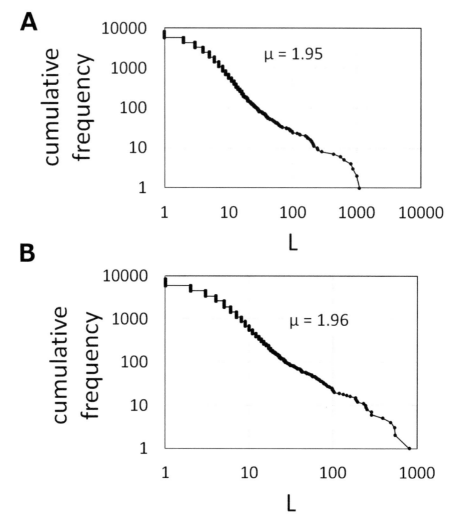

Fig. 4. The relationship between the step length and its cumulative frequency. A. *x*-axis. B- *y*-axis. Here, *count* = 5.

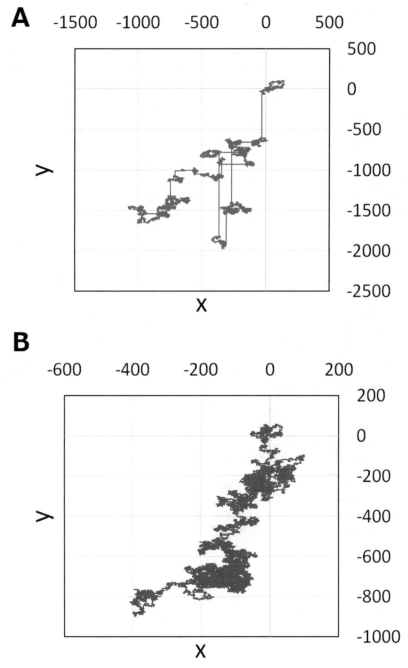

Fig. 5. An example of trajectories of walker. Here, *count* = 10 (A), *count* = 20 (B).

4 Conclusion

In this paper, we developed an agent-based model where a single walker selected one of the nearest cells at each time step according to its directional rule. In the proposed DC model, the walker sometimes coordinates its rule so as to maintain a self-avoiding walk [8–10]. The inconsistency between the directional movement at the current moment and the past one gives a rule coordination to the walker. Thus, the walker is likely to revise its rule when it regards that it cannot move in a certain direction stably. We found that the walker exhibits a power-law tailed movements with the optimal value. Moreover, the walker was likely to produce not a Lévy walk but a Brownian-like walk as it could access to more past information.

Our future work may be to check the memory-less effect or the perturbation analysis using the proposed model.

References

1. Viswanathan, G.M., et al.: Statistical physics of random searches. Braz. J. Phys. **31**, 102–108 (2001)
2. Bartumeus, F., Luz, M.G.E., Viswanathan, G.M., Catalan, J.: Animal search strategies: a quantitative random-walk analysis. Ecology **86**, 3078–3087 (2005)
3. Viswanathan, G.M.: et al. Optimizing the success of random searches. Nature **401**, 911–914 (1999)
4. Sakiyama, T.: A recipe for an optimal power law tailed walk. Chaos Interdisc. J. Nonlinear Sci. **31**, 023128 (2021). https://doi.org/10.1063/5.0038077
5. Sakiyama, T., Gunji, Y.P.: Emergence of an optimal search strategy from a simple random walk. J. R. Soc. Interface. **10**(86), 20130486 (2013)
6. Sakiyama, T., Gunji, Y.P.: Emergent weak home-range behaviour without spatial memory. R. Soc. Open Sci. **3**(6), 160214 (2016)
7. Sakiyama, T., Gunji, Y.P.: Optimal random search using limited spatial memory. Royal Soc. Open Sci. **5**(3), 171057 (2018)
8. Shlesinger, M.S.: Weierstrassian levy flights and self-avoiding random walks. J. Chem. Phys. **78**, 416–420 (1982)
9. Flory, P.: Principles of Polymer Chemistry, 672. Cornell University Press, Ithaca (1953)
10. Fisher, M.E.: Shape of a self-avoiding walk or polymer chain. J. Chem. Phys. **44**, 616 (1966)

Open Learning and Inclusive Education Through Information and Communication Technology

Actions to Promote Knowledge about Accessibility and Inclusion in Higher Education in Brazil

Cibele Cesario da Silva Spigel[1] (ID), Cibelle Albuquerque de la Higuera Amato[1] (ID),
and Valéria Farinazzo Martins[1,2(✉)] (ID)

[1] Programa de Pós-Graduação em Distúrbios do Desenvolvimento, São Paulo, Brazil
valfarinazzo@hotmail.com
[2] Programa de Pós-Graduação em Computação Aplicada, São Paulo, Brazil

Abstract. In the Brazilian context, public policies focus on inclusive education for elementary and high school students. Fortunately, these students are already starting to reach universities, but most teachers are not prepared to meet the demands of these students efficiently. This paper presents some strategies to promote knowledge to higher education teachers about accessibility and inclusion in a classroom context. For that, two modules of an online course were prepared, besides an e-book with contents in dialogue with each other. Two data collection instruments were used to understand the profile of the target audience and their perception of the course.

Keywords: Accessibility · Higher Education · Accessible digital teaching material

1 Introduction

Supported by public policies, not always applicable in their integrity, and the maturation of legislation in favor of inclusive education, from the 1960s to the present day, students with disabilities have increasingly reached higher education. However, in Brazil, unlike basic education teachers, higher education teachers are not always subjected to the assumptions of inclusive education in their initial and continuing training [1].

This lack of training often leaves teachers with no direction when faced with a student with a disability in the classroom regarding their actions and the pedagogical structure. This can trigger an increase in the evasion of this student since he is faced with the lack of accessibility, with teaching materials that are not adapted to his limitation, and the teacher is unprepared to deal with this public [2].

Considering the lack of preparation of higher education teachers on inclusive education, this paper aims to present concrete actions that can be applied to promote knowledge about accessibility and inclusion in higher education, reducing the gap mentioned above.

For this, we seek to understand how Digital Information and Communication Technologies (DICT), accessibility recommendations and Universal Design for Learning

A. Rocha et al. (Eds.): WorldCIST 2023, LNNS 802, pp. 431–440, 2024.
https://doi.org/10.1007/978-3-031-45651-0_43

(UDL) guidelines [3], as well as thinking about content planning from the perspective of the ADDIE model [4], can help teachers to build pedagogical practices with accessibility.

This paper is organized as follows. In Sect. 2, Theoretical Background, the concepts of Disabilities, Accessibility and Adaptation of Material and related works are presented. In Sect. 3, the Research Methodology is described. Section 4 contains the results of the research. Finally, Sect. 5 concludes this work.

2   Theoretical Background

2.1   Disabilities

Disability is part of the human condition. For some people, it is a condition imposed at birth that lasts throughout their lives. Due to accidents or illnesses, others may acquire temporary or non-temporary disabilities. Aging also places increasing limits on the body and mind [5].

Decree 5,296 of 2004 of the Brazilian legislation [6] defines the main types of disability present in society:

- Physical Disability: complete or partial alteration of one or more segments of the human body, leading to impairment of physical function, presenting itself in the form of paraplegia, paraparesis, monoplegia, monoparesis, tetraplegia, tetraparesis, triplegia, triparesis, hemiplegia, hemiparesis, ostomy, amputation or absence of a limb, cerebral palsy, dwarfism, limbs with congenital or acquired deformity, except for aesthetic deformities and those that do not cause difficulties in performing functions.
- Hearing impairment: bilateral partial or total loss of forty-one decibels (dB) or more, measured by audiogram at 500 Hz, 1,000 Hz, 2,000 Hz and 3,000 Hz.
- Visual impairment: blindness, in which the visual acuity is equal to or less than 0.05 in the best eye, with the best optical correction; low vision, which means visual acuity between 0.3 and 0.05 in the best eye, with the best optical correction; cases in which the sum of the measurement of the visual field in both eyes is equal to or less than 60°; or the simultaneous occurrence of any of the conditions above.
- Cognitive Impairment: intellectual functioning significantly lower than average, manifested before the age of eighteen and limitations associated with two or more areas of adaptive skills, such as communication, personal care, social skills, use of community resources, health and safety, academic skills, leisure and work.
- Multiple Disability: association of two or more disabilities.

2.2   Accessibility and Adaptation of Material

According to Brazilian Technical Standards Association (Associação Brasileira de Normas Técnicas [ABNT]), Standard 9050, accessibility is the "Possibility and condition of reach, perception and understanding for the safe and autonomous use of buildings, space, furniture, urban equipment and elements" [7]. For being focused on physical spaces, this standard specifies the patterns of height, lowered pavement, external circulation, and guideline, which are always based on universal design. However, extracting points to be used specifically for adapting digital learning material is possible. Some of these recommendations are presented below according to the type of disability.

- Motor, intellectual and hearing impairment:

  - Visual Range: When demonstrating something during a laboratory class, for example, ensure that what is being shown is within the visual range of students with motor, intellectual or hearing impairments.

- Visual impairment and low vision:

  - Sound range: When writing something on the blackboard, verbalize it so the student can know the content being worked on.

- Hearing and intellectual disability:

  - Information: Information should be complete, precise and clear so that the student with hearing or intellectual impairment understands what is being taught.

- Hearing and intellectual impairment and low vision:

  - Visual aids: Use text messages, symbols, contrasts, and figures to illustrate content that is more complex to be understood.

- Low vision and intellectual disability:

  - Contrast: Contrast highlights elements among themselves through light-dark and dark-light composition to catch the observer's attention.
  - Symbols and figures: Pay attention to a) solid and well-defined outlines; b) simplicity in shapes and few details.

- Hearing impairment:

  - Libras interpreter: Make sure that the interpreter is well positioned in the classroom space, both for the student (so that he/she can see him/her clearly) and for you (so that he/she can accurately translate your lesson).

- Visual impairment:

  - Sound Aids: Use them to enable comprehension by addition.

- Intellectual and hearing impairment, ADHD and ASD:

  - Material writing: It should be: a) objective; b) when tactile, it should contain essential information in high relief in Braille; c) contain complete sentence, in the order: subject, verb, and predicate; d) be in the active form and not passive; e) be in the affirmative and not negative form; f) emphasize the sequence of actions; g) avoid vertical texts.

- Low vision, intellectual disability, ASD:

  - Material typography: Use sans serif font, uppercase and lowercase letters. Avoid italic, decorated, handwritten, shaded, three-dimensional-looking, or distorted fonts. Avoid gloss and matte. Avoid vertical text.
  - Fonts: Preferably use the following fonts: Arial, Verdana, Helvetica, Univers and Folio.
  - Size of letters, figures and symbols: Letters, numbers, symbols and figures sizes must be proportional to the reading distance.

2.3   Related Work

When searching the literature, we found related works on inclusive education in higher education, teacher training and adaptation of teaching materials.

Marino [8] verified in his research the ease of adapting material by pedagogy professionals into accessible educational material for older people and how a digital platform supports the creation of accessible courses. For this, accessibility recommendations were created based on the W3C, and two courses were given to the participants: one on creating inclusive teaching material and the other on adapting this material, both supported by the principles of Universal Design for Learning (UDL) and with the use of ICT. Unlike the work by [8], which had pedagogy teachers as participants, this research aimed to have higher education teachers as participants from any area, considering that they are the public with the least assistance and training in inclusive education.

Miranda, Mourão and Gediel [9] investigated how ICT could help deaf students at a higher education institution in the Forest Zone of Minas Gerais State (Zona da Mata Mineira). For this, a material was built in five phases, involving teachers when providing the material to be adapted; deaf people and interpreters to validate what was made; and the technical team, which produced audio guides and signed video lessons. The study showed that laws, declarations and decrees are not enough. It is necessary to put accessibility into practice, which involves several agents, not just the teacher. Like the work by [9], this research showed the potential of ICT. However, it was not restricted to teaching students with deafness, preferring to broaden this look to any disability.

Barbosa -Vioto and Vitaliano [10] disclosed a very interesting result when studying the perception of pedagogy graduates about inclusive education in teacher training: in addition to the importance of curricular restructuring, with interdisciplinarity among the disciplines on inclusive education and others, it is necessary to have the practice itself, with internships in inclusive classrooms. That is, contemplating guidelines for inclusive pedagogical practice in initial teacher training is not enough. The teacher needs to experience this practice.

This view reinforced the need to promote practical actions for teachers, not just theory. Therefore, when preparing the script for adapting the material, applications widely used by teachers were included: Word, PowerPoint and Excel, so that they could adjust to what they were already more familiar with.

Most of the surveyed studies are directed to teacher training for basic education. Those aimed at higher education teachers focus on a specific disability and point to the urgent need for more significant support for teachers to promote inclusive and quality education.

3 Research Methodology

This study aims to present practical actions offered to Brazilian higher education teachers through a university extension project to promote knowledge about the inclusion and accessibility of students who arrive at universities in the country. To this end, concepts about accessibility and inclusion and legislation are presented and discussed, as well as practices for adapting digital teaching material into accessible.

To achieve this goal, an e-book with general guidelines and accessibility recommendations was developed. Two modules of a course on inclusive education were taught: the first - Inclusive attitudes in the classroom: possible paths - deals with disability types, the seven accessibility dimensions [11] and the UDL [3], and the second - Inclusive attitudes in the classroom: making my material accessible - guides about adapting material using ICTs, plus accessibility recommendations. Two questionnaires (profile and post-course feedback) were used as a data collection instrument.

4 Results

4.1 Profile of Participants

The total number of enrolled participants was 129. However, there were several withdrawals or not accounting because they were not the target audience. Of the 16 participants, the majority (81%) were female. Regarding the age range, most (50%) are between 41–50 years, 12% are between 31–40 years, 25% are between 51–60 years, and 13% are over 60 years. As for the area of training, there is prevalence in the humanities, 56.25%; in exact sciences, 18.75%; in social and applied sciences, 12.5%; and in linguistics, letters and arts, 12.5%. Concerning teaching time, most teachers (50%) have 20 or more years of experience; 6% have between 5–10 years of experience; 25% between 11–15 years; 19% between 16–20 years.

4.2 Course Modules

An extension course was divided into two parts, one day each, totaling six hours. It was taught online due to the ease of bringing together people from different parts of the country, offered at two different times (during the week and on the weekend).

The first module of the course, "Inclusive attitudes in the classroom: possible paths," brought up the following topics for discussion:

- Inclusive society – a brief history: definition of disability; types of disability; disorders; stages of the history of disability (exclusion, segregation, integration and inclusion); description of inclusion; the concept of minorities; primary legal devices.
- After all, what is accessibility?: definition of accessibility; the seven dimensions of accessibility (architectural, attitudinal, communicational, instrumental, methodological, natural and programmatic) [11].

- Universal Design for Learning (UDL) [3]: evolution of the universal design ramp to the UDL; definition of UDL by ABNT; seven principles of UDL.
- Accessibility recommendations: based on the guidelines of the Mackenzie Inclusion Program Manual (MIPM); in the studies pointed out by [12], the ABNT 9050 [13] classification and the DSM-V (Diagnostic and Statistical Manual of Mental Disorders); and the accessibility recommendations of eMag (Electronic Government Accessibility Model) [14].

The second module, "Inclusive attitudes in the classroom: making my material accessible," addressed the following topics:

- Accessibility: the definition of accessibility was resumed, as well as the seven dimensions [11], three of which would be worked on in the course – methodological, communicational and attitudinal.
- General guidelines: the accessibility recommendations were explained, for any material, regardless of whether or not it is in Word, PowerPoint and Excel. This part of the course showed in detail how to make a video accessible, from its insertion on YouTube to the generation of subtitles and transcription, ending with the insertion of Libras using VLibras. Many teachers were unaware of the Federal Government website and its various possibilities for mobile, web and developers. Privacy settings were shown when uploading a video on Youtube. Then, it was demonstrated how to insert subtitles and transcription without using Youtube for those who would not like to use the platform. This part of the class had a lot of interaction because most teachers did not know the difference between a serif and sans serif font and how the choice of colors for contrast influences the better understanding of people with and without disabilities. That is, simple attitudes of making the material could make any material accessible. Several examples of non-accessible material were also presented, and how the choice of accessible colors and fonts would already solve the problem. The detailed process of making a video lecture accessible was a little more challenging for participants who were unfamiliar with the technology.
- PowerPoint: the third part of the course module started showing how to activate the "Accessibility check" in the application and the differences between errors, warnings and tips. Later, it was taught how to insert alternative text into visual elements such as images, shapes, SmartArt, photos, etc. It was explained that even though screen readers are now very advanced, if a description is not inserted in PowerPoint visual elements, it will not be possible for a person with low vision or blindness to know what they are referring to. It was also shown how to make tables accessible, insert hidden or not titles in the slides and configure the reading order of the material.
- Word: this part of the course module was dedicated to one of the most used text applications: Word. As I had shown in the previous section how to activate the "Accessibility Check", the participants were already a little more familiar with the resource in this part. As with PowerPoint presentations, a description of the visual elements must be inserted. They were shown the whole process and then taught how to make the titles and headings accessible using the "Styles" feature, the tables, fonts, and contrasts, as well as maintaining accessibility by exporting the file as .pdf.
- Excel: this part was dedicated to Excel spreadsheets. A little more familiar with the accessibility features applied in the previous files, this application, which is often

used to generate graphs and spreadsheets, was left for last. Therefore, this part of the course started with accessibility graphics, explaining in detail how to use different textures as an alternative for people who cannot identify the colors. The importance of renaming all tabs in the worksheets was also explained so that the screen reader could read them without forgetting to reinforce the accessibility recommendations already described in the previous sections: font size, color and type; contrast colors; clear and direct texts etc.

4.3 E-book

For teachers to have at hand all the content worked in the course, an e-book was prepared to serve as a reference manual and made available at the end. The e-book "Inclusive attitudes in the classroom: making my material accessible" ("Atitudes inclusivas em sala de aula: tornando meu material acessível" [Portuguese version]) was prepared according to the accessibility recommendations of ABNT, eMag, IPM manual and in the notes of [12], as well as the site on accessibility at Microsoft. As teachers may have difficulty with technology if they do not understand the step-by-step instructions in the e-book, another alternative is to watch the explanatory videos (maximum five minutes) at the beginning of each subsection of the chapters.

   After the Summary section, where the operation of the book is explained, the e-book itself begins, bringing the following subjects (which are in line with what was presented in the course modules):

- "To start with…": Brief history of disability in Brazil and main legal provisions; the concept of minorities; accessibility; types of disability, with a table showing the definition of each one, how to act and the barriers; universal design for learning (ULD).
- "General guidelines": Accessible fonts, colors, contrast, visuals, and videos. This section was created thinking about adapting any material, not just the types listed in the e-book. In the "visual elements" section, how to insert a description in a visual element on Instagram was discussed, for example, thinking about teachers who use social networks as an extension of their classes, a practice that is increasingly common today. In this section, there are also steps on how to make a video accessible, showing how to insert it on Youtube to add subtitles and download it in.srt format, which will later be used to insert Libras in the video through the free VLibras platform.
- "Word": chapter presenting visual elements: description; hyperlink and screen-tips; color; contrast; built-in titles and styles; table; accessible .pdf; and accessibility checklist.
- "PowerPoint": chapter featuring alternative text in visual elements; hyperlink and screen tips; accessible color and design templates; contrast; titles; table; reading sense; font; accessible .pdf; accessibility checklist.
- "Excel": chapter showing alternative text in visual elements; hyperlink and screen tips; color; contrast; worksheet title; table; font; accessible .pdf; accessibility checklist.
- "Teams": chapter presenting definition; contrast and high contrast; caption and transcript.

Accessibility is shown in detail in the four Microsoft applications (Word, PowerPoint, Excel and Teams), maintaining the same structure and organization so that the teacher can quickly and practically find what he needs.

The e-book can be accessed at the link: https://issuu.com/cibele.spigel/docs/atitudes_inclusivas_ebook. Figure 1 presents some images from the book.

Fig. 1. E-book images

4.4 Some Feedback from Teachers

Regarding inclusive education, it was asked if the participants had already had contact with students with disabilities. Of the total, 44% answered "Sometimes,"; 44%, "Frequently"; and 13%, "Rarely". Asked if they had students with disabilities in their classroom in the last 10 years, the majority answered yes (93.75%).

Regarding the types of disability with which they have had contact, ASD and deafness lead the cases, 20% and 23%, respectively; then comes blindness, with 15%; motor disability and ADHD, with 10%. Other types of disabilities were reported in fewer numbers. When asked whether they had already received some training on inclusive education, most participants responded yes, 62.50%. This may reflect the need always to seek more knowledge on the subject and the need for more practical courses.

When asked if they had ever encountered difficulties in adapting material before the course, only 25% answered no, predominantly 43.75% of the participants answering "yes, very often" and 31.25%, "yes, sometimes", signaling the importance and need to hold courses on adapting material for teachers.

When questioned about the viability or not of adapting the material after the course modules had been taken, the rates were: 0% of the participants answered "unfeasible for daily use"; 62.5% answered "difficult, but possible"; 37.5% answered "quite reasonable"; and 0% "easy".

When asked if the course allowed them to do so: 100% of the participants stated that it was possible to know in practice how to adapt a material; 93.75% declared that it was possible to learn about the concepts of inclusion and accessibility, and only 6.25% said they were not able to adapt a material on their own. This suggests that the teachers could take advantage of the contents learned.

5 Conclusions

This work presented actions to promote a more excellent knowledge of higher education teachers regarding the concepts of accessibility and inclusion, culminating with practices of adapting digital didactic material into accessible. To achieve the objectives of this project, two modules of a course on inclusive education and adaptation of material using ICT were taught, and an e-book was developed with accessibility recommendations, both in theory and in practice.

As contributions of this work, the construction of the e-book and the course modules were built so that the contents of both could "talk" to each other. The accessibility recommendations found on the W3C and eMAG websites are targeted at creating web pages and applications and not toward teaching material. Therefore, we tried to develop proposals combining the contents of these two websites with those of ABNT Standard 9050 (Sect. 2.2) and Microsoft. Once these recommendations were finished, each one was tested, and examples were inserted in the e-book and shown in the courses so that teachers could understand each of them practically. In addition, videos were created to assist in the understanding of all the topics.

Creating a listening space for participants during the course was vital for them to feel welcomed in the face of something still recent in Brazilian higher education: inclusion. At that moment, the need to open more spaces for listening was perceived so that the exchange of experiences could help everyone.

After completing this research, it is clear that these types of actions for teachers must happen continuously. With this in mind, it would be interesting, after presenting the tools for adapting the material, to check later whether the teachers have been able to apply ICT for adapting the material in their daily lives, as well as whether after learning about the types of disability and its barriers have contributed to reducing their insecurity in the classroom.

Furthermore, it would be interesting to perform this research with a more significant number of participants, bearing in mind that the number of students with disabilities will increase, increasing the probability of having this public in the classroom, regardless of the course.

As a limitation of this work, although the number of applicants was high, more than 100, those who took the courses and had the expected profile reached only 16, resulting in a reduced and convenient sampling.

The Office version used for adapting material was Office 365, which is currently available (the year 2022). Some teachers may have used other arrangements as the course was online and not reported it. Version differences are very few, but there are. During the studies, only one teacher reported a difference in resource use. At the time, she was shown another way to adapt the material to circumvent the difference.

Since the course modules were offered online, it may have happened that many teachers averse to this modality or even averse to technology may not have participated.

Had the course been conducted face-to-face, the number of teachers would have been higher.

Acknowledgements. This work was supported by the Coordination for the Improvement of Higher Education Personnel – Brazil (Coordenação de Aperfeiçoamento de Pessoal de nível superior [CAPES]) – Excellence Program – (Programa de Excelência [Proex] 1133/2019.

References

1. ABED – Associação Brasileira de Educação a Distância. Censo EAD.BR: relatório analítico da aprendizagem a distância no Brasil 2019/2020. Curitiba: Intersaberes, 2021. Disponível em: http://abed.org.br/arquivos/CENSO_EAD_2019_PORTUGUES.pdf. Acesso em: 16 dez. 2021
2. Mendes et al.: A formação dos professores especializados segundo os pesquisadores do observatório nacional de educação especial. Educação e Fronteiras, Dourados, v. 5, n. 13, p. 84–95, maio/ago. 2015
3. Glass, D., Meyer, A., Rose, D.: Universal design for learning and the arts. Harv. Educ. Rev. **83**(1), 98–119 (2013)
4. Filatro, A.: Design instrucional na prática. Pearson, São Paulo (2008)
5. WHO - World Health Organization. World report on disability 2011. World Health Organization (2011)
6. Brasil, A.D.I.: Decreto nº 5.296, de 2 de dezembro de 2004. Diário Oficial da União (2004)
7. ABNT. Sobre. 2022. Disponível em: http://www.abnt.org.br/institucional/sobre Acesso em: 23 maio 2022
8. Marino, M.C.H.: Acessibilidade em material didático digital para idosos: estudo de caso em uma plataforma educacional digital. Dissertação (Mestrado em Distúrbios do Desenvolvimento) – Ciências Biológicas e da Saúde, Universidade Presbiteriana Mackenzie, São Paulo (2020)
9. Miranda, I., Mourão, V.L.A., Gediel, A.L.B.: As tecnologias da informação e comunicação (TIC) e os desafios da inclusão: a criação de aulas sinalizadas no contexto do ensino superior. Periferia: Educação, Cultura & Comunicação, v. 9, n. 1, jan./jun. 2017
10. Barbosa-Vioto, J., Vitaliano, C.R.: Educação inclusiva e formação docente: percepções de formandos em pedagogia. magis, Revista Internacional de Investigación en Educación **5**(11), 353–373 (2013)
11. Romeu Kazumi, S.: Inclusão: acessibilidade no lazer, trabalho e educação. Revista Nacional de Reabilitação (Reação), São Paulo, Ano XII, mar./abr. pp. 10–16 (2009)
12. Amato, C., Brunoni, D. (Org), Boggio, P.S. (Org.) Distúrbios do Desenvolvimento - Estudos Interdisciplinares. 1. ed. São Paulo: Memnon, v. 1. 506 p. (2018)
13. ABNT. Norma 9050. Acessibilidade a edificações, mobiliário, espaços e equipamentos urbanos (2015)
14. Brasil. Ministério do Planejamento, Orçamento e Gestão. Secretaria de Logística e Tecnologia da Informação. Gov.br e-mag versão 3.1: modelo de acessibilidade em governo eletrônico. Brasília, DF, 2014. Disponível em: http://www.governoeletronico.gov.br/eixos-de-atuacao/gestao/acessibilidade/emag-modelo-de-acessibilidade-em-governo-eletronico. Acesso em 23 jul. 2016

A Digital Ecosystem with Pedagogical Aspects to Support the Creation of Accessible Educational Resources

Maria Amelia Eliseo[(✉)] [ID], Valéria Farinazzo Martins [ID],
Cibelle Albuquerque de la Higuera Amato Amato [ID], and Ismar Frango Silveira [ID]

Universidade Presbiteriana Mackenzie, São Paulo, SP 01239-001, Brazil
{mariaamelia.eliseo,valeria.farinazzo,cibelle.amato,
ismar.silveira}@mackenzie.br

Abstract. The lack of support tools for teachers and pedagogical content creators in the creation and implementation of accessible digital educational resources encouraged the realization of the Smart Ecosystem for Learning and Inclusion Project - SELI Project, which involved researchers from different countries in Europe, Latin America, and the Caribbean. Based on the difficulties regarding the Digital Exclusion of the countries involved in this project, an authoring tool, called SELI Learning Platform, was implemented to assist and guide teachers in planning, structuring, and building accessible courses. In this context, this paper presents some keys aspects of pedagogical planning for the creation of accessible educational digital resources on the SELI Learning Platform. Eight pilot studies of the development of accessible courses using this ecosystem are implemented to check its efficiency, and feasibility and point out some improvements. How instructional design guide, one of the improvements, based on the ADDIE framework, and accessibility aspects were incorporated into the ecosystem are discussed. The results of the usability evaluation with eight teachers of these pilot studies are presented, with their strengths and weaknesses.

Keywords: Accessible Digital Educational Resources · Smart Digital Ecosystem · Pedagogical Platform · Authoring Tool

1 Introduction

The creation and implementation of accessible digital educational resources has always been a challenge in the computer-supported learning environments. The lack of supporting tools is a barrier for teachers and content creators for integrating pedagogical aspects with accessibility features to attend the most diverse student profiles. Considering this scenario, a Digital Ecosystem for designing, producing, and sharing accessible learning contents was envisioned - thus the SELI Project (Smart Ecosystem for Learning and Inclusion Project) was born, inspired by the main difficulties found in different countries regarding the Digital Divide in Europe and Latin America and Caribbean (Martins et al. 2019; Tomczyk et al. 2019).

A. Rocha et al. (Eds.): WorldCIST 2023, LNNS 802, pp. 441–451, 2024.
https://doi.org/10.1007/978-3-031-45651-0_44

During the development of this project, an authoring tool, called SELI Learning Platform, was implemented to assist and guide teachers in planning, structuring, and building accessible courses. Teachers can create their courses in an accessible way, considering their students' limitations and disabilities. After completing the course implementation, it can be published and made available to students. Thus, SELI learning platform is part of a learning ecosystem that offer authoring services, Content Management System (CMS) and a Learning Management System (LMS) services and learning analytics service (Eliseo et al. 2020; Oyelere et al. 2020). In addition, it offers some course models, based on Pedagogical Patterns suggesting to the teacher a methodology in pedagogical practice. It suggests different ways of presenting didactic concepts and tasks according to the type of content to be presented to the student.

Using the SELI learning platform, some pilot courses were built to check the efficiency and feasibility of using the tool both on the teacher's side when using the instructional design and accessibility resources offered, and on the part of the student when studying with a course built to be accessible. The pilot courses were designed, in some way, to capacity teachers in the use of ICTs with focus on accessibility.

At first the pilot courses started their development following the ADDIE framework (Peterson 2003), an instructional systems design (ISD). In this phase, eight pilot courses were implemented by the partner countries. Table 1 shows the pilot courses implemented by each country.

Table 1. Pilot courses implemented.

Country	Course Title
Bolivia	Problem-solving by computational approach
Turkey	Digital Storytelling training course with teachers
	Inclusive Education for Physical Education Teacher Education
Brazil	Accessible Educational Materials for the Elderly
Poland	Prevention of cyberbullying
	Digital inclusion
Uruguay	ICT tools by teachers to teach new media and English as a second language
Dominican Rep	FlippedClassroom with the H5P Tool

As pilot courses were being planned and implemented on the SELI learning platform, improvements in the tool's functionalities were being made. One of them was the implementation of the instructional design guide, based on the ADDIE framework described in this article.

This paper presents some aspects of pedagogical planning implemented in a digital platform for the creation of accessible educational digital resources. It describes the implementation of the instructional design guide, based on the ADDIE framework, as a resource to assist the teacher in creating these resources. A usability evaluation was carried out by eight teachers, from different countries, who used the SELI learning platform. The intention of this evaluation was to verify the ease of use, ease of learning, usefulness, and satisfaction related to this new functionality. It is not the intention of this paper to describe the use of the SELI learning platform, but rather to show the theoretical background of the resources that help to build the courses accessible on the platform.

2   Instructional Design and ADDIE Framework

There are innumerous tasks regarding course design that enlarges the complexity spectrum form teachers. Such tasks, which vary from planning to applying different learning materials, techniques, and activities, gains a wider level of challenge in the context of remote courses. To facilitate the organization of these activities, drawing upon a learner-centered approach, instructional design (Filatro 2008) generates knowledge pertaining to the principles and methodologies of instruction that are most appropriate for various teaching contexts, thereby fostering an optimal learning experience. (Merrill and Twitchell 1994). According to McGriff (2000), by being grounded in a learner-centered approach, instructional design encompasses a structured methodology that facilitates the identification of problems or learning needs. Subsequently, it orchestrates a comprehensive process encompassing problem recognition, solution design, implementation, and evaluation, thereby ensuring an effective and tailored approach to address the specific issues identified within diverse teaching contexts.

Peterson (2003) and Filatro (2008) introduce ADDIE (acronym for Analyze, Design, Develop, Implement and Evaluate) framework, an instructional design approach that integrates an iterative process comprising vital steps crucial for the development of a highly efficient course or class. ADDIE follows a cyclic, evolutionary nature, persisting throughout the instructional planning and implementation process. It operates as a systematic framework, employing a defined methodology and sequential steps to navigate complex scenarios encountered during the development of educational products and learning resources. This process serves as a facilitator for constructing knowledge and skills within guided learning environments.

Thus, ADDIE process encompasses a systematic and iterative approach to instructional design, progressing through the stages of Analysis, Design, Development, Implementation, and Evaluation. The initial stage, analysis, involves a comprehensive examination of the learning needs, including the identification of learning gaps or problems that necessitate instructional intervention. It includes conducting a profound analysis of the target audience, learning objectives, and contextual factors to ensure the subsequent design aligns effectively with the previously identified needs. The following stage is the Design, when instructional designers conceptualize and outline comprehensive solutions that aligns with the needs that were elicited during the previous stage. Then the Development stage starts, involving the effective creation of learning resources. In this stage, subject matter experts collaborate with instructional designers to produce learning materials that align with the design specifications outlined in the previous stage. Following, the Implementation stage takes place, when learning materials are deployed and made available through some learning environment or by other means. In the last stage, Evaluation, the effectiveness of instructional solutions pass by an assessment process, preferably guided by data gathered during the usage of learning resources by students. These data, which includes feedback from instructors and students, can be used to improve some previous stage, allowing the incorporation of such improvements in future iterations of the instructional design process (Branch 2009). The systematization of the teaching learning process with the instructional design supported by the ADDIE framework ensures the development of learning activities and establishes the principles, objectives, and teaching methods most appropriate to the needs of students. In summary, by following these stages, instructional designers can effectively address complex educational challenges and create impactful learning experiences.

These concepts of Instructional Design and ADDIE framework were used as requirements in the implementation of the SELI learning platform to guide teachers to plan their courses.

2.1   Instructional Design Implemented in SELI Learning Platform

SELI learning platform offers the teacher the option of building accessible courses freely or guided by the tool. In the guided form, the teacher will follow the instruction design guide, based on ADDIE Framework. The analysis phase consists of understanding the educational problem, covering the survey of educational needs, the characterization of students and the verification of Constraints.

In this analysis phase, title, audience, inclusion goals, learning objectives, outcomes, and constraint and the pedagogical considerations are defined. The learning objectives was implemented based on Bloom's taxonomy that divides these goals into three domains: cognitive domain, affective domain and psychomotor domain (Krathwohl 2002; Ferraz and Belhot 2010). By acknowledging all domains in a holistic way, instructional designers can develop comprehensive learning experiences.

To help the teacher establish these objectives, the SELI learning platform offers suggestions for these action verbs, such as to build, to combine, to design, etc. (Fig. 1), based on Bloom's taxonomy (Krathwohl 2002; Ferraz and Belhot 2010). When choosing the verb, the teacher must complete the sentence clearly showing the objectives of the educational resource that is being built.

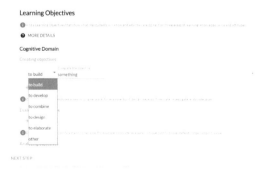

Fig. 1.   Learning objectives in the analysis phase.

In addition to the learning objectives, the teacher must define the learning outcomes, indicating what the students will be able to accomplish with the acquired knowledge.

Learning outcomes address content objectives, skill objectives and values or attitudes objectives. As well as learning objectives, learning outcomes are described using action verbs and should use specific language, and clearly indicate expectations for student performance. Generally, the phrases that indicate the learning outcomes begin as follows: "By the end of this course, students will be able..." completed by the action verb according to the category (content, skills, and values/attitudes objectives). In this item, the SELI learning platform also offers action verb suggestions in each category to guide teachers.

The course structure is defined in the design phase. For each subject, the instructional design process entails a comprehensive delineation of learning objectives, assessment instruments, exercises, content, subject matter analysis, lesson planning, and media selection. This systematic approach ensures that all components of the instructional design align cohesively, supporting the intended learning outcomes and addressing the identified needs of the learners. (Filatro 2008).

In the development phase, the didactic content is to develop and build. In SELI learning platform, during this phase, the instructional design tool can provide guidance to educators in aligning their teaching resources with accessibility standards. This aspect ensures that the created materials are inclusive and cater to diverse learner needs, promoting equal access to educational content. Figure 2 shows an example of the program screen, where the teacher chose to use images and videos for the course content, in addition to the textual content.

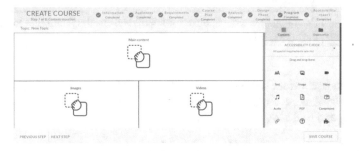

Fig. 2. Program screen to Development phase.

After the development of learning materials, they are made available to students. If the accessibility suggestions were followed during the construction of the resource, the student will receive didactic material that may be adapted to their needs according to their limitations and deficiencies.

Finally, it comes that las ADDIE stage, the evaluation. This stage not necessarily occurs only after all stages, in a waterfall style, but it can be applied together with all other stages, improving their results, and allowing quick responses to eventual problems. Regarding accessibility issues, it is important to evaluate the conformity to standards at each iteration's steps.

3   Accessibility Requirements in SELI Learning Platform

The accessibility requirements were made according to the WCAG guidelines (Web Content Accessibility Guidelines), with the purpose of making the SELI learning platform accessible to students with cognitive, hearing, visual and physical disabilities, in addition to people who have a diversity of these disabilities, such as elderly. For this purpose, in the teacher's view, the course creation was designed so that they can choose an accessibility goal and then include several alternatives for such audiences (Fig. 3). In Table 2 there are the main alternatives for each type of component.

Table 2. Alternative approaches for each component.

Component	Alternative approach
Images	Textual alternative of short description and long description
Texts	Must be produced to allow the use of screen readers
PDF files	Guidelines for the teacher to produce an accessible PDF and register the accessibility alternatives present in it
Videos	Textual alternative of short description and transcription, option to include captions, audio description, sign language and warning of seizure risk
Audio files	Textual alternative of short description and transcription (Sign Language to be supported)
Quizzes	Disabling or extending time restrictions and enabling messages to warn that the time is running out

In the student's view, some requirements were made so all the pages they can view are accessible. Among them, using screen readers and keyboard navigation, adjusting line spacing and font size, among others.

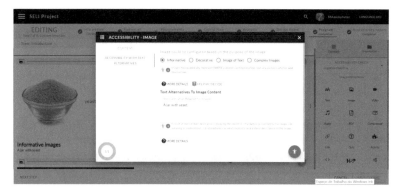

Fig. 3. Image accessibility feature.

4   Usability Evaluation

During the SELI Learning Platform implementation process, teachers from partner countries used the tool to create and make available their accessible courses, as previously shown in Table 1. After finishing their courses, eight teachers performed an usability evaluation and answered the USE Questionnaire (Lund 2001). The questionnaire was sent to teachers by e-mail and answered via Google forms to measure the usability of the instructional design guide functionality. The questionnaire was constructed as seven-point Likert rating scales and users were asked to rate agreement with the statements, ranging from strongly disagree to strongly agree. The USE Questionnaire consists of thirty questions divided in four usability dimensions: Usefulness (eight questions), Ease of Use (eleven questions), Ease of Learning (four questions), and Satisfaction (seven questions). Table 3 shows the composition of the USE questionnaire, highlighting the questions corresponding to each dimension. In addition to the USE Questionnaire questions, three open questions were included: "List the most negative aspect(s)", "List the most positive aspect(s)" and "Leave your comments/suggestions".

Table 3. Composition of the USE questionnaire.

Dimension	Questions	Total Questions
Usefulness	(1) It helps me be more effective. (2) It helps me be more productive. (3) It is useful. (4) It gives me more control over the activities in my life. (5) It makes the things I want to accomplish easier to get done. (6) It saves me time when I use it. (7) It meets my needs. (8) It does everything I would expect it to do	8
Ease of Use	(9) It is easy to use. (10) It is simple to use. (11) It is user friendly. (12) It requires the fewest steps possible to accomplish what I want to do with it. (13) It is flexible. (14) Using it is effortless. (15) I can use it without written instructions. (16) I don't notice any inconsistencies as I use it. (17) Both occasional and regular users would like it. (18) I can recover from mistakes quickly and easily. (19) I can use it successfully every time	11
Ease of Learning	(20) I learned to use it quickly. (21) I easily remember how to use it. (22) It is easy to learn to use it. (23) I quickly became skillful with it	4
Satisfaction	(24) I am satisfied with it. (25) I would recommend it to a friend. (26) It is fun to use. (27) It works the way I want it to work. (28) It is wonderful. (29) I feel I need to have it. (30) It is pleasant to use	7

5   Results

In this analysis, responses 1 to 3 is going to somewhat disagree, disagree or strongly disagree were considered negative responses, indicating an unacceptable assessment on that item. Response 4 is neutral, neither agree or disagree and response 5 to 7 is going to somewhat agree, agree or strongly agree were considered positive responses, indicating an acceptable assessment on that item.

In summary, as shown in Fig. 4 regarding usefulness, the questions (1), (2), (5) and (7), 75% of respondents agreed that the tool was efficient and productive, it makes the things they want to accomplish easier to get done and it meets their needs; 87.5% agree that is (3) useful and (8) it does everything they would expect it to do (although none of the respondents gave a score of 7 for this question). In question (4) and (6), 62.5% agree that the tool gives them more control over the activities of their routine and it saves them time when they use it, although 12.5% were neutral on these issues. With these percentages and with a mean score of 4.953125 of 7 (Table 4), we concluded that the SELI Learning Platform is useful.

Related to ease of use (Fig. 4), the questions (9) to (13) received 75% of agree and this percentual of respondents believe that the tool is ease and simple to use, user friendly, it requires the fewest steps possible to accomplish what they want to do with it and flexible. Already 87.5% consider that the SELI learning platform is (14) effortless in its use. Relating to questions (15) and (17), 75% agree that they can use it without written instructions and both occasional and regular users would like it, but 12.5% were neutral in this issue. In the questions (16), (18) and (19), 62.5% didn't notice any inconsistencies as they use it, they could recover from mistakes quickly and easily and could use it successfully every time, but 12.5% were neutral about these issues. With these percentages and with mean score of 4.875 of 7 (Table 4), we concluded that the SELI Learning Platform is easy to use, although there was a smaller percentage in relation to questions (16), (18) and (19).

In the four questions referring to Ease of Learning dimension, 87.5% of respondents agreed that the tool is quick to learn, ease to remember how to use it, ease to learn to use it and they quickly became skillful with it (Fig. 4). It demonstrates that the SELI Learning Platform interface helps users to learn and use it. The mean score of this dimension was 5.6875 of 7 (Table 4).

Analyzing the responses of satisfaction dimension (Fig. 4), 87.5% of respondents were satisfied (24) with the tool. Considering the questions (25), (26), (27) e (30), 75% of respondents would recommend the tool to a friend, they thought that it is fun to use, it works the way they wanted it to work and is considered pleasant to use. To the question (28), 62.5% believe that it is wonderful and 25% were neutral in this issue. And about the question (29), only 50% of respondents felt the need to have the tool. These responses demonstrate that in the Satisfaction aspect, the tool still needs to improve to become attractive and satisfactory, although the mean score was 4.982142857 of 7 (Table 4). Regarding the open questions, the following negative and positive aspects and comments/suggestions are indicated as follows.

Some negative aspects that were pointed out involve complaints about the user friendness due to some technical problems in both Latin American and European mirror installations of the platform. Comments and suggestions have arisen from the group, regarding

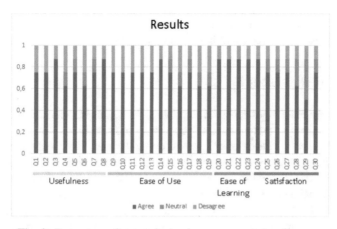

Fig. 4. Percentage of respondents who agree, neutral or disagree.

Table 4. Mean score of each dimension of USE Questionnaire.

Dimension	Mean Score
Usefulness	4.953125
Ease of Use	4.875
Ease of Learning	5.6875
Satisfaction	4.982142857

the lack of stability of some mirrors. These problems have been addressed and some improvements were incorporated later in a new version of the platform.

The main positive aspects that have emerged are the fact of being an innovative platform, with the purpose of "including everyone", as mentioned by one the teachers. Aspects of the user interface, such as modern look-and-feel, quick configurability, configurable accessibility for people with special educational needs were mentioned. Many respondents pointed out that intuitive resources are easily loaded with simple operations, being easy to use for both students and teachers.

Most respondents expressed negative points related to technical problems they faced. In this item, it should be noted that the tool was tested in the development process, so technical errors happen at this stage and the tests helped to verify these points to be improved. Regarding the positive aspects, 100% responded to this question, highlighting the ease in creating teaching materials along with the resources offered for this task, in addition to the issue of inclusion. Based on the comments/suggestions, we inferred that most respondents found the tool interesting. However, there was a percentage of respondents who did not feel comfortable using the tool, which can be seen in both the responses to closed and open questions. This denotes that, despite the positive aspects pointed out by the respondents, there are still points that must be analyzed in the use of the tool. For example, providing users with a shorter, more streamlined option for the course planning process in the Analysis phase. It is important to mention that teachers are not always able and interested in this long process.

6 Final Considerations

There are many challenges regarding inclusion in digital learning environments, that vary from technical aspects - lack of supporting tools, difficulty to address different demands for specific audiences and so - to cultural facets: from system developers to content creators, it is necessary to stimulate a culture of inclusion. Accessibility must not be considered just a feature more, but instead a core part of the design of every system. This attitude must be reflected also in the content creation process.

Teachers and content creators tend to be more open to the need to support inclusion in their productions, maybe because of the proximity they usually have with their audience, being able to go through the needs and expectations of those students that need to be included.

The main goal of the SELI Project was to empower teachers and content creators to include accessibility from the beginning in their educational products. This paper has shown some initial qualitative tests that were conducted with a reduced group of teachers, which was a key action to implement improvements in the authoring tool, whose results could be experienced in the Brazilian mirror available at SELI.

Further works include to evolve the Ecosystem itself, mainly the authoring tool to improve stability and usability aspects, targeting the optimization of the user experience. More tests are to be performed with specific audiences that were not yet covered, aiming to enhance accessibility and even to discover new requirements for newer versions of the Ecosystem.

Acknowledgments. Thanks to Coordenação de Aperfeiçoamento de Pessoal de Nível Superior (CAPES, Coordination for the Improvement of Higher Education Personnel) - Brazil - Program of Excellence - Proex 0653/2018 and ERANET-LAC project - European Union's Seventh Framework Programme - Project Smart Ecosystem for Learning and Inclusion - ERANet17/ICT-0076SELI to support this project.

References

Branch, R.M.: Instructional Design: The ADDIE Approach, vol. 722. Springer, New York (2009). https://doi.org/10.1007/978-0-387-09506-6

Eliseo, M.A., et al.: Framework to creation of inclusive and didactic digital material for elderly. In 2020 15th Iberian Conference on Information Systems and Technologies (CISTI), pp. 1–6. IEEE, June 2020. https://doi.org/10.23919/CISTI49556.2020.9140993

Ferraz, A.P.D.C.M., Belhot, R.V.: Taxonomia de Bloom: revisão teórica e apresentação das adequações do instrumento para definição de objetivos instrucionais. Gestão and Produção **17**(2), 421–431 (2010). https://doi.org/10.1590/S0104-530X2010000200015

Filatro, A.: Design instrucional na prática. Pearson Education do Brasil, São Paulo (2008)

Krathwohl, D.R.: A revision of Bloom's taxonomy: an overview. Theor. Pract. **41**(4), 212–218 (2002)

Lund, A.M.: Measuring usability with the USE questionnaire. Usability Interface **8**(2), 3–6 (2001)

Martins, V., et al.: A blockchain microsites-based ecosystem for learning and inclusion. In: Brazilian Symposium on Computers in Education (Simpósio Brasileiro de Informática na Educação-SBIE), vol. 30, no. 1, p. 229, November 2019. https://doi.org/10.5753/cbie.sbie.201 9.229

McGriff, S.: Instructional System Design (ISD): Using the ADDIE Model. Instructional Systems, College of Education, Penn State University (2000)

Merrill, M.D., Twitchell, D.: Instructional design theory. Educ. Technol. (1994)

Oyelere, S.S., et al.: Digital storytelling and blockchain as pedagogy and technology to support the development of an inclusive smart learning ecosystem. In: Rocha, Á., Adeli, H., Reis, L., Costanzo, S., Orovic, I., Moreira, F. (eds.) Trends and Innovations in Information Systems and Technologies. WorldCIST 2020. Advances in Intelligent Systems and Computing, vol. 1161, pp. 397–408. Springer, Cham (2020). https://doi.org/10.1007/978-3-030-45697-9_39

Peterson, C.: Bringing ADDIE to life: instructional design at its best. J. Educ. Multimedia Hypermedia **12**(3), 227–241 (2003). Norfolk, VA: Association for the Advancement of Computing in Education (AACE). Accessed https://www.learntechlib.org/primary/p/2074/

Tomczyk, Ł., et al.: Digital divide in Latin America and Europe: main characteristics in selected countries. In: 2019 14th Iberian Conference on Information Systems and Technologies (CISTI), pp. 1–6. IEEE, June 2019. https://doi.org/10.23919/CISTI.2019.8760821

Author Index

Printed in the United States
by Baker & Taylor Publisher Services